STUDENT RESOURCES
to accompany

PRECALCULUS
with Calculus Previews

FIFTH EDITION

by Dennis G. Zill and Jacqueline M. Dewar

Warren S. Wright
Loyola Marymount University

Carol D. Wright

JONES & BARTLETT
LEARNING

World Headquarters
Jones & Bartlett Learning
5 Wall Street
Burlington, MA 01803
978-443-5000
info@jblearning.com
www.jblearning.com

Jones & Bartlett Learning books and products are available through most bookstores and online booksellers. To contact Jones & Bartlett Learning directly, call 800-832-0034, fax 978-443-8000, or visit our website, www.jblearning.com.

Substantial discounts on bulk quantities of Jones & Bartlett Learning publications are available to corporations, professional associations, and other qualified organizations. For details and specific discount information, contact the special sales department at Jones & Bartlett Learning via the above contact information or send an email to specialsales@jblearning.com.

Copyright © 2013 by Jones & Bartlett Learning, LLC, an Ascend Learning Company

All rights reserved. No part of the material protected by this copyright may be reproduced or utilized in any form, electronic or mechanical, including photocopying, recording, or by any information storage and retrieval system, without written permission from the copyright owner.

Production Credits
Publisher: Cathleen Sether
Senior Acquisitions Editor: Timothy Anderson
Managing Editor: Amy Bloom
Production Editor: Tiffany Sliter
Senior Marketing Manager: Andrea DeFronzo
Cover and Title Page Design: Kristin E. Parker
Composition: Northeast Compositors, Inc.
Cover and Title Page Image: Whirlpool galaxy: © Steve Nagy/Design Pics Inc./Alamy Images; Coma Cluster of galaxies: © Stocktrek Images, Inc./Alamy Images

Complete Solutions Manual to accompany *Precalculus with Calculus Previews, Fifth Edition* by Dennis G. Zill and Jacqueline M. Dewar ISBN: 978-1-4496-4912-8

ISBN: 978-1-4496-8636-9

6048

Printed in the United States of America
16 15 14 13 12   10 9 8 7 6 5 4 3 2 1

Contents

Preface

This manual is intended to accompany *Precalculus with Calculus Previews, Fifth Edition*, by Dennis G. Zill and Jacqueline M. Dewar. It consists of five parts, described below.

Topics in Algebra This part consists of short discussions of appropriate topics from a prerequisite algebra course (such as synthetic division), as well as topics intended to assist the student in becoming a more effective problem solver (such as implicit conditions in a word problem). See the Table of Contents for a complete list of topics covered.

Use of a Calculator This part is not intended to be a comprehensive manual on the use of a graphing calculator in a precalculus course. While much of the material discussed will be pertinent to any graphing calculator, the references in this manual are to the TI-84 family of calculators. After a few brief comments on the use of the TI-84 calculator, the focus in this manual will be on how to use the calculator to either assist in the solution of selected problems in the text or to check that your solution is correct or at least reasonable.

Basic Skills This is a list for each section in the text of the skills needed to solve the more manipulative problems in the section.

Selected Solutions Here, complete solutions to every third problem (starting with Problem 3) in each section of the text are given. Solutions are not provided for **Calculator Problems** or **Calculator/Computer Problems** or for the **Discussion Problems** or problems labeled **For Discussion**. Solutions for problems labeled as some type of **Application** are however given.

Final Examination Answers As indicated by the title of this part, the answers are provided for each of the 92 problems on the final examination on pages 545–549 in the text.

Part I

Topics in Algebra

1 | Multiplication by an Unknown in an Inequality

By (i) in the **Properties of Inequalities** in Section 1.1 of the text, there is no problem adding an expression containing the variable to both sides of an inequality. This is also the case with subtraction. For example, in $3x > 2x - 4$, it is fine to subtract $2x$ from both sides and conclude that $3x - 2x > 2x - 4 - 2x$ or $x > -4$.

In an inequality like $x(x-1)/(x-2) \leq 0$ it is tempting to multiply both sides by $x - 2$ resulting in $x(x-1) \leq 0$. *This is incorrect* because, by (iii) in the **Properties of Inequalities**, $x - 2$ will be negative for some values of x, which will result in the inequality changing from $\leq$ to $\geq$. Thus, you should never multiply both sides of an inequality by an expression containing a variable that could cause the expression to be negative. On the other hand, if you are confident that an expression is always positive, there is no problem multiplying or dividing by that expression in an inequality. For example, $(x + 1)/(x^2 + 4) > 0$ is equivalent to $x + 1 > 0$, because both sides of the inequality can be multiplied by $x^2 + 4$, which is positive for all values of x.

2 | Division by Zero When Solving an Equation

Of course, you know that you should never divide by zero. Sometimes, however, you may unwittingly do so. For example, if you attempt to solve the equation

$$(x - 3)^2 = 4(x - 3)$$

by dividing both sides by $x - 3$ obtaining $x - 3 = 4$, you may have divided by zero. In fact, in this case you have divided by zero, since the resulting equation, $x - 3 = 4$, has only the solution $x = 7$. This is a solution of the original equation, but another solution, $x = 3$, has been lost. This is because when $x = 3$, $x - 3 = 0$, and you have divided by $x - 3$. A way to solve the equation without dividing by zero involves factoring and is shown below

$$(x - 3)^2 = 4(x - 3)$$
$$(x - 3)^2 - 4(x - 3) = 0$$
$$(x - 3)(x - 3 - 4) = 0$$
$$(x - 3)(x - 7) = 0$$

Thus, $x = 3$ and $x = 7$ are solutions of the original equation.

3 Implicit Conditions in a Word Problem

Frequently, a word problem will contain a condition or conditions that is/are not explicitly stated as part of the problem, but is/are implicit in the nature of the problem. For example, if t, representing time, is the unknown in a problem, then it will usually be the case that an implicit condition is $t \geq 0$. This means that if, say, $t = -4$ arises in the solution of an equation or inequality, it can and should be ignored and only nonnegative values of t should be considered.

As another, more detailed example, consider the following geometric problem:

Suppose the area of the ring between a pair of concentric circles is to be at most $9\pi \, \text{cm}^2$. Find the possible dimensions for the radius of the inner circle if the radius of the outer circle is fixed at 5 cm. (See the figure below.)

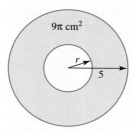

The solution of this problem involves the inequalities

$$\pi(25) - \pi r^2 < 9\pi$$
$$(25 - r^2 - 9)\pi < 0$$
$$16 - r^2 < 0$$
$$(4 + r)(4 - r) < 0,$$

which leads to a sign chart involving $r = -4$ and $r = 4$. Although it is not explicitly stated, $r > 0$ is an implicit condition in this problem. Thus, we can ignore $r = -4$ and consider only positive values of r.

$$
\begin{array}{rccc}
r + 4 & + & + & + \\
r - 4 & - & 0 & + \\
(r + 4)(r - 4) & - & 0 & +
\end{array}
$$

From the graph we conclude that r must be greater than 4 for the area of the ring to be at most $9 \, \text{cm}^2$. But, wait a minute! If we read the problem more carefully, we see that another implicit condition is $r < 5$, so we should have stopped the line diagram above at $r = 5$, as shown below.

The correct solution of the problem is $4 < r < 5$.

Once you have a solution you should reread the problem, being on the lookout for implicit conditions that you may have initially overlooked.

4 Extraneous Roots

When solving an algebraic equation, a number that appears to be a solution may in fact not satisfy the original equation and hence not be a solution. We call such a number an **extraneous solution** or **extraneous root**. We will discuss two instances that can give rise to extraneous solutions. The first instance occurs when both sides of an equation are multiplied by an expression that is zero for some value or values of the unknown. In this case, it is *essential* to check any solutions that have been obtained—they may or may not be solutions of the original equation. For example, multiplying both sides of the equation

$$2 - \frac{1}{x+1} = \frac{x}{x+1}$$

by $x + 1$ yields

$$(x+1)\left(2 - \frac{1}{x+1}\right) = (x+1)\left(\frac{x}{x+1}\right)$$

$$2x + 2 - 1 = x.$$

The solution of the last equation is $x = -1$, but this is not a solution of the original equation because substitution into the original equation results in division by 0, which is not permissible.

Extraneous solutions will not be obtained when both sides of an equation are multiplied by an expression that can never be zero. For example, when solving

$$\frac{x}{x^2 + 1} = \frac{1}{2},$$

both sides can be multiplied by $2(x^2 + 1)$ resulting in

$$2x = x^2 + 1$$

$$x^2 - 2x + 1 = 0$$

$$(x - 1)^2 = 0,$$

so $x = 1$. The number 1 cannot be an extraneous solution because both sides of the original equation were multiplied by the expression $2(x^2 + 1)$, which can never be 0. This is not intended to discourage you from checking your answer. This should always be done.

The second instance that can result in extraneous solutions occurs when both sides of an equation are squared. For example, squaring both sides of the equation

$$\sqrt{3x + 4} = x - 2$$

results in

$$3x + 4 = (x - 2)^2$$

$$3x + 4 = x^2 - 4x + 4$$

$$x^2 - 7x = 0$$

$$x(x - 7) = 0.$$

Thus, it appears that $x = 7$ and $x = 0$ are solutions of the equation. In fact, substitution of $x = 7$ into the original equation yields the true statement $5 = 5$. However, substitution of $x = 0$ into the original equation yields $2 = -2$, which is false. In this case, $x = 0$ is an extraneous solution and the only solution of the original equation is $x = 7$.

5 | Factorial Notation

Factorial notation is used to represent a descending product of nonnegative integers. The symbol $n!$ is defined for any nonnegative integer n by

$$1! = 1,$$
$$2! = 2 \cdot 1 = 2,$$
$$3! = 3 \cdot 2 \cdot 1 = 6,$$
$$4! = 4 \cdot 3 \cdot 2 \cdot 1 = 24,$$

and so on. In general

$$n! = n(n-1)(n-2) \cdots 3 \cdot 2 \cdot 1,$$

where n is a positive integer. There are a number of reasons that justify the additional definition of

$$0! = 1.$$

We will see one of these reasons in the section on the Binomial Theorem.

EXAMPLE   Simplify $7!/4!$.

Solution   Using the definition of factorial we have

$$\frac{7!}{4!} = \frac{7 \cdot 6 \cdot 5 \cdot 4!}{4!} = 7 \cdot 6 \cdot 5 = 210. \qquad \square$$

EXAMPLE   Simplify

$$\frac{n!(n+1)}{(n-1)!}.$$

Solution   Using the definition of factorial we can write the numerator as

$$n!(n+1) = (n+1)n! = (n+1)n(n-1) \cdots 3 \cdot 2 \cdot 1 = (n+1)n(n-1)!.$$

Then,

$$\frac{n!(n+1)}{(n-1)!} = \frac{(n+1)n(n-1)!}{(n-1)!} = (n+1)n. \qquad \square$$

6 | Pascal's Triangle

Pascal's triangle is a triangular array of positive integers. We will see an application of Pascal's triangle in the next section on the Binomial Theorem. The first five rows of the triangle are shown below.

$$
\begin{array}{ccccccccc}
 & & & & 1 & & & & \\
 & & & 1 & & 1 & & & \\
 & & 1 & & 2 & & 1 & & \\
 & 1 & & 3 & & 3 & & 1 & \\
1 & & 4 & & 6 & & 4 & & 1 \\
\end{array}
$$

Observe that the outer sides of the triangle consist entirely of 1's, and that each number in the interior of the triangular array is the sum of the two numbers above to the left and to the right of that number. For example, the first 4 in the fifth row is the sum of 1 and 3, the two numbers above the 4.

EXAMPLE Find the next row in Pascal's triangle.

Solution The diagram below shows how to obtain the sixth row of Pascal's triangle by using the fifth row.

$$
\begin{array}{ccccccccc}
& 1 & & 4 & & 6 & & 4 & & 1 \\
1 & & 5 & & 10 & & 10 & & 5 & & 1
\end{array}
$$

☐

7 | Binomial Theorem

When $(a+b)^n$ is expanded, or multiplied out, for an arbitrary positive integer n, the exponents of a and b follow a definite pattern. For example, from

$$(a+b)^2 = a^2 + 2ab + b^2,$$
$$(a+b)^3 = a^3 + 3a^2b + 3ab^2 + b^3,$$
$$(a+b)^4 = a^4 + 4a^3b + 6a^2b^2 + 4ab^3 + b^4,$$

we see that the exponents of a decrease by 1, starting with the first term, whereas the exponents of b increase by 1, starting with the second term. To extend this pattern, we consider the first and last terms to be multiplied by b^0 and a^0, respectively; that is,

$$(a+b)^4 = a^4b^0 + 4a^3b^1 + 6a^2b^2 + 4a^1b^3 + a^0b^4.$$

We note that the sum of the exponents in each term of the expansion of $(a+b)^4$ is 4.

The following two examples illustrate the use of the product formulas given at the beginning of this section.

EXAMPLE Expand $(x^2 + 2k)^2$.

Solution Identifying $a = x^2$ and $b = 2k$, it follows from the formula for $(a+b)^2$ at the start of this section that

$$(x^2 + 2k)^2 = (x^2)^2 + 2(x^2)(2k) + (2k)^2 = x^4 + 4x^2k + 4k^2.$$

☐

EXAMPLE Expand $(\sqrt{x} - 4y^2)^3$.

Solution Identifying $a = \sqrt{x}$ and $b = -4y^2$, it follows from the formula for $(a+b)^3$ at the start of this section that

$$\left(\sqrt{x} - 4y^2\right)^3 = \left(\sqrt{x}\right)^3 + 3\left(\sqrt{x}\right)^2\left(-4y^2\right) + 3\left(\sqrt{x}\right)\left(-4y^2\right)^2 + \left(-4y^2\right)^3$$
$$= x^{3/2} - 12xy^2 + 48x^{1/2}y^4 - 64y^6.$$

☐

In the last example we expressed $\left(\sqrt{x}\right)^3$ in the form $x^{3/2}$ and $\sqrt{x}$ in the form $x^{1/2}$. That is, we switched from the use of radical notation to the use of fractional exponents. This is generally good practice, especially in calculus, because fractional exponents are usually easier to manipulate than expressions containing radicals.

As discussed above, the exponents in a binomial expansion follow a pattern. The same is true of the coefficients. In fact, in the case of the coefficients, this pattern is given by Pascal's triangle, discussed in the preceding section of this manual. For example,

$$(a+b)^5 = a^5 + 5a^4b + 10a^3b^2 + 10a^2b^3 + 5ab^4 + b^5,$$

and the coefficients constitute the sixth row of Pascal's triangle.

■ **EXAMPLE**   Expand $(x - y)^7$.

Solution We begin by expanding Pascal's triangle through the eighth row:

$$
\begin{array}{ccccccccccccccc}
&&&&&&& 1 &&&&&&& \\
&&&&&& 1 && 1 &&&&&& \\
&&&&& 1 && 2 && 1 &&&&& \\
&&&& 1 && 3 && 3 && 1 &&&& \\
&&& 1 && 4 && 6 && 4 && 1 &&& \\
&& 1 && 5 && 10 && 10 && 5 && 1 && \\
& 1 && 6 && 15 && 20 && 15 && 6 && 1 & \\
1 && 7 && 21 && 35 && 35 && 21 && 7 && 1
\end{array}
$$

Now, identifying a with x and b with $-y$, it follows that

$$
\begin{aligned}
(x - y)^7 &= x^7 + 7x^6(-y) + 21x^5(-y)^2 + 35x^4(-y)^3 + 35x^3(-y)^4 \\
&\qquad\qquad + 21x^2(-y)^5 + 7x(-y)^6 + (-y)^7 \\
&= x^7 - 7x^6 y + 21x^5 y^2 - 35x^4 y^3 + 35x^3 y^4 - 21x^2 y^5 + 7xy^6 - y^7.
\end{aligned}
$$ □

While Pascal's triangle is easy to use when expanding a binomial expression of the form $(a + b)^n$ for relatively small values of n, it would be very time consuming to generate enough rows of Pascal's triangle to expand $(a + b)^{14}$. In this case it is more convenient to use the general formula for the expansion of $(a + b)^n$ given by the **Binomial Theorem**.

BINOMIAL THEOREM   For any positive integer n,

$$
\begin{aligned}
(a + b)^n &= a^n + \frac{n}{1!}\, a^{n-1}b + \frac{n(n-1)}{2!}\, a^{n-2}b^2 \\
&\quad + \cdots + \frac{n(n-1)\cdots(n-k+1)}{k!}\, a^{n-k}b^k + \cdots + b^n.
\end{aligned}
$$ ■

We can write the Binomial Theorem in a more compact form using summation notation (discussed in Section 3.6 in the text) and without the ellipses ($\cdots$) by using the fact that for any integer $0 \le k \le n$,

$$
\begin{aligned}
n(n-1)\cdots(n-k-1) &= \frac{n(n-1)\cdots(n-k+1)(n-k)(n-k-1)\cdots 3\cdot 2\cdot 1}{(n-k)(n-k-1)\cdots 3\cdot 2\cdot 1} \\
&= \frac{n!}{(n-k)!}.
\end{aligned}
$$

Using this fact, we then have

$$
\begin{aligned}
(a + b)^n &= \frac{n!}{0!(n-0)!}\, a^{n-0}b^0 + \frac{n!}{1!(n-1)!}\, a^{n-1}b^1 \\
&\quad + \cdots + \frac{n!}{k!(n-k)!}\, a^{n-k}b^k + \cdots + \frac{n!}{n!(n-n)!} a^{n-n}b^{n-0} \\
&= \sum_{k=0}^{n} \frac{n!}{k!(n-k)!}\, a^{n-k}b^k.
\end{aligned}
$$

Note that at this point we require that $0! = 1$.

The Binomial Theorem provides an alternative to Pascal's triangle for finding the coefficients of an expression of the form $(a + b)^n$.

EXAMPLE   Use the Binomial Theorem to expand $(a + b)^4$.

Solution From the Binomial Theorem we have

$$(a + b)^4 = a^4 + \frac{4}{1!} a^{4-1}b + \frac{4(3)}{2!} a^{4-2}b^2 + \frac{4(3)(2)}{3!} a^{4-3}b^3 + \frac{4(3)(2)(1)}{4!} b^4$$
$$= a^4 + 4a^3b + \frac{12}{2} a^2b^2 + \frac{24}{6} ab^3 + \frac{24}{24} b^4$$
$$= a^4 + 4a^3b + 6a^2b^2 + 4ab^3 + b^4.$$

□

EXAMPLE   Find the sixth term of the expansion of $(x^2 - 2y)^7$.

Solution Letting $n = 7$ in the summation notation form of the Binomial Theorem we have

$$(a + b)^7 = \sum_{k=0}^{7} \frac{7!}{k!(7 - k)!} a^{7-k}b^k.$$

Since the sixth term corresponds to $k = 5$ (because k starts with 0 instead of 1), the sixth term is

$$\frac{7!}{5!(7 - 5)!} a^{7-5}b^5 = \frac{7!}{5!(2!)} a^2b^5 = \frac{7(6)}{2(1)} a^2b^5 = 21a^2b^5.$$

Now, identifying $a = x^2$ and $b = -2y$, we see that the sixth term is

$$21(x^2)^2(-2y)^5 = 21x^4(-32y^5) = -672x^4y^5.$$

□

8 Factoring

Factoring polynomials is the reverse of multiplying polynomials. That is, given a polynomial like $5x^3 + 6x^2 - 29x - 6$, we would like to be able to express it as a product of simpler polynomials. For example,

$$5x^3 + 6x^2 - 29x - 6 = (5x + 1)(x - 2)(x + 3).$$

Factoring can be very useful when solving equations or evaluating limits.

8.1 Common Factors and Grouping

In general, the first step in factoring any algebraic expression is to determine whether the terms have a common factor.

EXAMPLE   Factor $6x^4y^4 - 4x^2y^2 + 10xy^3 - 2xy^2$.

Solution Since $2xy^2$ is a common factor of the terms, we have

$$6x^4y^4 - 4x^2y^2 + 10xy^3 - 2xy^2 = 2xy^2(3x^3y^2) - 2xy^2(2x) + 2xy^2(5y) - 2xy^2$$
$$= 2xy^2(3x^3y^2 - 2x + 5y - 1).$$

□

When the terms of an expression do not have a common factor, it may still be possible to factor by grouping the terms in an appropriate manner.

◾ **EXAMPLE**   Factor $x^2 + 2xy - x - 2y$.

Solution Grouping the first two terms and the last two terms gives

$$x^2 + 2xy - x - 2y = (x^2 + 2xy) + (-x - 2y) = x(x + 2y) + (-1)(x + 2y).$$

We observe the common factor $x + 2y$ and complete the factorization as

$$x^2 + 2xy - x - 2y = (x - 1)(x + 2y).$$ ◻

8.2 | Use of Factorization Formulas |

By reversing several product formulas we obtain the following important factorization formulas.

Factorization Formulas

(i)   **Square of a Sum:** $X^2 + 2XY + Y^2 = (X + Y)^2$

(ii) **Square of a Difference:** $X^2 - 2XY + Y^2 = (X - Y)^2$

(iii) **Difference of Two Squares:** $X^2 - Y^2 = (X - Y)(X + Y)$

(iv) **Sum of Two Cubes:** $X^3 + Y^3 = (X + Y)(X^2 - XY + Y^2)$

(v)   **Difference of Two Cubes:** $X^3 - Y^3 = (X - Y)(X^2 + XY + Y^2)$

We have used capital letters in these formulas to clarify our work when we apply the formulas. Thus, in the next example, X represents $4x^2y$ and Y represents 5.

◾ **EXAMPLE**   Factor $16x^4y^2 - 25$.

Solution This is the difference of two squares. Thus, from **(iii)**, with $X = 4x^2y$ and $Y = 5$, we have

$$16x^4y^2 - 25 = (4x^2y)^2 - (5)^2 = (4x^2y - 5)(4x^2y + 5).$$ ◻

◾ **EXAMPLE**   Factor $8a^3 + 27b^6$.

Solution This is the sum of two cubes. Thus, from **(iv)**, with $X = 2a$ and $Y = 3b^2$, we have

$$8a^3 + 27b^6 = (2a)^3 + (3b^2)^3 = (2a + 3b^2)[(2a)^2 - (2a)(3b^2) + (3b^2)^2]$$
$$= (2a + 3b^2)(4a^2 - 6ab^2 + 9b^4).$$ ◻

Observe in the above factorization formulas that there is no formula for the sum of two squares. An expression of the form $X^2 + Y^2$ *never* factors in the real number system. On the other hand, formulas **(iii)**–**(v)** indicate that the difference of two squares and the sum and difference of two cubes always factor, provided that we do not restrict the coefficients to the set of integers. For example, using **(iii)** to factor $x^2 - 5$, we identify $X = x$ and $Y = \sqrt{5}$ so that

$$x^2 - 5 = x^2 - (\sqrt{5})^2 = (x - \sqrt{5})(x + \sqrt{5}).$$

For the remainder of this section, however, we will seek only polynomial factors with *integer* coefficients.

8.3 Factoring Quadratic Polynomials

It is sometimes possible to factor the quadratic polynomial $ax^2 + bx + c$, where a, b, and c are integers, as

$$(Ax + B)(Cx + D),$$

where A, B, C, and D are also integers. Initially, to simplify our discussion we assume that the quadratic polynomial has as its leading coefficient $a = 1$. If $x^2 + bx + c$ has a factorization using integer coefficients, then it will be of the form

$$(x + B)(x + D),$$

where B and D are integers. Finding the product and comparing coefficients,

$$(x + B)(x + D) = x^2 + \overbrace{(B + D)}^{B+D=b}x + \underbrace{BD = x^2 + bx + c,}_{BD=c}$$

we see that

$$B + D = b \quad \text{and} \quad BD = c.$$

Thus, to factor $x^2 + bx + c$ with integer coefficients, we list all possible factorizations of c as a product of two integers B and D. We then check which, if any, of the sums $B + D$ equals b.

■ **EXAMPLE**   Factor $x^2 - 9x + 18$.

Solution With $b = -9$ and $c = 18$, we look for integers B and D such that

$$B + D = -9 \quad \text{and} \quad BD = 18.$$

We can write 18 as a product BD in the following ways:

$$1(18), \quad 2(9), \quad 3(6), \quad (-1)(-18), \quad (-2)(-9), \quad \text{or} \quad (-3)(-6).$$

Since -9 is the sum of -3 and -6, the factorization is

$$x^2 - 9x + 18 = (x - 3)(x - 6). \qquad \square$$

Once you have obtained a factorization, you should *always* check your work by multiplying out the factors.

As we see in the next example, it is always possible that a quadratic expression will not factor.

■ **EXAMPLE**   Factor $x^2 + 3x - 2$.

Solution The number -2 can be written as a product of integers in two ways: $(-1)(2)$ and $(1)(-2)$. Since neither $-1 + 2$ nor $1 + (-2)$ is equal to 3, the expression $x^2 + 3x - 2$ cannot be factored using integer coefficients. $\qquad \square$

It is more complicated to factor the general quadratic polynomial $ax^2 + bx + c$, when $a \neq 1$, since, in this case, we must consider factors of a as well as of c. Finding the product and comparing the coefficients in

$$(Ax + B)(Cx + D) = \overbrace{ACx^2}^{AC=a} + \underbrace{(AD + BC)}_{AD+BC=b}x + \overbrace{BD = ax^2 + bx + c,}^{BD=c}$$

we see that $ax^2 + bx + c$ factors as $(Ax + B)(Cx + D)$ if

$$AC = a, \qquad AD + BC = b, \qquad \text{and} \qquad BD = c.$$

■ EXAMPLE   Factor $2x^2 + 11x - 6$.

Solution The factors will be

$$(2x + \underline{})(1x + \underline{}),$$

where the blanks are to be filled with a pair of integers B and D whose product BD equals -6. Possible pairs are

$$\begin{array}{llll}
1 \text{ and } -6, & -1 \text{ and } 6, & 3 \text{ and } -2, & -3 \text{ and } 2, \\
-6 \text{ and } 1, & 6 \text{ and } -1, & -2 \text{ and } 3, & 2 \text{ and } -3.
\end{array}$$

Now we must check to see if one of the pairs gives 11 as the value of $AD + BC$ (the coefficient of the middle term), where $A = 2$ and $C = 1$. We find

$$2(6) + (-1)(1) = 11,$$

so $2x^2 + 11x - 6 = (2x - 1)(x + 6)$.   □

This general method can be applied to expressions of the form $ax^2 + bxy + cy^2$, where a, b, and c are integers.

■ EXAMPLE   Factor $15x^2 + 17xy + 4y^2$.

Solution The possible factors have the form

$$(5x + \underline{}y)(3x + \underline{}y) \qquad \text{or} \qquad (15x + \underline{}y)(1x + \underline{}y).$$

There is no need to consider the cases

$$(-5x + \underline{}y)(-3x + \underline{}y) \qquad \text{or} \qquad (-15x + \underline{}y)(-x + \underline{}y)$$

because we can factor out -1 from each term and $(-1)(-1) = 1$, so these two cases are equivalent to the first two cases. Now, the blanks must be filled with a pair of integers whose product is 4. Possible pairs are

$$1 \text{ and } 4, \qquad -1 \text{ and } -4, \qquad 2 \text{ and } 2, \qquad -2 \text{ and } -2.$$

We check pairs of numbers to see which combination, if any, gives a coefficient of 17 for the middle term. We find

$$15x^2 + 17xy + 4y^2 = (5x + 4y)(3x + y).$$ □

■ EXAMPLE   Factor $2x^4 + 11x^2 + 12$.

Solution Letting $X = x^2$, we can regard this expression as a quadratic polynomial in the variable X,

$$2X^2 + 11X + 12.$$

We then factor this quadratic polynomial. Since all of the coefficients of $2X^2 + 11X + 12$ are positive, we can assume that the factors will have the form

$$(X + \underline{})(2X + \underline{}),$$

where the blanks are to be filled with a pair of positive integers whose product is 12. The possible pairs are

$$1 \text{ and } 12, \quad 2 \text{ and } 6, \quad 3 \text{ and } 4, \quad 12 \text{ and } 1, \quad 6 \text{ and } 2, \quad 4 \text{ and } 3.$$

We check each pair to see which combination, if any, gives a coefficient of 11 for the middle term. We find

$$2X^2 + 11X + 12 = (X + 4)(2X + 3).$$

Substituting x^2 for X gives

$$2x^4 + 11x^2 + 12 = (x^2 + 4)(2x^2 + 3). \qquad \square$$

In the preceding example you should consider the possibility that $x^2 + 4$ or $2x^2 + 3$ factor. In this case, since each is a sum of squares, neither factors.

9 Rationalizing the Numerator

When we remove radicals from the numerator or denominator of a fraction, we say that we are **rationalizing**. In algebra, we usually rationalize the denominator, but in calculus it is sometimes important to rationalize the numerator. The procedure of rationalizing involves multiplying the fraction by 1, written in a special way. For example, to rationalize $2/\sqrt{3}$ we multiply by $\sqrt{3}/\sqrt{3}$:

$$\frac{2}{\sqrt{3}} = \frac{2}{\sqrt{3}}\left(\frac{\sqrt{3}}{\sqrt{3}}\right) = \frac{2\sqrt{3}}{3}.$$

If a fraction contains an expression such as $2 + \sqrt{x}$, we make use of the fact that the product of $2 + \sqrt{x}$ and its *conjugate* $2 - \sqrt{x}$ contains no radicals:

$$\frac{x}{2 + \sqrt{x}} = \frac{x}{2 + \sqrt{x}}\left(\frac{2 - \sqrt{x}}{2 - \sqrt{x}}\right) = \frac{2x - x\sqrt{x}}{4 - x}.$$

We proceed in a similar fashion to rationalize the numerator of a fraction containing radicals:

$$\frac{\sqrt{x-2} - \sqrt{y}}{x + y} = \frac{\sqrt{x-2} - \sqrt{y}}{x + y}\left(\frac{\sqrt{x-2} + \sqrt{y}}{\sqrt{x-2} + \sqrt{y}}\right) = \frac{x - 2 - y}{(x + y)(\sqrt{x-2} + \sqrt{y})}.$$

10 Rational Expressions

A quotient of two polynomials is called a **rational expression**. For example,

$$\frac{2x^2 + 5}{x + 1} \qquad \text{and} \qquad \frac{3}{2x^3 - x + 8}$$

are rational expressions. In solving problems, we must often combine rational expressions and then simplify the results. To do this we use the following properties of numbers, which are valid provided each denominator is nonzero.

Properties of Fractions	
(i) **Cancellation:**	$\dfrac{ac}{bc} = \dfrac{a}{b}$
(ii) **Addition or Subtraction:**	$\dfrac{a}{b} \pm \dfrac{c}{b} = \dfrac{a \pm c}{b}$
(iii) **Multiplication:**	$\dfrac{a}{b} \cdot \dfrac{c}{d} = \dfrac{ac}{bd}$
(iv) **Division:**	$\dfrac{a}{b} \div \dfrac{c}{d} = \dfrac{a}{b} \cdot \dfrac{d}{c}$

◼ **EXAMPLE**   Simplify

$$\frac{2x^2 - x - 1}{x^2 - 1}.$$

Solution We factor the numerator and denominator and cancel common factors (noted in gray type) using cancellation property **(i)**:

$$\frac{2x^2 - x - 1}{x^2 - 1} = \frac{(2x + 1)(x - 1)}{(x + 1)(x - 1)} = \frac{2x + 1}{x + 1}.$$ □

Note that in the above example the cancellation of the common factor $x - 1$ is valid only for those values of x such that $x - 1$ is nonzero; that is, for $x \neq 1$. However, since the expression $(2x^2 - x - 1)/(x^2 - 1)$ is not defined for $x = 1$, our simplification is valid for all real numbers in the domain of the variable x in the *original* expression. We emphasize that the equation

$$\frac{2x^2 - x - 1}{x^2 - 1} = \frac{2x + 1}{x + 1}$$

is not valid for $x = 1$, even though the right-hand side, $(2x + 1)/(x + 1)$, is defined for $x = 1$. Considerations of this sort are important when solving equations involving rational expressions.

Henceforth, we will assume without further comment that variables are restricted to values for which all denominators in an equation are nonzero.

◼ **EXAMPLE**   Simplify

$$\frac{4x^2 + 11x - 3}{2 - 5x - 12x^2}.$$

Solution We factor the numerator and denominator and cancel common factors using cancellation property **(i)**:

$$\begin{aligned}
\frac{4x^2 + 11x - 3}{2 - 5x - 12x^2} &= \frac{(4x - 1)(x + 3)}{(1 - 4x)(2 + 3x)} \\
&= \frac{(4x - 1)(x + 3)}{-(4x - 1)(2 + 3x)} \\
&= -\frac{x + 3}{3x + 2}.
\end{aligned}$$ □

11 $\boxed{\text{Least Common Denominator}}$

In order to add or subtract rational expressions, we proceed just as we do when adding or subtracting fractions. We first find a common denominator and then apply **(ii)** in the **Properties of Fractions** above. Although any common denominator will do, less work is involved if we use the **least common denominator** (LCD). This is found by factoring each denominator completely and forming a product of the distinct factors, using each factor with the highest exponent with which it occurs in any single denominator.

◼ **EXAMPLE**   Find the LCD of

$$\frac{1}{x^4 - x^2}, \qquad \frac{x + 2}{x^2 + 2x + 1}, \qquad \text{and} \qquad \frac{1}{x}.$$

Solution Factoring the denominators in the rational expressions, we obtain

$$\frac{1}{x^2(x - 1)(x + 1)}, \qquad \frac{x + 2}{(x + 1)^2}, \qquad \text{and} \qquad \frac{1}{x}.$$

The distinct factors of the denominators are x, $x-1$, and $x+1$. We use each factor with the highest exponent with which it occurs in any single denominator. Thus, the LCD is

$$x^2(x-1)(x+1)^2. \qquad \square$$

■ **EXAMPLE** Combine and simplify

$$\frac{x}{x^2-4} + \frac{1}{x^2+4x+4}.$$

Solution In factored form, the denominators are $(x-2)(x+2)$ and $(x+2)^2$. Thus, the LCD is $(x-2)(x+2)^2$. We use **(i)** in the **Properties of Fractions** above in reverse order to rewrite each rational expression with the LCD as a common denominator:

$$\frac{x}{x^2-4} = \frac{x}{(x-2)(x+2)} = \frac{x(x+2)}{(x-2)(x+2)(x+2)}$$

$$\frac{1}{x^2+4x+4} = \frac{1}{(x+2)^2} = \frac{1(x-2)}{(x+2)^2(x-2)}.$$

Then, using **(ii)**, we add and simplify:

$$\frac{x}{x^2-4} + \frac{1}{x^2+4x+4} = \frac{x(x+2)}{(x-2)(x+2)^2} + \frac{x-2}{(x-2)(x+2)^2}$$

$$= \frac{x(x+2)+x-2}{(x-2)(x+2)^2}$$

$$= \frac{x^2+2x+x-2}{(x-2)(x+2)^2}$$

$$= \frac{x^2+3x-2}{(x-2)(x+2)^2}. \qquad \square$$

To multiply or divide rational expressions, we apply **(iii)** or **(iv)** and then simplify.

■ **EXAMPLE** Combine and simplify

$$\frac{x}{5x^2+21x+4} \cdot \frac{25x^2+10x+1}{3x^2+x}.$$

Solution We factor, cancel, and then use **(iii)** in the **Properties of Fractions** above to combine:

$$\frac{x}{5x^2+21x+4} \cdot \frac{25x^2+10x+1}{3x^2+x} = \frac{x(25x^2+10x+1)}{(5x^2+21x+4)(3x^2+x)}$$

$$= \frac{x(5x+1)(5x+1)}{(5x+1)(x+4)x(3x+1)}$$

$$= \frac{5x+1}{(x+4)(3x+1)}. \qquad \square$$

12   Complex Fractions

A **complex fraction** is a fractional expression in which the numerator or denominator, or both, is itself a fraction. The examples below illustrate two techniques that can be used to simplify complex fractions.

■ EXAMPLE   Simplify

$$\frac{\dfrac{1}{x} - \dfrac{x}{x+1}}{1 + \dfrac{1}{x}}.$$

Solution (Method I)   First we obtain a single rational expression for the numerator and the denominator:

$$\frac{1}{x} - \frac{x}{x+1} = \frac{1(x+1)}{x(x+1)} - \frac{x \cdot x}{x(x+1)} = \frac{x+1-x^2}{x(x+1)} = \frac{-x^2+x+1}{x(x+1)}$$

and

$$1 + \frac{1}{x} = \frac{x}{x} + \frac{1}{x} = \frac{x+1}{x}.$$

Thus,

$$\frac{\dfrac{1}{x} - \dfrac{x}{x+1}}{1 + \dfrac{1}{x}} = \frac{\dfrac{-x^2+x+1}{x(x+1)}}{\dfrac{x+1}{x}}.$$

Now we apply **(iv)** in the **Properties of Fractions** above to this quotient to obtain

$$\frac{\dfrac{-x^2+x+1}{x(x+1)}}{\dfrac{x+1}{x}} = \frac{-x^2+x+1}{x(x+1)} \cdot \frac{x}{x+1} = \frac{-x^2+x+1}{(x+1)^2}. \qquad \square$$

An alternative method of simplifying a complex fraction assumes that the original fraction has no complex fractions in either its numerator or denominator. That is, the method described below would not apply to the complex fraction

$$\frac{\dfrac{x}{x-1} - \dfrac{\dfrac{1}{x}+x}{x+1}}{\dfrac{1}{x} + \dfrac{x}{x-1}}$$

because the numerator itself contains the complex fraction

$$\frac{\dfrac{1}{x}+x}{x+1}.$$

To describe this method for simplifying complex fractions we introduce the notion of the **least common multiple** of polynomials. This is closely related to the notion of LCD. The least common multiple of a set of polynomials is obtained by factoring each polynomial completely and forming a product of the distinct factors, using each factor with the highest exponent with which it occurs in any factorization. We note that the LCD of a set of rational expressions is the least common multiple of the denominators of the expressions.

■ **EXAMPLE**   Simplify

$$\frac{\dfrac{1}{x} - \dfrac{x}{x+1}}{1 + \dfrac{1}{x}}.$$

Solution (Method II)   The denominators in the numerator of the original fraction are x and $x+1$, and the denominator in the denominator of the original fraction is simply x. The least common multiple of these so-called *secondary denominators* is $x(x+1)$. We now multiply the original fraction by 1 written in the form $x(x+1)/x(x+1)$ and then simplify the result:

$$\frac{\dfrac{1}{x} - \dfrac{x}{x+1}}{1 + \dfrac{1}{x}} = \frac{\dfrac{1}{x} - \dfrac{x}{x+1}}{1 + \dfrac{1}{x}} \; \frac{x(x+1)}{x(x+1)} = \frac{(x+1) - x^2}{x(x+1) + (x+1)}$$

$$= \frac{-x^2 + x + 1}{(x+1)(x+1)} = \frac{-x^2 + x + 1}{(x+1)^2}. \qquad \square$$

The techniques discussed above can often be applied to expressions containing negative exponents, as we see in the following example.

■ **EXAMPLE**   Simplify $(a^{-1} + b^{-1})^{-1}$.

Solution (Method I)   We first replace all negative exponents by the equivalent quotients and then use properties of fractions to simplify the resulting expression:

$$(a^{-1} + b^{-1})^{-1} = \frac{1}{a^{-1} + b^{-1}} = \frac{1}{\dfrac{1}{a} + \dfrac{1}{b}} = \frac{1}{\dfrac{b+a}{ab}} = \frac{ab}{b+a}. \qquad \square$$

Solution (Method II)   We replace all negative exponents by the equivalent quotients and then multiply by ab/ab, the least common multiple of the secondary denominators:

$$(a^{-1} + b^{-1})^{-1} = \frac{1}{a^{-1} + b^{-1}} = \frac{1}{\dfrac{1}{a} + \dfrac{1}{b}} \; \frac{ab}{ab} = \frac{ab}{b+a}. \qquad \square$$

Our final example in this section illustrates a common rational expression occurring in calculus.

■ **EXAMPLE**   Simplify

$$\frac{\dfrac{1}{x+h} - \dfrac{1}{x}}{h}.$$

Solution (Method I)   First we obtain a single rational expression for the numerator and then simplify the resulting fraction:

$$\frac{\dfrac{1}{x+h} - \dfrac{1}{x}}{h} = \frac{\dfrac{x - (x+h)}{(x+h)x}}{h} = \frac{\dfrac{-h}{x(x+h)}}{h} = \frac{-h}{x(x+h)h} = \frac{-1}{x(x+h)}. \qquad \square$$

Solution (Method II)   The least common multiple of the secondary denominators, $x+h$ and x, is $x(x+h)$. We multiply the numerator and denominator by this least common multiple and simplify:

$$\frac{\dfrac{1}{x+h}-\dfrac{1}{x}}{h} = \frac{\dfrac{1}{x+h}-\dfrac{1}{x}}{h}\,\frac{x(x+h)}{x(x+h)} = \frac{x-(x+h)}{hx(x+h)} = \frac{-h}{hx(x+h)} = \frac{-1}{x(x+h)}\,. \qquad \square$$

13 │ Even and Odd Functions │

As discussed in the text, a function f is even if it is symmetric with respect to the y-axis, and is odd if it is symmetric with respect to the origin. Using the tests for symmetry given in Section 2.2 of the text, we have the following:

Tests for Even and Odd Functions

(i)   **Even Function:**
   A function $f(x)$ is even if $f(-x) = f(x)$ for every x in the domain of f.

(ii)   **Odd Function:**
   A function $f(x)$ is odd if $f(-x) = -f(x)$ for every x in the domain of f.

It should be easy to convince yourself graphically that the only function f that is both even and odd is $f(x) = 0$. It is also easy to verify this analytically:

Suppose that $f(x)$ is both even, so that $f(-x) = f(x)$, and odd, so that $f(-x) = -f(x)$. Then $f(x) = -f(x)$, since both are equal to $f(-x)$, and, adding $f(x)$ to both sides of the last equation, we have $2f(x) = 0$ or $f(x) = 0$.

14 │ Graphing a Quadratic Function by Hand │

In Exercises 2.4 of the text you are asked in several of the problems to sketch the graph of a quadratic function. As discussed in the text, the graph is a parabola and it is sufficient to plot only two points on the graph, providing one of them is the vertex of the parabola. Obtaining an accurate graph, however, is more difficult when there are no x-intercepts and the y-intercept is very close to the vertex. The example below illustrates a method for dealing with a situation such as this.

■ EXAMPLE   Find the intercepts and vertex of the parabola determined by

$$f(x) = 8x^2 + 4x + \frac{5}{2}\,.$$

Solution To find the x-intercepts we first determine if $f(x)$ factors easily. We begin by factoring out $\frac{1}{2}$ and noting that a factorization of $f(x)$ must have the form

$$f(x) = \frac{1}{2}(16x^2 + 8x + 5) = \frac{1}{2}(\underline{}x + 1)(\underline{}x + 5).$$

The blanks in the above "factorization" must be one of the pairs 4, 4; 2, 8; 8, 2; 16, 1; or 1, 16. (Note that it is not necessary to consider negative coefficients. Why?) Since none of these pairs works, the expression for $f(x)$ does not easily factor—that is, with integer coefficients after the $\frac{1}{2}$ has been factored out. We next set $f(x) = 8x^2 + 4x + \frac{5}{2} = 0$ and use the quadratic formula. This gives

$$x = \frac{-4 \pm \sqrt{16 - 4(8)(\frac{5}{2})}}{2(8)} = \frac{-4 \pm \sqrt{16 - 80}}{16},$$

which are not real numbers. Thus, the graph of $f(x)$ has no x-intercepts.

To find the y-intercept, *which a quadratic function always has*, compute $f(0) = \frac{5}{2}$.

To find the vertex of the parabola we obtain the standard form of the equation by completing the square after first factoring 8 from the two terms containing x:

$$f(x) = 8x^2 + 4x + \frac{5}{2}$$

$$= 8\left(x^2 + \frac{1}{2}x + \phantom{\frac{1}{16}}\right) + \frac{5}{2}$$

$$= 8\left(x^2 + \frac{1}{2}x + \frac{1}{16}\right) + \frac{5}{2} - 8\left(\frac{1}{16}\right)$$

$$= 8\left(x + \frac{1}{4}\right)^2 + 2.$$

In this case, the vertex, $(-\frac{1}{4}, 2)$, and the y-intercept, $(0, \frac{5}{2})$, are close together, making it difficult to sketch an accurate graph by hand. To solve this problem, we simply choose a value for x that is farther away from $x = -\frac{1}{4}$, say $x = 1$, and plot the corresponding point on the graph of $f(x)$. Using $f(1) = 8(1)^2 + 4(1) + \frac{5}{2} = 14.5$ we see that the graph also passes through the point $(1, 14.5)$. (Note that it must also then pass through $(-\frac{1}{4} - \frac{5}{4}, 14.5) = (-\frac{3}{2}, 14.5)$. Why?) The graph of $f(x)$ is shown below to the right of a plot of the vertex and y-intercept.

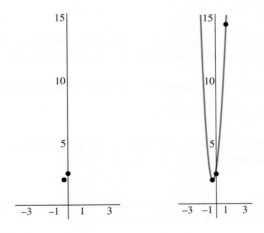

See Section 12 in Part II, **Use of a Calculator**, in this manual for a further discussion on graphing quadratic functions.

15 Domain of a Composition

As discussed in Section 2.6 of this text, the domain of the composition of two functions f and g is the set of all numbers x in the domain of g such that $g(x)$ is in the domain of f. This is somewhat complicated to apply and it is tempting to find the domain of $f \circ g$ by computing a formula for $(f \circ g)(x)$ and then using this formula to determine the domain of $(f \circ g)(x)$. *Be careful*; this method does not always work, as shown in the example below.

■ EXAMPLE   Find the domain of $f \circ g$, given

$$f(x) = \frac{3x+2}{x+1} \quad \text{and} \quad g(x) = \frac{x}{x-3}.$$

Solution The domain of g is $\{x \mid x \neq 3\}$ so the domain of $f \circ g$ cannot contain $x = 3$. Now, we must also have $g(x)$ in the domain of f. Since the domain of f is $\{x \mid x \neq -1\}$, we must have $g(x) = x/(x-3) \neq -1$. Solving $x/(x-3) = -1$ we have

$$\frac{x}{x-3} = -1$$
$$x = -1(x-3) = -x + 3$$
$$2x = 3$$
$$x = \frac{3}{2}.$$

Thus, $x = \frac{3}{2}$ is not in the domain of $f \circ g$. The domain of $f \circ g$ is then $\{x \mid x \neq 3, \frac{3}{2}\}$, or in interval notation, $(-\infty, \frac{3}{2}) \cup (\frac{3}{2}, 3) \cup (3, \infty)$. □

In this example we could have computed $(f \circ g)(x)$:

$$(f \circ g)(x) = f\big(g(x)\big) = f\left(\frac{x}{x-3}\right) = \frac{3\left(\dfrac{x}{x-3}\right) + 2}{\dfrac{x}{x-3} + 1} = \frac{3\left(\dfrac{x}{x-3}\right) + 2}{\dfrac{x}{x-3} + 1}\left(\dfrac{x-3}{x-3}\right)$$

$$= \frac{3x + 2(x-3)}{x + (x-3)} = \frac{5x - 6}{2x - 3}.$$

The formula for $(f \circ g)(x)$ is $(5x-6)/(2x-3)$, which excludes only $x = \frac{3}{2}$. However, as we saw in the example above, the domain of $f \circ g$ must also exclude $x = 3$. This fact can be lost when using the formula for $f \circ g$ to find the domain of $f \circ g$.

The preceding discussion suggests an alternative method for finding the domain of the composition of two functions.

> The domain of the composition $f \circ g$ of two functions, f and g, is the intersection of the domains of the formulas for $g(x)$ and $(f \circ g)(x)$.

This method is especially appropriate when a formula for $f \circ g$ must already be found as part of an exercise. We illustrate this technique with the following example.

■ EXAMPLE   Find the domain of $f \circ g$, given

$$f(x) = \frac{1}{x^2 - 1} \quad \text{and} \quad g(x) = \sqrt{x+3}.$$

Solution The domain of g is $\{x \mid x + 3 \geq 0\} = \{x \mid x \geq -3\}$. We compute

$$(f \circ g)(x) = f\big(g(x)\big) = f\big(\sqrt{x+3}\big) = \frac{1}{\big(\sqrt{x+3}\big)^2 - 1} = \frac{1}{x + 3 - 1} = \frac{1}{x + 2}.$$

The domain of the formula for $f \circ g$ is $\{x \mid x \neq -2\}$, so the domain of $f \circ g$ is

$$\{x \mid x \geq -3\} \cap \{x \mid x \neq -2\} = [-3, -2) \cup (-2, \infty).$$ $\square$

16 Synthetic Division

Polynomial long division, when the divisor has the form $x - c$, for any real number c, can be simplified using a technique called **synthetic division**. For example, when $f(x) = 3x^3 + 5x^2 - 4x + 2$ is divided by $d(x) = x + 2$ we see by regular long division

$$
\begin{array}{r}
\boxed{3}\,x^2 \;\boxed{-1}\,x \;\boxed{-2} \\
x - \boxed{-2}\,\overline{)\;\boxed{3}\,x^3 + \boxed{5}\,x^2\;\boxed{-4}\,x + \boxed{2}} \\
\underline{3\,x^3 + 6x^2} \\
\boxed{-1}\,x^2 - 4\,x + 2 \\
\underline{-1\,x^2 - 2\,x} \\
\boxed{-2}\,x + 2 \\
\underline{-2\,x - 4} \\
\boxed{6}
\end{array}
$$

that the quotient is $q(x) = 3x^2 - x - 2$ and the remainder is $r = 6$. We see in this process that much of the writing is extraneous and all of the pertinent information is contained in the boxed numbers, with various styles of boxes used to distinguish the numbers involved in the divisor $\boxed{c}$, the dividend $\boxed{d}$, the quotient $\boxed{q}$, and the remainder $\boxed{r}$. The procedure of synthetic division for dividing $f(x)$, a polynomial of degree $n > 0$, by $x - c$ is summarized as follows.

Method of Synthetic Division

(i) Write c followed by the coefficients of $f(x)$. Be sure to include any coefficients of $f(x)$ that are 0, including the constant term. Note that the form of the divisor is $x - c$, so that if $f(x)$ is divided by $x + 3$, for example, the value of c is -3.

(ii) Bring down the first coefficient of $f(x)$ to the third row.

(iii) Multiply this number by c and write the product directly under the second coefficient of $f(x)$. Then add the two numbers in this column and write the sum beneath them in the third row.

(iv) Multiply this sum by c and write the product in the second row of the next column. Then add the two numbers in this column and write the sum beneath them in the third row.

(v) Repeat the preceding step as many times as necessary.

(vi) The last number in the third row is the constant remainder r; the numbers preceding it in the third row are the coefficients of $q(x)$, the quotient polynomial of degree $n - 1$.

EXAMPLE   Use synthetic division to find the quotient and remainder when

$$f(x) = 3x^3 + 5x^2 - 4x + 2 \quad \text{is divided by} \quad d(x) = x + 2.$$

Solution Identifying $c = -2$ we have

$$
\begin{array}{r|rrrr}
-2 & 3 & 5 & -4 & 2 \\
   &   & -6 & 2 & 4 \\
\hline
   & 3 & -1 & -2 & \boxed{6 = r}
\end{array}
$$

The coefficients of the quotient are the first three numbers in the third row, so $q(x) = 3x^2 - x - 2$. The remainder is the far right entry in the third row, $r = 6$. □

   As mentioned in the text, one of the applications of synthetic division is the evaluation of a polynomial $f(x)$ at a number c. As the following example shows, this is not always the most effective way to compute $f(c)$.

■ **EXAMPLE** Evaluate $f(-1)$ if $f(x) = 2x^5 + 3x - 4$.

Solution Since $f(-1)$ is the remainder when $f(x) = 2x^5 + 3x - 4$ is divided by $x + 1$, its value could be found using synthetic division by $x + 1$. In this case, however, it is easier to simply plug -1 into the formula for $f(x)$:

$$
f(-1) = 2(-1)^5 + 3(-1) - 4 = -2 - 3 - 4 = -9.
$$

Contrasting this with the use of synthetic division

$$
\begin{array}{r|rrrrrr}
-1 & 2 & 0 & 0 & 0 & 3 & -4 \\
   &   & -2 & 2 & -2 & 2 & -5 \\
\hline
   & 2 & -2 & 2 & -2 & 5 & \boxed{-9 = r}
\end{array}
$$

which also gives $f(-1) = -9$ (of course), we see that direct evaluation of the function at $x = -1$ involves fewer computations. □

17 | Complex Numbers

Complex numbers naturally arise when considering the solutions of polynomial equations. For example, the simple polynomial equation $x^2 + 1 = 0$ has no real zeros because there is no real number whose square is -1. However, if we define a new number, denoted by i, having the property that $i^2 = -1$, or $i = \sqrt{-1}$, we then see that $x = i$ is a zero of $x^2 + 1$. In general, any expression of the form $z = a + bi$, where a and b are real numbers and $i^2 = -1$, is called a **complex number**. The numbers a and b are called the **real** and **imaginary** parts of z, respectively. A complex number of the form $0 + bi$, $b \neq 0$, is said to be a **pure imaginary number**. Choosing $b = 0$ in $z = a + bi$ we see that every real number is a complex number. Thus, the set of real numbers is a subset of the set of complex numbers.

■ **EXAMPLE**

(a) The complex number $z = 4 - 5i = 4 + (-5)i$ has real part 4 and imaginary part -5.
(b) $z = 10i = 0 + 10i$ is a pure imaginary number.
(c) $z = -6 + 0i = -6$ is a real number. (It is also a complex number.) □

17.1 | Equality

To define equality of two complex numbers we require that their real parts be equal and their imaginary parts be equal. That is, if $z_1 = a + bi$ and $z_2 = c + di$, then $z_1 = z_2$ if and only if $a = c$ and $b = d$.

17.2   Addition and Subtraction

To add and subtract complex numbers it is helpful to think of them as linear polynomials in the variable i. The complex numbers are then added or subtracted as though they were polynomials.

EXAMPLE   If $z_1 = 5 - 6i$ and $z_2 = 2 + 4i$, then find $z_1 + z_2$ and $z_1 - z_2$.

Solution Thinking of z_1 and z_2 as polynomials and combining like terms we have

$$z_1 + z_2 = (5 - 6i) + (2 + 4i) = (5 + 2) + (-6 + 4)i = 7 - 2i$$

and

$$z_1 - z_2 = (5 - 6i) - (2 + 4i) = (5 - 2) + (-6 - 4)i = 3 - 10i. \qquad \square$$

We formally define addition and subtraction of $z_1 = a + bi$ and $z_2 = c + di$ by

$$z_1 + z_2 = (a + bi) + (c + di) = (a + c) + (b + d)i$$

and

$$z_1 - z_2 = (a + bi) - (c + di) = (a - c) + (b - d)i.$$

17.3   Complex Conjugate

Consider the quadratic equation $x^2 - 8x + 25 = 0$. Since the left-hand side of this equation does not factor with integer coefficients, we use the quadratic formula and the fact that $\sqrt{-1} = i$ to solve the equation:

$$x = \frac{-(-8) \pm \sqrt{(-8)^2 - 4(1)(25)}}{2(1)} = \frac{8 \pm \sqrt{64 - 100}}{2}$$

$$= \frac{8 \pm \sqrt{-36}}{2} = \frac{8 \pm \sqrt{36}\sqrt{-1}}{2} = \frac{8 \pm 6i}{2} = 4 \pm 3i.$$

In general, whenever the use of the quadratic formula to solve a quadratic equation with real coefficients leads to an expression involving $\sqrt{-1} = i$, say $a + bi$, a second solution will always be $a - bi$. We say that $a - bi$ is the **complex conjugate** of $z = a + bi$ and denote it by $\bar{z} = \overline{a + bi}$. For example, the complex conjugate, or simply **conjugate**, of $z = 2 - 5i$ is

$$\bar{z} = \overline{2 - 5i} = 2 - (-5)i = 2 + 5i.$$

As indicated by the Conjugate Zeros Theorem in the text, if $f(x)$ is a polynomial function of degree $n > 1$, and z is a complex zero of $f(x)$, then the conjugate $\bar{z}$ is also a zero of $f(x)$. This theorem is illustrated in the solution above of the quadratic equation $x^2 - 8x + 25 = 0$. (This is the equation that is solved to find the zeros of the quadratic polynomial function $f(x) = x^2 - 8x + 25$.)

17.4   Multiplication and Division

As with addition and subtraction of complex numbers, a formula for the product of two complex numbers $z_1 = a + bi$ and $z_2 = c + di$ can be found by treating the complex numbers as linear polynomials in i and multiplying them using $i^2 = -1$. Rather than memorizing a formula for this, it is better to simply carry out this process each time using the specific complex numbers.

EXAMPLE   If $z_1 = 5 - 6i$ and $z_2 = 2 + 4i$, find the product $z_1 z_2$.

Solution We use the distributive, commutative, and associative properties. Then

$$z_1 z_2 = (5 - 6i)(2 + 4i) = 5(2) + [5(4) + (-6)(2)]i + (-6)(4)i^2$$

$$= 10 + (20 - 12)i - 24(-1) = 34 + 8i. \qquad \square$$

To divide complex numbers and represent the quotient in the form $a+bi$ we use the fact that the product of a complex number and its conjugate is a real number. To see this let $z = a + bi$ and multiply:

$$z\bar{z} = (a + bi)(a - bi) = a^2 - (bi)^2 = a^2 - b^2 i^2 = a^2 - b^2(-1) = a^2 + b^2.$$

The following example illustrates how to find the quotient of two complex numbers.

■ EXAMPLE   If $z_1 = 3 - 2i$ and $z_2 = 4 + 5i$, find z_1/z_2.

Solution To compute the quotient we multiply both the numerator and denominator of z_1/z_2 by the conjugate of z_2:

$$\frac{z_1}{z_2} = \frac{3 - 2i}{4 + 5i} = \frac{3 - 2i}{4 + 5i}\left(\frac{4 - 5i}{4 - 5i}\right) = \frac{12 + [3(-5) + (-2)(4)]i + (-2)(-5)i^2}{4^2 + 5^2}$$

$$= \frac{12 - 23i + 10(-1)}{16 + 25} = \frac{2 - 23i}{41} = \frac{2}{41} - \frac{23}{41}i. \qquad \square$$

18 | Checking Your Answer

When solving a problem, especially one that involves a numerical solution, it is *always* a good idea to check your result. This can generally be done in a variety of ways, a few of which are discussed below.

Rework the Problem. This is perhaps the most obvious way to check your result. Two fairly obvious drawbacks, however, are that it can be time consuming, and that an error made the first time through the problem may well be repeated the second time.

Reasonableness. Is the answer at least about what you expected it to be; does it make sense as a solution? For example, if you want to find the x-intercepts of the graph of $y = 2x^2 - x - 7$ and you conclude from the quadratic formula that there are none, you can check this result by noting that the discriminant, $(-1)^2 - 4(2)(-7) = 57$, is positive and therefore the equation $y = 2x^2 - x - 7$ has real solutions. Thus, the graph of $y = 2x^2 - x - 7$ must, in fact, have two distinct x-intercepts, and your original contention that there are no x-intercepts must be incorrect.

Another example of an unreasonable answer is a very small or very large numerical result when you expect the answer to be something between, say, 20 and 100. Or maybe you get a negative result when you know that the answer must be positive. Sometimes a calculator can be used to convert an exact value like $1/(\sqrt{95} - 3\pi)$ into a decimal approximation that can easily be seen to be positive or negative, or large or small. While the reasonableness criterion will not give you any idea of what the actual answer is, it is very quick and easy to apply.

Use a Calculator to Draw a Graph. The problem may not require it, but it is generally easy enough to obtain the graph of a function and at least determine if your answer is close to the one indicated by the graph.

For example, consider the trigonometric problem of finding all values of x in the interval $[0, 2\pi]$ for which $\sin x = \cot x$. We use $\cot x = \cos x/\sin x$, the Pythagorean identity $\sin^2 x + \cos^2 x = 1$, and solve for $\cos x$:

$$\sin x = \cot x = \frac{\cos x}{\sin x}$$

$$\sin^2 x = \cos x$$

$$1 - \cos^2 x = \cos x$$

$$\cos^2 x + \cos x - 1 = 0$$

$$\cos x = \frac{-1 \pm \sqrt{1 + 4}}{2} = -\frac{1}{2} \pm \frac{1}{2}\sqrt{5}.$$

Since $-\frac{1}{2} - \frac{1}{2}\sqrt{5} < -1$, we have only $\cos x = -\frac{1}{2} + \frac{1}{2}\sqrt{5}$. Since $-\frac{1}{2} + \frac{1}{2}\sqrt{5} > 0$ and $\cos x$ is positive in the first and fourth quadrants,

$$x = \cos^{-1}\left(\frac{1}{2}\sqrt{5} - \frac{1}{2}\right) \quad \text{or} \quad x = 2\pi - \cos^{-1}\left(\frac{1}{2}\sqrt{5} - \frac{1}{2}\right).$$

Using a calculator or computer, we find $x = 0.904557$ and $x = 5.37863$, respectively. These results can be checked by plotting $y = \sin x$ and $y = \cot x$ on $[0, 2\pi]$ and observing that they do in fact intersect at about $x = 0.9$ and $x = 5.4$. (See the figure below.) While this does not absolutely guarantee that our answer is exactly correct, it does show that at least we are very close.

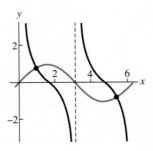

19 | Extraneous Roots of Logarithmic Equations

When solving a logarithmic equation, it is possible to obtain an extraneous solution. This means that it is essential that you check your answers when solving an equation involving logarithms.

■ EXAMPLE   Solve $\ln x + \ln(x + 3) = \ln(2x)$.

Solution   Using the properties of logarithms and the fact that $\ln x$ is a one-to-one function, we have
$$\ln x + \ln(x + 3) = \ln 2x$$
$$\ln x(x + 3) = \ln 2x$$
$$x(x + 3) = 2x$$
$$x^2 + 3x = 2x$$
$$x^2 + x = 0$$
$$x(x + 1) = 0.$$

Thus, $x = 0$ or $x = -1$. When we check this result, however, we note that the original equation contains $\ln x$, so x must be positive. Thus, neither of our "solutions" checks, and we conclude that the equation has no solutions.   □

Part II

Use of a Calculator

1 Introduction

This part is not intended to be a comprehensive manual on the use of a graphing calculator in a precalculus course. While much of the material discussed will be pertinent to any graphing calculator, the references in this manual will be to the TI-84 family of calculators. A number of students entering college already have a TI-83 calculator that they used in their high school mathematics classes. The TI-83 is similar to the TI-84 and most of what is discussed below applies to the TI-83 as well. After a few brief comments on the use of the TI-84 calculator, the focus in this manual will be on how to use the calculator to assist in the solution of a number of the problems in the text. While the TI-84 calculator contains many sophisticated routines that could be used, the focus in this manual will be on using techniques that require a minimum amount of expertise with the calculator. Before deciding on the extent to which you plan to use a calculator in your precalculus course, it is very important that you keep two things in mind:

1. Your instructor may not allow the use of calculators on quizzes or tests. If this is the case, you should definitely avoid becoming dependent on the calculator as a source of solutions to problems. At most, you should use the calculator as a way to check the plausibility of your answer to a problem.

2. Even if you are allowed free use of a calculator, it is important to keep in mind that many of the problems in the text are frequently designed to have "nice" solutions. That is, the answers will generally involve integers or relatively simple fractions. In actual practice, real-world problems will generally have solutions that are more complicated and can best be obtained and approximated with a calculator or computer. You should keep in mind that answers obtained on a calculator are generally only approximations to the exact answer, and that the best result is an exact solution that is then given an approximate (and usually more usable) value. To illustrate this consider a jet that is moving at twice the speed of sound. In this case we say that the jet has a velocity of Mach two. This value is related to the angle of the conical sound wave created by the jet moving at a high rate of speed through the atmosphere. This angle can then be used to find the Mach number of the plane. In a problem in the text you are asked to find the Mach number corresponding to an angle of $30°$. The exact value of the corresponding Mach number is $2/\sqrt{2-\sqrt{3}}$. This value gives little feeling for the actual speed of the jet. However, after using a calculator to convert this number to the approximate value 3.86, it is easily seen that the jet is traveling at nearly four times the speed of sound; a much more intuitively appealing result.

2 Mode Settings

When you first turn on the calculator it is a good idea to check the settings in the calculator. To do this press the MODE key to see something like

Generally, you will want to use the default settings that are highlighted in the screenshot shown above. Note especially the RADIAN and DEGREE settings. These are particularly important in Chapters 4 and 5 of the text.

3 Clearing the Calculator Screen

To clear the screen when you are not in a graphics mode simply position the cursor on a blank line and press the CLEAR key. If the cursor is not on a blank line, pressing the CLEAR key will generally erase only that line. In this case, if there are still entries on the screen and you want to erase them, press the CLEAR key again.

When in the Y= window, to clear a function definition, position the cursor at the end of the function and press the CLEAR key.

Suppose a graphics object, like a line connecting $(1, 3)$ and $(-5, -4)$, has been entered using the DRAW menu. Then, later, you want to graph the function $y = x^2$. In this case you are likely to see the following screen:

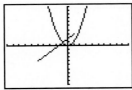

You probably didn't expect to see the line in the graph. To clear the line, press the 2ND DRAW keys and select the first option, 1 : ClrDraw.

When the CLEAR key doesn't work, try using the 2ND QUIT keys to clear a graphics screen.

4 Basic Calculations and Memory

In the TI-84 calculator, expressions like $(2\pi - 3)(5 + \sqrt{2})^3$ can be computed as shown:

If the result of this operation is to be used again, do *not* write the number 866.4143622 on a piece of paper and then key it in (or even worse, a rounded version of it like 866.41) at the appropriate spot. For example, if you later want to find the reciprocal of the square root of this number, then you should store it in a memory register and call it up when needed. This process is demonstrated below.

The fourth line in the window, which stores the result of the calculation in the variable B, is obtained by pressing the STO   ALPHA   B keys. To clear a value assigned to a variable, simply store the number 0 in that variable.

5   Functions

To enter a function that is to be evaluated or graphed, press the Y= key and enter the function (or functions) that is (are) to be used. In the screenshot below, the function $y = x^3 - 2x + 4$ is entered.

5.1   Function Evaluation

If a function has been entered as Y1, then it can be evaluated at any number, say $x = -3.2$, using the 2ND   TBLSET and 2ND   TABLE keys. The following sequence of screenshots demonstrates how to do this.

 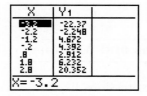

In this case, we see that when $x = -3.2$, $x^3 - 2x + 4$ equals -22.37. When entering a negative number like -3.2 be sure to use the (-) key in the bottom row on the calculator, as opposed to the subtraction key − just to the right of the 6 key. This is a very common mistake, and should generally be the first possibility you consider whenever the calculator gives you an error message. Also, once you have input the number in the **TABLESETUP** screen, be sure to press the ENTER key.

5.2 |Graphing a Function|

To graph Y1 use the |WINDOW| key to set the viewing rectangle. For example, if you set Xmin=-10, Xmax=10, Ymin=-10, and Ymax=10, and then press the |GRAPH| key, you obtain the following two screens:

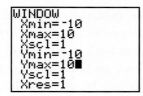 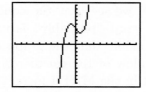

To change the viewing rectangle simply press the |WINDOW| key and reset the boundaries of the window. Based on the screen shown above, you might want to reset Xmin to be -4 and Xmax to be 3. In this case the following screen is obtained:

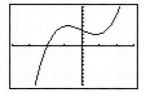

5.3 |An Alternative Method of Function Evaluation|

An alternative way to evaluate a function that has been entered in Y1 uses the |2ND| |CALC| key combination. Suppose you want to evaluate $f(-3.4)$ when

$$ f(x) = \frac{\sqrt{x^2 - 3}}{x + 4} . $$

First, enter this function as Y1 in the |Y=| window. Next, be sure that the graphing window includes the value -3.4. Then, in the |2ND| |CALC| window, choose 1 : value. Enter the value for x, in this case -3.4, and press |ENTER|. (Did you remember to use the |(-)| key instead of the subtraction key?) In the lower right-hand portion of the screen you will see the value of $f(x)$; in this case Y = 4.8762463. You should see something like the following sequence of screens:

 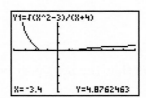

6 |Solution of an Equation|

To find the solution or solutions of an equation such as $x^3 - x^2 - 12x + 7 = 0$, first graph the function to get a sense of how many solutions there are and approximately where they are. The graph of Y2 = $x^3 - x^2 - 12x + 7 = 0$ is shown below with Xmin $= -10$, Xmax $= 10$, Ymin $= -10$, and Ymax $= 10$.

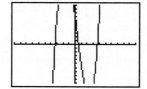

From the graph above, we see that the equation probably has three solutions near $x = -3$, $x = 1$, and $x = 4$. The word "probably" is used because, without knowledge of the behavior of cubic functions, it is possible that the graph comes back down (or up on the left) and crosses the x-axis outside the window of the calculator. Now, to find the solutions of the equation, use the EQUATION SOLVER in the MATH menu. This can be accessed by pressing the $\boxed{\text{MATH}}$ key followed by the $\boxed{0}$ key. Where the screen reads eqn : 0 = type in the equation. (If the screen does not say EQUATIONSOLVER at the top, press the up cursor.) You should see the following screen:

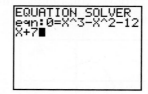

Now press $\boxed{\text{ENTER}}$ and assign a value to X that is close to a solution. For example, to find the smallest positive solution of the equation in this example you could assign x = 1 followed by $\boxed{\text{ALPHA}}$ $\boxed{\text{SOLVE}}$. You should see the following sequence of two screens:

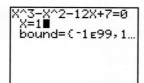 

The smallest positive solution is $x = .57166826799$ rounded to 10 decimal places. To find the other positive solution press $\boxed{4}$ followed by $\boxed{\text{SOLVE}}$.

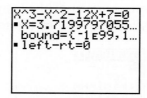

This shows that the next positive root is $x = 3.7199797055$.

7 $\boxed{\text{Solution of an Inequality}}$

To find the solution set of an inequality such as $x^2 + 3x - 10 \geq 0$, first graph Y1 $= x^2 + 3x - 10$ and then solve the equation $x^2 + 3x - 10 = 0$ as described in **6: Solution of an Equation** on the previous page. This gives $x = -5$ and $x = 2$. Now, pressing the $\boxed{\text{GRAPH}}$ key again shows that $x^2 + 3x - 10 \geq 0$ (that is, lies above the x-axis) when $x \leq -5$ and $x \geq 2$. You should see the following sequence of calculator screens:

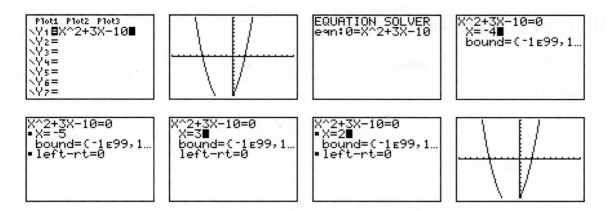

While the above technique could be used to approximate the solution of an inequality, it should only be used to check the answer that you arrived at using the sign chart technique described in the text.

8 Solution of an Inequality Involving an Absolute Value

To find an approximation to the solution of an inequality involving an absolute value such as $|2x - 8| \geq 12$, first write the inequality as $|2x - 8| - 12 \geq 0$. Then graph $\texttt{Y1} = \texttt{abs}(2x - 8) - 12$ and find the solutions of $|2x - 8| = 12$ by estimating where the graph intersects the x-axis. The solution set of the inequality will then be the union of the intervals where the graph is above the x-axis.

9 Graphing a Circle

Option 9 in the DRAW menu is used to draw a circle. In this case, three inputs are needed—the first two are the coordinates of the center and the third is the radius of the circle. For example, to draw a circle centered at $(1, 2)$ with radius 3 use the following sequence of keystrokes: $\boxed{\text{2ND}}$ $\boxed{\text{DRAW}}$ $\boxed{9}$ $\boxed{1}$ $\boxed{2}$ $\boxed{3}$. You will probably see a screen that looks something like the following:

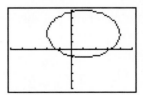

To make this circle actually look like a circle rather than an oval, use the $\boxed{\text{ZOOM}}$ key followed by $5 : \texttt{ZSquare}$. To see the resulting circle you may need to press the $\boxed{\text{CLEAR}}$ key followed by the $\boxed{\text{ENTER}}$ key. The screen should now show the following:

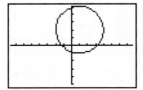

10 Limit of a Function

While you should generally not use your calculator to find the limit of a function, it can be useful to check that the limit you have obtained using hand computations is plausible. For example, suppose you want to find

$$\lim_{x \to -1} \frac{2x + 2}{x^2 - x - 2}.$$

By hand, you compute

$$\frac{2x + 2}{x^2 - x - 2} = \frac{2(x + 1)}{(x - 2)(x + 1)} = \frac{2}{x - 2}, \quad x \neq -1,$$

so that

$$\lim_{x \to -1} \frac{2x + 2}{x^2 - x - 2} = \lim_{x \to -1} \frac{2}{x - 2} = \frac{2}{-1 - 2} = -\frac{2}{3}.$$

There are two ways you can use your calculator to check the plausibility of this result. First, graph the function

$$Y1 = \frac{2x + 2}{x^2 - x - 2},$$

and then use the $\boxed{\text{TRACE}}$ key to position the cursor over $x = -1$ (or as close as you can get to $x = -1$). The corresponding y-coordinate will be displayed in the lower right-hand portion of the screen, as shown below:

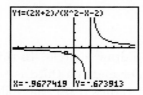

Since $Y = -.673913$ is close to $-\frac{2}{3}$, you have reason to believe that your hand computations are correct.

Alternatively, you could evaluate the function (as discussed in Section 5.1 of this part of the manual)

$$Y1 = \frac{2x + 2}{x^2 - x - 2},$$

at a value very close to $x = -1$, say, at $x = -1.01$. In this case, you will see a screen like the one below:

X	Y1	
-1.01	-.6645	
-.01	-.995	
.99	-1.98	
1.99	-200	
2.99	2.0202	
3.99	1.005	
4.99	.6689	
X=-1.01		

In this case, when $x = -1.01$ the corresponding y value is $-.6645$. Since this number is close to $-\frac{2}{3}$, you have reason to believe that your hand computations are correct.

11 Factorials

Factorials play a very important role in probability, which explains its location in the MATH window of the calculator. To find, say 6!, press $\boxed{6}$ $\boxed{\text{MATH}}$. You will see the following screen:

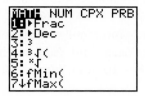

Now press the right cursor three times to highlight PRB followed by $\boxed{4}$ $\boxed{\text{ENTER}}$. This will give the value of 6!.

To test your ability with factorials on the calculator, you might want to find values of n for which the calculator switches to scientific notation, and values for which it overflows. You should also determine what the calculator thinks 0! is, and what it thinks of $(-4)!$. In the latter case, be sure to include the parentheses around the -4 before trying to find the factorial, and remember to use the $\boxed{(\text{-})}$ key rather than the subtraction key $\boxed{-}$.

12 Graphing Quadratic Functions

As discussed in Section 2.4 of the text, the graph of a quadratic function is always a parabola opening upward or downward. When the coefficients of the function are integers or simple fractions you should be able to sketch a graph of the function by hand. In this case, a calculator may be used to confirm the results of your work. However, when the function does not have simple coefficients, it can be very tedious to find the vertex and the intercepts by hand.

Suppose, for example, that you want to approximate the intercepts and vertex of the parabola determined by $f(x) = 2.83x^2 - 4.32x - 31.09$. To get a sense of the intercepts and vertex of the parabola you could graph the function. To do this, enter the function as Y1 using the $\boxed{\text{Y=}}$ key and make sure that all other functions, Y1, Y2, Y3, . . . , are clear. Then press the $\boxed{\text{GRAPH}}$ key. Depending on the settings in the $\boxed{\text{WINDOW}}$ screen you will see a screen like

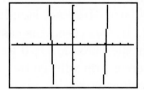

Knowing that the graph is a parabola, you can zoom in on the x-intercepts to determine their locations with reasonable accuracy. To do this, first press the $\boxed{\text{TRACE}}$ key and then use the left and right cursor keys, $\boxed{\blacktriangleleft}$ and $\boxed{\blacktriangleright}$, to place the cursor near an x-intercept. (Initially, the cursor may be off the screen, but its x- and y-coordinates are shown at the bottom of the screen, so you can move it into the viewing window.) Having done this, you should see a screen like the following:

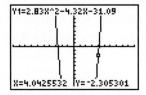

Now, press the $\boxed{\text{ZOOM}}$ $\boxed{2}$ $\boxed{\text{ENTER}}$ key sequence to zoom in on the part of the graph around the cursor. At this point, the cursor is no longer attached to the graph and can be moved using any of the four cursor keys, making it easier to position the cursor more directly on top of the x-intercept. After doing this, press the $\boxed{\text{ZOOM}}$ $\boxed{2}$ $\boxed{\text{ENTER}}$ key sequence to zoom in again. Successively repositioning the cursor and noting the x-values (the y-value should remain 0), you soon see that, to two decimal places, this x-intercept is 4.16.

To zoom back out to find the other x-intercept you can either use the $\boxed{\text{ZOOM}}$ $\boxed{3}$ $\boxed{\text{ENTER}}$ key sequence, or you can reset Xmin, Xmax, Ymin, and Ymax in the $\boxed{\text{WINDOW}}$ screen. For example, if you set Xmin $= -10$, Xmax $= 10$, Ymin $= -10$, and Ymax $= 10$ and then press the $\boxed{\text{GRAPH}}$ key you will see the following screen:

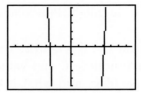

Repeating the process above to find the other x-intercept, you should determine that it is approximately $x = 2.637$. (If in the above process you lose the y-axis from the screen, it can always be recovered by resetting the Ymin and Ymax values in the $\boxed{\text{WINDOW}}$ menu.)

Notice in the picture above (where Xmin $= -10$, Xmax $= 10$, Ymin $= -10$, and Ymax $= 10$) that both the vertex and the y-intercept of the parabola are below the bottom of the calculator screen. To get a more complete view of the parabola, you could change Ymin using the $\boxed{\text{WINDOW}}$ menu to, say, Ymin $= -20$. This is better, but still not enough, so try again. Eventually, Ymin $= -40$ will work and you should see the following screen:

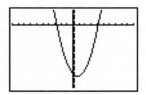

At this point, you can see the y-intercept and could approximate its coordinates using the $\boxed{\text{TRACE}}$ and $\boxed{\text{ZOOM}}$ keys as above when approximating the x-coordinates, but this is an overuse of the graphing capabilities of the calculator. Instead, simply use one of the techniques described above in Section 5.1 or 5.3 to find $f(0)$. In this case, it is easier to use Section 5.3, so in the $\boxed{\text{2ND}}$ $\boxed{\text{CALC}}$ window choose 1 : value. Enter the value for x, 0 for the y-intercept, and press $\boxed{\text{ENTER}}$. In the lower right-hand portion of the screen you will see the value of $f(x)$, in this case Y $= -31.09$. You should see something like the following screen:

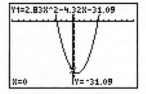

Next, to find the vertex of the parabola, you can use the $\boxed{\text{TRACE}}$ and $\boxed{\text{ZOOM}}$ keys as described above for finding the x-intercepts. You should find that the vertex is approximately at $(0.76, -32.74)$.

13 ⟦Constructing a Table⟧

The table-building capability of the calculator can be used to generate the data in a table. We illustrate this by considering part (d) of Problem 51 in Section 2.4 of the text. We begin by entering the function $R(D) = kD(P - D)$ as Y1 in the ⟦Y=⟧ window. To do this we let $k = 0.00003$, $P = 10,000$, and identify D with X. You should see the following screen:

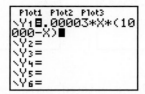

Next, use the ⟦2ND⟧ ⟦TBLSET⟧ key combination to change the Indpnt : setting from Auto to Ask. (The TblStart and ΔTbl setting do not matter.) You should now see the following window with Ask flashing after you press the ⟦ENTER⟧ key:

Now press the ⟦2ND⟧ ⟦TABLE⟧ key combination and enter 125 after X = at the bottom of the window, followed by the ⟦ENTER⟧ key. You should see 125 in the X column of the table and 37.031 in the Y1 column of the table. Round 37.031 to the nearest integer since it represents a number of persons, add it to 125, and enter the result 162 after X =. Press the ⟦ENTER⟧ key, and you should see

X	Y₁	
125	37.031	
162	47.813	

X=162

Continuing in this fashion, you will generate the following table containing the number of infected individuals in the Y1 column:

X	Y₁	
125	37.031	
162	47.813	
210	61.677	
272	79.38	
351	101.6	
453	129.74	

X=

14 | Converting a Decimal to a Fraction

Selecting 1:▶Frac in the MATH menu will sometimes convert a decimal to a fraction. The following screens illustrate this:

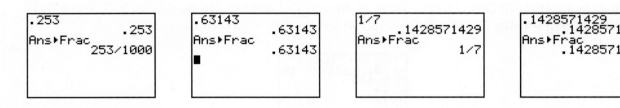

Notes: (1) After the decimal number is entered, press the MATH key, followed by 1 ENTER to obtain the fraction.

(2) As the second screen shows, not every decimal can be converted to a fraction. You can try experimenting with various decimals to see whether you can determine those that will be converted to fractions.

(3) If the decimal arises as the quotient of two integers, as shown above in the third screen, the decimal result *may* be converted back to the fraction. On the other hand, when the same decimal is entered directly, it *may not* be converted to a fraction. See the fourth screen above.

15 | Arithmetic of Complex Numbers

It is a relatively simple task to have the calculator perform the four basic arithmetic operations of addition, subtraction, multiplication, and division on complex numbers. To do this, you first need to put the calculator in rectangular complex mode. Press the MODE key, scroll down to **REAL**, and then scroll right to $a + bi$. Finally, press the ENTER key.

To find $(2 - 4i) - (1 + 3i)$, first clear the calculator screen (see Section 3 in this manual); then the following sequence of keystrokes will cause the calculator to display the difference of the two complex numbers: (2 − 4 2ND i) − (1 + 3 2ND i) ENTER.

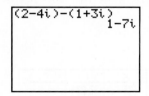

A quotient of complex numbers with integer real and imaginary parts will usually result in a complex number with fractional (meaning noninteger) real and imaginary parts. The calculator will display the resulting quotient as a complex number with decimal real and imaginary parts. In this case, decimal to fraction conversion (see Section 14 of this manual) can be used. The following screen illustrates this in the computation of $(5 + 7i)/(4 - 2i)$:

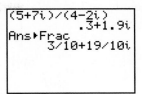

16 | Rational Zeros of a Polynomial with Integer Coefficients

We want to find the rational zeros of a polynomial $f(x) = a_n x^n + a_{n-1} x^{n-1} + \cdots + a_1 x + a_0$, where $n \geq 1$ and a_i is an integer for $0 \leq i \leq n$. As discussed in the text, all rational zeros of $f(x)$ must have the form p/s where p is an integer factor of a_0 and s is an integer factor of a_n. With the calculator it is a simple matter to directly check each possible rational zero. Using the $\boxed{\text{Y=}}$ key enter the function as Y1. In the screen below we are trying to find the rational zeros of $f(x) = 6x^4 + 11x^3 + 14x^2 - 7x - 6$.

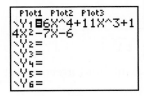

Now press the $\boxed{\text{TRACE}}$ key. The set of possible rational zeros is ± 1, $\pm \frac{1}{2}$, $\pm \frac{1}{3}$, $\pm \frac{1}{6}$, ± 2, $\pm \frac{2}{3}$, ± 3, $\pm \frac{3}{2}$, and ± 6. To test -1 simply press $\boxed{(-)}\,\boxed{1}\,\boxed{\text{ENTER}}$ and read Y = 10. Thus, $f(1) = 10$, so 1 is not a zero of $f(x)$. Next, press $\boxed{1}\,\boxed{\text{ENTER}}$ and read Y = 18. Continuing in this manner we find that $-\frac{1}{2}$ and $\frac{2}{3}$ are the rational zeros of $f(x)$. Some sample screens are shown below:

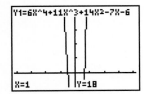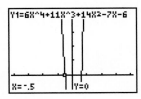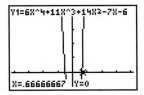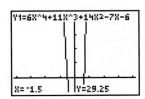

You should be aware that the technique described above is very mechanical and will easily find the rational zeros of $f(x)$. It will not, however, find any irrational zeros, nor will it find any nonlinear factors of $f(x)$. Most significantly, the technique will not contribute to your understanding of the concepts involved in the theory of rational zeros of polynomials.

17 | Calculation of the Other Trigonometric Functions

The keyboard of the calculator contains no keys for directly computing the cotangent, secant, cosecant, inverse cotangent, inverse secant, and inverse cosecant functions. To compute the cot, sec, and csc functions, we use the identities

$$\cot x = \frac{1}{\tan x}, \quad \sec x = \frac{1}{\cos x}, \quad \text{and} \quad \csc x = \frac{1}{\sin x}.$$

For example, to find $\cot 137°$, first be sure that the calculator is set to degree mode by pressing $\boxed{\text{MODE}}$, then scroll down to DEGREE and press $\boxed{\text{ENTER}}$ to select degree mode:

After using $\boxed{\text{CLEAR}}$ to return to the main window, use the keystrokes $\boxed{\text{TAN}}$ $\boxed{1}$ $\boxed{3}$ $\boxed{7}$ $\boxed{)}$ $\boxed{\text{X}^{-1}}$ to see that $\cot 137° = -1.07236871$. The screen output is shown below:

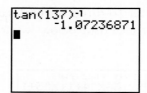

Alternatively, you could use $\boxed{1}$ $\boxed{\div}$ $\boxed{\text{TAN}}$ $\boxed{1}$ $\boxed{3}$ $\boxed{7}$ $\boxed{)}$ to obtain the same result.

To compute the arccotangent, arcsecant, and arccosecant functions, use the identities

$$\operatorname{arccot} x = \frac{\pi}{2} - \arctan x,$$

$$\operatorname{arcsec} x = \arccos\left(\frac{1}{x}\right),$$

$$\operatorname{arccsc} x = \arcsin\left(\frac{1}{x}\right).$$

Part III

Basic Skills

1 Introduction

Each section in the text contains problems that focus primarily on manipulative skills. This part of the manual lists those skills. It does not make reference to the other types of problems contained in most sections of the text that require a deeper conceptual understanding of the material.

Chapter 1
Inequalities, Equations, and Graphs

1.1 The Real Line

- Solve a linear inequality using the properties of inequalities

- Solve a nonlinear inequality using the sign chart method

- Express the solution of an inequality in interval notation

- Graph the solution set of an inequality

- Solve word problems involving inequalities

1.2 Absolute Value

- Find the absolute value of an expression containing an unknown number, x, subject to a constraint on x

- Find the distance between two numbers on the real line

- Find the midpoint of two numbers on the real line

- Solve an equation involving an absolute value

- Solve an inequality involving an absolute value

- Find an inequality involving an absolute value having a given interval or union of intervals as its solution set

1.3 The Rectangular Coordinate System

- Locate the quadrant in which a point lies based on the algebraic signs of its coordinates

- Find the distance between two points in the plane

- Find the midpoint of the line segment joining two points in the plane

1.4 Circles and Graphs

- Given the equation of a circle, find its center and radius (completing the square if necessary)

- Determine which of the four possible semicircles is defined by a given equation solved for x or y

- Find the x- and y-intercepts of an equation

- Determine if the graph of an equation possesses symmetry with respect to the x-axis, the y-axis, or the origin

1.5 Calculus Preview – Algebra and Limits

- Simplify an algebraic expression by factoring and adding rational expressions

- Find the limit of an algebraic expression by simplifying it

- Be able to use a binomial expansion to simplify a given expression

Chapter 2
Functions

2.1 Functions and Graphs

- Find the domain of a function given a formula for the function

- Find the domain of a function given the graph of the function

- Find the value of a function at a point in its domain given a formula for the function

- Find the value of a function at a point in its domain given the graph of the function

- Use the vertical line test to determine if a given graph is the graph of a function

- Find the zeros of a given function

- Find the x-intercepts of a given function

- Find the y-intercept of a given function

- Evaluate the factorial function at a positive integer

2.2 Symmetry and Transformations

- Determine if a function is even or odd

- Given a partial graph of a function that is known to be even or odd, find the entire graph

- Given the graph of a function, sketch the graph of a translation, reflection, stretching, or compression of the function

2.3 Linear Functions

- Find the slope of the line through two given points

- Given the equation of a line, find the slope and intercepts of the line

- Find the equation of the line through a given point with a given slope

- Find the equation of the line through two given points

- Find the equation of a line that is parallel to a given line

- Find the equation of a line that is perpendicular to a given line

- Determine if two given lines are parallel to each other

- Determine if two given lines are perpendicular to each other

- Find the point of intersection of two given lines

2.4 Quadratic Functions

- Complete the square to express a quadratic function in standard form

- Find the vertex of the graph of a quadratic function

- Find the x- and y-intercepts of the graph of a quadratic function

- Find the maximum or minimum value of a quadratic function

- Identify the transformations on the graph of $y = x^2$ used to obtain the graph of a quadratic function

2.5 Piecewise-Defined Functions

- Evaluate a piecewise-defined function at a given number

- Find the intercepts of a piecewise-defined function

- Sketch the graph of a piecewise-defined function

- Sketch the graph of functions defined using the greatest integer function

2.6 Combining Functions

- Given formulas for $f(x)$ and $g(x)$, find formulas for $(f+g)(x), (f-g)(x), (fg)(x)$, and $(f/g)(x)$

- Find the composition of two or more functions

- Find the domain of the composition of two functions

- Given a function f that is the composition of functions f and g, find $f(x)$ and $g(x)$

2.7 Inverse Functions

- Use the horizontal line test to determine if a function is one-to-one

- Use the definition to show that a given function is, or is not, one-to-one

- When appropriate, show that a function $f(x)$ is not one-to-one by finding $x_1 \neq x_2$ such that $f(x_1) = f(x_2)$

- Given the domain and range of a one-to-one function, find the domain and range of its inverse function

- Given a formula for a one-to-one function, find a formula for its inverse function

- Given the graph of a one-to-one function, sketch the graph of its inverse function

2.8 Building a Function from Words

- Translate words into an appropriate function

- Find the objective function associated with a word problem

- Find the domain of an objective function

2.9 Calculus Preview – The Tangent Line Problem

- Compute the difference quotient, $[f(a + h) - f(a)]/h$, for a given function at a given value of a

- Find the equation of the tangent line to a given function at a given point

- Compute the derivative of a given function

- Use the derivative of a given function to compute the slope of the tangent line to the graph of the function at a given x-coordinate on the graph of the function

- Compute the difference quotient, $[f(x) - f(a)]/(x - a)$, for a given function

- Use the difference quotient, $[f(x) - f(a)]/(x - a)$, to find the derivative of $f(x)$ and $x = a$

Chapter 3
Polynomial and Rational Functions

3.1 Polynomial Functions

- Determine if a polynomial function is even, odd, or neither

- Graph a polynomial function using its zeros and degree

3.2 Division of Polynomial Functions

- Use long division to find the quotient and remainder of two polynomial functions

- Use the Remainder Theorem to find the remainder when a polynomial is divided by a linear polynomial

- Use polynomial long division to find the value of a polynomial at a number

- Use synthetic division to find the quotient and remainder when a polynomial is divided by a linear polynomial

- Use synthetic division and the Remainder Theorem to evaluate a polynomial function at a given number

3.3 Zeros and Factors of Polynomial Functions

- Find the zeros of a polynomial function and give the complete factorization of the polynomial

- Determine if a given number is a zero of a polynomial

- Use synthetic division to determine if a given linear polynomial is a factor of a given polynomial

- Use polynomial long division to determine if a given polynomial is a factor of another polynomial

- Determine if a given complex number is a zero of a given polynomial

- Given a complex zero of a polynomial, find all other zeros and completely factor the polynomial

- Given the zeros, with their multiplicities, of a polynomial function, find the polynomial function

- Completely factor a polynomial to determine the zeros and their multiplicities of the polynomial

3.4 Real Zeros of Polynomial Functions

- Determine all *possible* rational zeros of a polynomial function having integer coefficients

- Determine all rational zeros of a polynomial function having integer coefficients

- Find all real zeros of a polynomial function with rational coefficients (when possible) and then factor the polynomial using only real numbers

3.5 Approximating Real Zeros

- Apply the Bisection Method to approximate the zeros of a polynomial equation

3.6 Rational Functions

- Find the vertical asymptotes of a rational function

- Find the horizontal asymptote of a rational function

- Find the slant asymptote of a rational function whose numerator has degree one greater than the degree of the denominator

- Determine where the graph of a rational function crosses any horizontal or slant asymptotes

- Find any holes in the graph of a rational function

- Graph a rational function

3.7 Calculus Preview – The Area Problem

- Assuming that the graph of a function $f(x)$ is never negative on an interval $[a, b]$, approximate the area bounded by the lines $x = a$, $x = b$, the graph of the function, and the x-axis

Chapter 4
Trigonometric Functions

4.1 Angles and Their Measurement

- Draw an angle given in either radian or degree form

- Convert an angle given in radians to degrees

- Convert an angle given in degrees to radians

- Find an angle between $0°$ and $360°$ that is coterminal with a given angle

- Find an angle between 0 and 2π radians that is coterminal with a given angle

- Find angles (when possible) that are complementary and supplementary to a given angle

- Find the arc length subtended by a given central angle in a circle with a given radius

4.2 The Sine and Cosine Functions

- Given the sine of an angle t in a specified quadrant, find the cosine of the angle

- Given the cosine of an angle t in a specified quadrant, find the sine of the angle

- Find the reference angle of a given angle

- Find the exact values of the sine and cosine of all angles having reference angle 0, $\pi/6$, $\pi/4$, $\pi/3$, or $\pi/2$

4.3 Graphs of Sine and Cosine Functions

- Sketch one cycle of the graph of a function involving $\sin x$ or $\cos x$ using shifting, stretching, compressing, and reflecting

- Given one cycle of a sine or cosine graph having period 2π, match it with an equation of the form $y = A \sin x + D$ or $y = A \cos x + D$

- Find the x-intercepts of a sine or cosine graph

- Given one cycle of a sine or cosine graph, match it with an equation of the form $y = A \sin x$ or $y = A \cos x$

- Find the amplitude, period, and phase shift of a function of the form $y = A \sin(Bx + C)$ or $y = A \cos(Bx + C)$

4.4 Other Trigonometric Functions

- Find the exact value of any of the six trigonometric functions of an angle with reference angle 0, $\pi/6$, $\pi/4$, $\pi/3$, or $\pi/2$

- Given the value of any of the six trigonometric functions of an angle t in a specified quadrant, find the values of the other five trigonometric functions of that angle

- Find the period, x-intercepts, and vertical asymptotes of any of the four functions of the forms $y = A\tan(Bx + C)$, $y = A\cot(Bx + C)$, $y = A\sec(Bx + C)$, and $y = A\csc(Bx + C)$

4.5 Verifying Trigonometric Identities

- Simplify a trigonometric expression

- Verify a trigonometric identity

- Use a substitution involving the square root of a quadratic expression in x (such as $\sqrt{a^2 + x^2}$) to rewrite a simple trigonometric expression (such as $x = a\tan\theta$) without radicals

4.6 Sum and Difference Formulas

- Use a sum or difference formula to find the exact value of the sine, cosine, or tangent of a given angle

- Use a double-angle formula to simplify an expression

- Use a double-angle formula to find the exact value of the sine, cosine, or tangent of a given angle

- Use a half-angle formula to find the exact value of the sine, cosine, or tangent of a given angle

4.7 Product-to-Sum and Sum-to-Product Formulas

- Use a product-to-sum formula to write a product of trigonometric expressions as a sum of sines or cosines

- Use a product-to-sum formula to find the exact value of an appropriate trigonometric expression

- Use a sum-to-product formula to write a sum or difference of trigonometric expressions as a product of sines, cosines, or one sine and one cosine function

- Use a sum-to-product formula to find the exact value of an appropriate trigonometric expression

4.8 Inverse Trigonometric Functions

- Find the value of an inverse trigonometric function of a real number

- Find the value of a trigonometric function of an inverse trigonometric function

- Find the value of an inverse trigonometric function of a trigonometric function

- Express $f(g^{-1}(x))$, where f is one of the six basic trigonometric functions and g is one of the six basic inverse trigonometric functions, in terms of x without any trigonometric functions

- Sketch the graphs of the six basic inverse trigonometric function

4.9 Trigonometric Equations

- Find all solutions of a given trigonometric equation in which the unknown represents a real number

- Find all solutions of a given trigonometric equation in which the unknown represents an angle measured in degrees

- Use a sum-to-product formula to find all solutions of an appropriate trigonometric equation

- Determine graphically if a trigonometric equation has any real solutions

- Find the first three positive x-intercepts of a function involving trigonometric functions

- Use inverse trigonometric functions to solve a trigonometric equation

4.10 Simple Harmonic Motion

- Reduce a given expression to one of the form $A\sin(Bx + \phi)$

4.11 Calculus Preview – The Limit Concept Revisited

- Evaluate the limit of a given trigonometric function at a given number

- Find the amplitude, period, and frequency of motion of a given spring/mass system

Chapter 5
Triangle Trigonometry

5.1 Right Triangle Trigonometry

- Given two sides of a right triangle, find the six trigonometric functions of an angle of the triangle

- Given two sides, or one angle and one side, of a right triangle, find the remaining side(s) and angle(s) of the triangle

5.2 Applications of Right Triangles

- Solve word problems involving right triangles

5.3 Law of Sines

- Know when the Law of Sines can lead to two possible triangles

- Given three of the six parts of a triangle, find the remaining parts using the Law of Sines

- Solve word problems using the Law of Sines

- Know when it is appropriate to use the Law of Sines

5.4 Law of Cosines

- Given three of the six parts of a triangle, find the remaining parts using the Law of Cosines

- Solve word problems using the Law of Cosines

- Know when it is appropriate to use the Law of Cosines, rather than the Law of Sines

5.5 Calculus Preview – Vectors and Dot Product

- Graph a vector, identifying the magnitude and smallest positive direction angle

- Perform basic arithmetic operations on scalar multiples of vectors

- Represent a vector in component form or as a linear combination of the basis vectors **i** and **j**

- Find a unit vector in the same (or opposite) direction as a given vector

- Normalize a vector

- Express a vector in trigonometric form

- Find the resultant force associated with a pair of forces

- Find the speed associated with a velocity vector

- Find the dot product of two vectors given in component form

- Find the dot product of two vectors given their lengths and the angle between them

- Determine if two vectors are orthogonal

- Find the component of one vector on another

- Find the projection of one vector on another

- Find the work done by a force acting at an angle to the direction of motion

Chapter 6
Exponential and Logarithmic Functions

6.1 Exponential Functions

- Sketch the graph of an exponential function

- Find the value of b such that the graph of $f(x) = b^x$ passes through a given point

- Determine the range of a given exponential function

- Use a graph to solve an exponential inequality

- Determine if a given exponential function is even or odd or neither

6.2 Logarithmic Functions

- Write a given exponential expression as an equivalent logarithmic expression

- Write a given logarithmic expression as an equivalent exponential expression

- Find the exact value of a given expression involving a logarithm

- Sketch the graph of a logarithmic function

- Find the value of b such that the graph of $f(x) = \log_b x$ passes through a given point

- Find the domain of a given logarithmic function

- Find the asymptote of a given logarithmic function

- Use a graph to solve a given inequality involving logarithms

- Use the laws of logarithms to rewrite a given logarithmic expression as a single logarithm

- Use the laws of logarithms to write a given expression involving products, quotients and powers as an expression that contains sums and differences of logarithms

- Find the exact solution of an equation that involves logarithms
- Use the natural logarithm to solve an equation in which the unknown is in the exponent

6.3 | Exponential and Logarithmic Equations |

- Find the exact value of the solution of an exponential equation
- Find the exact value of the solution of a logarithmic equation
- Find the x and y-intercepts of the graphs of exponential and logarithmic functions
- Be able to recognize extraneous solutions

6.4 | Exponential and Logarithmic Models |

- Solve word problems involving population growth
- Solve word problems involving radioactive decay and half-life
- Solve word problems involving carbon dating
- Solve word problems involving Newton's law of cooling/warming
- Solve word problems involving compound interest
- Solve word problems involving the Richter scale
- Solve word problems involving the pH of a chemical solution

6.5 | Calculus Preview – The Hyperbolic Functions |

- Find the derivative of a simple logarithmic function
- Find the derivative of a simple exponential function
- Know the definitions of the hyperbolic functions
- Know the graphs of the hyperbolic functions
- Verify identities involving the hyperbolic functions

Chapter 7
Conic Sections

7.1 The Parabola

- Find the vertex, focus, directrix, and axis of the parabola determined by a given equation

- Given the equation of a parabola, sketch its graph

- Given information about the vertex, focus, and directrix of a parabola, find the equation of the parabola

- Find the x- and y-intercepts of a parabola

- Determine the focal width of a given parabola

7.2 The Ellipse

- Find the center, foci, vertices, endpoints of the minor axis, and eccentricity of the ellipse determined by a given equation

- Given the equation of an ellipse, sketch its graph

- Given information about the center, vertices, foci, and endpoints of the minor axis, find the equation of the ellipse

- Determine the focal width of a given ellipse

7.3 The Hyperbola

- Find the center, foci, vertices, asymptotes, and eccentricity of the hyperbola determined by a given equation

- Given the equation of a hyperbola, sketch its graph

- Given information about the center, vertices, foci, and asymptotes, find the equation of the hyperbola

- Determine the focal width of a given hyperbola

7.4 Rotation of Axes

- Given an angle of rotation, find the $x'y'$-coordinates of a given xy-point

- Given an angle of rotation, find the xy-coordinates of a given $x'y'$-point

- Eliminate the xy-term in a quadratic expression, and identify and graph the conic

- Use the discriminant to determine the type of conic section represented by a quadratic expression

7.5 | Calculus Preview – 3-Space |

- Graph points and planes in 3-space

- Graph spheres in 3-space

- Find the distance between two points in 3-space

- Find the distance between a point and a plane that is parallel to one of the coordinate axes

- Find the midpoint between two points in 3-space

- Perform simple vector operations in 3-space

- Find the cross product of two vectors in 3-space

Chapter 8
Polar Coordinates

8.1 | The Polar Coordinate System |

- Plot the point with given polar coordinates

- Be able to find multiple polar representations of the same polar point

- Find the rectangular coordinates of a point given in polar coordinates

- Find the polar coordinates of a point given in rectangular coordinates

- Find a polar equation that has the same graph as a given rectangular equation

8.2 | Graphs of Polar Equations |

- Recognize the graphs and equations of a cardioid, a limaçon with an interior loop, a dimpled limaçon, a convex limaçon, a rose curve with an even number of petals, a rose curve with an odd number of petals, a circle centered on a coordinate axis that passes through the origin, and a lemniscate

- Identify by name the graph of a given polar equation

- Find the points of intersection of the graphs of a given pair of polar equations

8.3   Conic Sections in Polar Coordinates

- Given the polar equation of a conic, determine its eccentricity

- Given the polar equation of a conic, sketch its graph

- Given the polar equation of a conic, find the rectangular equation

- Find the polar equation of a conic, given its eccentricity and directrix

- Find the polar equation of a parabola with focus at the origin, given the polar coordinates of its vertex

8.4   Calculus Preview – Parametric Equations

- Sketch the curve that has a given set of parametric equations

- Eliminate the parameter from a given set of parametric equations and obtain a rectangular equation having the same graph

- Compare the graphs of a rectangular equation and a set of parametric equations whose graph lies on the graph of the rectangular equation

- Recognize the role played by the domain of the parameter in the set of parametric equations of a circle centered at the origin

- Find the x- and y-intercepts of the graph determined by a set of parametric equations

Chapter 9
Systems of Equations and Inequalities

9.1   Systems of Linear Equations

- Solve a linear system by substitution

- Solve a linear system by elimination

- Determine if a linear system is consistent, with independent or dependent equations; or if it is inconsistent

9.2   Determinants and Cramer's Rule

- Evaluate 2×2 and 3×3 determinants

- Use Cramer's rule to solve a linear system of equations

9.3 Systems of Nonlinear Equations

- Use graphs to determine if a nonlinear system has a solution

- Find the solution of a nonlinear system of equations

9.4 Systems of Inequalities

- Graph the solution set of a system of inequalities

- Use the graph of a system of inequalities to determine the inequalities

9.5 Calculus Preview – Partial Fractions

- Find the partial fraction decomposition of a rational function whose numerator has smaller degree than the denominator

- Find the partial fraction decomposition of a rational function whose numerator has equal or larger degree than the denominator

Chapter 10
Sequences and Series

10.1 Sequences

- List the first n terms of a sequence whose nth term is given

- List the first n terms of a sequence that is defined recursively

- Distinguish between arithmetic and geometric sequences

- Determine the common difference of an arithmetic sequence

- Determine the common ratio of a geometric sequence

10.2 Series

- Find the sum of the first n terms of an arithmetic series

- Find the sum of the first n terms of a geometric series

- Use sigma notation to represent a given series

10.3   Mathematical Induction

- Use the Principle of Mathematical Induction to prove a given statement

10.4   The Binomial Theorem

- Evaluate the factorial of a nonnegative integer
- Evaluate a binomial coefficient
- Expand a power of a binomial expression using the Binomial Theorem
- Find the nth term in the expansion of a binomial expression

10.5   Principles of Counting

- Use a tree diagram to list all possible arrangements of m objects chosen from n objects
- Be able to use the Fundamental Counting Principle
- Be able to compute $C(n, r)$ and $P(n, r)$
- Know when to use a permutation, $P(n, r)$ and when to use a combination, $C(n, r)$

10.6   Introduction to Probability

- Find the sample space S of an experiment
- Be able to find the probability of an event E within a sample space S
- Use the notion of the complement of an event E
- Use the notion of mutually exclusive events
- Find the probability of the union of two events

10.7   Convergence of Sequences and Series

- Determine if a given sequence converges or diverges
- Express a repeating decimal as a quotient of integers
- Determine whether or not an infinite geometric series converges or diverges

Part IV
Selected Solutions

Chapter 1

Inequalities, Equations, and Graphs

1.1 | The Real Line

The graphs of the solutions sets in this section show the appropriate numbers in their proper order, but the distances between the numbers are sometimes not shown to the proper scale.

3. Since "nonnegative" means positive or 0, the statement is equivalent to $a + b \geq 0$.

6. $c - 1 \leq 5$

9. $[5, \infty)$

12. $(-5, -3]$

15. $-7 \leq x \leq 9$

18. $x \geq -5$

21. Using the properties of inequalities, we have

$$\frac{3}{2} x + 4 \leq 10$$
$$\frac{3}{2} x + 4 - 4 \leq 10 - 4 \qquad \leftarrow \text{ by } (ii)$$
$$\frac{3}{2} x \leq 6$$
$$\left(\frac{2}{3}\right) \frac{3}{2} x \leq \left(\frac{2}{3}\right) 6$$
$$x \leq 4.$$

In interval notation this is $(-\infty, 4]$, with graph

54

24. Using the properties of inequalities, we have

$$-(1 - x) \geq 2x - 1$$
$$-1 + x \geq 2x - 1$$
$$1 - 1 + x \geq 1 + 2x - 1 \qquad \leftarrow \text{ by } (i) \text{ of Theorem 1.1.1}$$
$$x \geq 2x$$
$$-x + x \geq -x + 2x \qquad \leftarrow \text{ by } (i) \text{ of Theorem 1.1.1}$$
$$0 \geq x.$$

To express the solution in interval notation we rewrite the last inequality as $x \leq 0$. In interval notation this is $(-\infty, 0]$ with graph

$$0$$

27. Using the properties of inequalities, we have

$$-\frac{20}{3} < \frac{2}{3}x < 4$$
$$\frac{3}{2}\left(-\frac{20}{3}\right) < \frac{3}{2}\left(\frac{2}{3}x\right) < \frac{3}{2}(4) \qquad \leftarrow \text{ by } (ii) \text{ of Theorem 1.1.1}$$
$$-10 < x < 6.$$

In interval notation this is $(-10, 6)$, with graph

$$-10 \qquad 6$$

30. Using the properties of inequalities, we have

$$3 < x + 4 \leq 10$$
$$3 - 4 < x + 4 - 4 \leq 10 - 4 \qquad \leftarrow \text{ by } (i) \text{ of Theorem 1.1.1}$$
$$-1 < x \leq 6.$$

In interval notation this is $(-1, 6]$, with graph

$$-1 \qquad 6$$

33. Using the properties of inequalities, we have

$$-1 \leq \frac{x - 4}{4} < \frac{1}{2}$$
$$4(-1) \leq 4\left(\frac{x - 4}{4}\right) < 4\left(\frac{1}{2}\right) \qquad \leftarrow \text{ by } (ii) \text{ of Theorem 1.1.1}$$
$$-4 \leq x - 4 < 2$$
$$-4 + 4 \leq x - 4 + 4 < 2 + 4 \qquad \leftarrow \text{ by } (i) \text{ of Theorem 1.1.1}$$
$$0 \leq x < 6.$$

In interval notation this is $[0, 6)$, with graph

36. First, rewrite the inequality with all nonzero terms to the left of the inequality symbol and then factor.

$$x^2 - 16 \geq 0$$
$$(x + 4)(x - 4) \geq 0.$$

Placing $x = -4$ and $x = 4$ on the number line determines three intervals. The sign chart below, in turn, determines the graph of the solution set.

$$
\begin{array}{ccccccc}
x + 4 & - & 0 & + & + & + \\
x - 4 & - & - & - & 0 & + \\
(x + 4)(x - 4) & + & 0 & - & 0 & +
\end{array}
$$

$$\xleftarrow{\qquad} \overset{]}{\underset{-4}{}} \qquad \overset{[}{\underset{4}{}} \xrightarrow{\qquad}$$

The solution set is $(-\infty, -4] \cup [4, \infty)$.

39. Factoring, we have

$$x^2 - 8x + 12 < 0$$
$$(x - 2)(x - 6) < 0.$$

Placing $x = 2$ and $x = 6$ on the number line determines three intervals. The sign chart below, in turn, determines the graph of the solution set.

$$
\begin{array}{ccccccc}
x - 2 & - & 0 & + & + & + \\
x - 6 & - & - & - & 0 & + \\
(x - 2)(x - 6) & + & 0 & - & 0 & +
\end{array}
$$

$$\underset{2}{(} \rule{2cm}{0.4pt} \underset{6}{)}$$

The solution set is $(2, 6)$.

42. First, rewrite the inequality with all nonzero terms to the left of the inequality symbol and then factor.

$$4x^2 > 9x + 9$$
$$4x^2 - 9x - 9 > 0$$
$$(4x + 3)(x - 3) > 0.$$

Placing $x = -\frac{3}{4}$ and $x = 3$ on the number line determines three intervals. The sign chart below, in turn, determines the graph of the solution set.

$4x + 3$	$-$	0	$+$	$+$	$+$
$x - 3$	$-$	$-$	$-$	0	$+$
$(4x + 3)(x - 3)$	$+$	0	$-$	0	$+$

The solution set is $(-\infty, -\frac{3}{4}) \cup (3, \infty)$.

45. Factoring we have

$$(x^2 - 1)(x^2 - 4) \le 0$$
$$(x + 1)(x - 1)(x + 2)(x - 2) \le 0.$$

Placing $x = -2$, $x = -1$, $x = 1$, and $x = 2$ on the number line determines five intervals. The sign chart below, in turn, determines the graph of the solution set.

$x + 1$	$-$	$-$	$-$	0	$+$	$+$	$+$	$+$	$+$
$x - 1$	$-$	$-$	$-$	$-$	$-$	0	$+$	$+$	$+$
$x + 2$	$-$	0	$+$	$+$	$+$	$+$	$+$	$+$	$+$
$x - 2$	$-$	$-$	$-$	$-$	$-$	$-$	$-$	0	$+$
$(x + 1)(x - 1)(x + 2)(x - 2)$	$+$	0	$-$	0	$+$	0	$-$	0	$+$

The solution set is $[-2, -1] \cup [1, 2]$.

48. Since $x^2 + 2$ is positive for all real values of x, the quotient $10/(x^2 + 2)$ is always positive. The solution set is $(-\infty, \infty)$ and the graph is

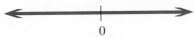

51. We begin by putting the terms over a common denominator:

$$\frac{x + 1}{x - 1} + 2 > 0$$
$$\frac{x + 1 + 2(x - 1)}{x - 1} > 0$$
$$\frac{3x - 1}{x - 1} > 0.$$

Placing $x = \frac{1}{3}$ and $x = 1$ on the number line determines three intervals. The sign chart below, in turn, determines the graph of the solution set.

$$
\begin{array}{ccccccc}
3x - 1 & - & 0 & + & + & + \\
x - 1 & - & - & - & 0 & + \\
(3x-1)/(x-1) & + & 0 & - & \text{undefined} & +
\end{array}
$$

The solution set is $(-\infty, \frac{1}{3}) \cup (1, \infty)$.

54. The left-hand side is already in the proper form. Placing $x = -1$, $x = 0$, and $x = 1$ on the number line determines four intervals. The sign chart below, in turn, determines the graph of the solution set.

$$
\begin{array}{cccccccc}
1 + x & - & 0 & + & + & + & + & + \\
1 - x & + & + & + & + & + & 0 & - \\
x & - & - & - & 0 & + & + & + \\
(1+x)(1-x)/x & + & 0 & - & \text{undefined} & + & 0 & -
\end{array}
$$

The solution set is $(-1, 0] \cup [1, \infty)$.

57. First, rewrite the left-hand side of the inequality as a simple fraction.

$$
\frac{2}{x+3} - \frac{1}{x+1} < 0
$$

$$
\frac{2(x+1) - (x+3)}{(x+3)(x+1)} > 0
$$

$$
\frac{x-1}{(x+3)(x+1)} > 0.
$$

Placing $x = -3$, $x = -1$, and $x = 1$ on the number line determines four intervals. The sign chart below, in turn, determines the graph of the solution set.

$$
\begin{array}{cccccccc}
x - 1 & - & - & - & - & - & 0 & + \\
x + 3 & - & 0 & + & + & + & + & + \\
x + 1 & - & - & - & 0 & + & + & + \\
(x-1)/(x+3)(x+1) & - & \text{undefined} & + & \text{undefined} & - & 0 & +
\end{array}
$$

The solution set is $(-\infty, -3) \cup (-1, 1)$.

60. The sides of the resulting rectangle are $x + 2$ and $x + 5$. An implicit condition in this case is $x > 0$. See the section on **Implicit Conditions in a Word Problem** in **Part I, Topics in Algebra,** of this manual. We want $(x + 5)(x + 2) < 130$. To solve this inequality, we move all nonzero terms to the left-hand side of the inequality, expand the left-hand side, and factor the result.

$$(x + 5)(x + 2) < 130$$
$$x^2 + 7x + 10 - 130 < 0$$
$$x^2 + 7x - 120 < 0$$
$$(x + 15)(x - 8) < 0$$

Since $x > 0$, we can ignore $x = -15$ and work with a number line that extends from 0 to ∞ and contains 8.

$x + 15$	$+$	$+$	$+$
$x - 8$	$-$	0	$+$
$(x + 15)(x - 8)$	$-$	0	$+$

$$\overset{\displaystyle(\!\!-\!\!-\!\!-\!\!-\!\!-\!\!-\!\!-\!\!-\!\!)\quad\quad}{\underset{0\qquad\qquad 8}{}}$$

The possible lengths of the original square are between 0 and 8 inches, with neither 0 nor 8 included.

Miscellaneous Applications

63. Let x be the width of the flower bed. Then its length is $2x$ and the area is $(2x)x = 2x^2$. An implicit condition in this case is $x > 0$. See the section on **Implicit Conditions in a Word Problem** in **Part I, Topics in Algebra,** of this manual. We want $2x^2 > 98$. Rewrite the inequality with all nonzero terms on the left-hand side of the inequality symbol and then factor.

$$2x^2 > 98$$
$$2x^2 - 98 > 0$$
$$2(x + 7)(x - 7) > 0$$

Since $x > 0$, we can ignore $x = -7$ and work with a number line that extends from 0 to ∞ and contains 7.

$x + 7$	$+$	$+$	$+$
$x - 7$	$-$	0	$+$
$2(x + 7)(x - 7)$	$-$	0	$+$

$$\overset{\displaystyle(\!\!-\!\!-\!\!-\!\!-\!\!-\!\!-\!\!)\longrightarrow}{\underset{0\qquad\qquad 7}{}}$$

The width of the flower bed must be greater than 7 m.

66. Letting $g = 32$, $v_0 = 72$, and $s_0 = 0$ we want to solve $-\frac{1}{2}(32)t^2 + 72t > 80$. An implicit condition in this case is $t \geq 0$. See the section on **Implicit Conditions in a Word Problem** in **Part I, Topics in Algebra**, of this manual. Rewrite the inequality with all nonzero terms on the left-hand side of the inequality symbol and then factor.

$$-\frac{1}{2}(32)t^2 + 72t > 80$$

$$-16t^2 + 72t - 80 > 0$$

$$16t^2 - 72t + 80 < 0 \qquad \leftarrow \text{ by } (iii) \text{ of Theorem 1.1.1}$$

$$8(2t^2 - 9t + 10) < 0$$

$$8(t - 2)(2t - 5) < 0$$

Using $t \geq 0$ we place $t = 0$, $t = 2$, and $t = \frac{5}{2}$ on the number line extending to the right from 0. The sign chart below is used to determine the solution set.

$$
\begin{array}{lcccccc}
t - 2 & - & - & 0 & + & + & + \\
2t - 5 & - & - & - & - & 0 & + \\
8(t-2)(2t-5) & + & + & 0 & - & 0 & + \\
\end{array}
$$

The rocket will be more that 80 feet above the ground between $t = 2$ and $t = \frac{5}{2}$ seconds.

1.2   Absolute Value

3. Since $\sqrt{63} < \sqrt{64} = 8$, $8 - \sqrt{63}$ is positive and so $|8 - \sqrt{63}| = 8 - \sqrt{63}$.

6. $\big||-3| - |10|\big| = |3 - 10| = |-7| = 7$

9. If $x < 6$, then $x - 6 < 0$ and $|x - 6| = -(x - 6) = 6 - x$.

12. Using the Properties of Absolute Values in the text, we have for $x \neq y$

$$
\begin{aligned}
\left|\frac{x-y}{y-x}\right| &= \frac{|x-y|}{|y-x|} \qquad \leftarrow \text{ by}(iv) \\
&= \frac{|x-y|}{-(x-y)|} \\
&= \frac{x-y}{x-y} \qquad \leftarrow \text{ by}(i) \\
&= 1.
\end{aligned}
$$

15. Since x is in $(3, 4]$, $3 < x \leq 4$, so $x - 2 > 0$ and $x - 5 < 0$. Thus,

$$|x - 2| + |x - 5| = x - 2 - (x - 5) = x - 2 - x + 5 = 3.$$

18. Since x is in $(0, 1)$, $0 < x < 1$, so $x + 1 > 0$ and $x - 3 < 0$. Thus,

$$|x + 1| - |x - 3| = x + 1 - [-(x - 3)] = x + 1 + x - 3 = 2x - 2.$$

21. $d(3, 7) = |7 - 3| = |4| = 4$

$$m = \frac{3 + 7}{2} = \frac{10}{2} = 5$$

24. $d\left(-\frac{1}{4}, \frac{7}{4}\right) = \left|\frac{7}{4} - \left(-\frac{1}{4}\right)\right| = \left|\frac{7}{4} + \frac{1}{4}\right| = \left|\frac{8}{4}\right| = |2| = 2$

$$m = \frac{-1/4 + 7/4}{2} = \frac{6/4}{2} = \frac{6}{4(2)} = \frac{6}{8} = \frac{3}{4}$$

27. Since $a = 4$ and $d(a, m) = \pi$, we have $m - a = \pi$ or $m = 4 + \pi$. Then

$$m = \frac{a + b}{2} = \frac{4 + b}{2} = 2 + \frac{1}{2}b = 4 + \pi,$$

so $\frac{1}{2}b = 2 + \pi$ and $b = 4 + 2\pi$.

30. The given equation is equivalent to

$$5v - 4 = 7 \qquad \text{or} \qquad 5v - 4 = -7.$$

From $5v - 4 = 7$ we obtain $5v = 7 + 4 = 11$, so $v = \frac{11}{5}$. From $5v - 4 = -7$ we obtain $5v = -7 + 4 = -3$, so $v = -\frac{3}{5}$. Therefore, the solutions are $\frac{11}{5}$ and $-\frac{3}{5}$.

33. The given equation is equivalent to

$$\frac{x}{x - 1} = 2 \qquad \text{or} \qquad \frac{x}{x - 1} = -2.$$

From $x/(x - 1) = 2$ we obtain $x = 2(x - 1) = 2x - 2$, so $x = 2$. From $x/(x - 1) = -2$ we obtain $x = -2(x - 1) = -2x + 2$, so $3x = 2$ and $x = \frac{2}{3}$. Therefore, the solutions are 2 and $\frac{2}{3}$.

36. The inequality $|3x| > 18$ is equivalent to

$$3x > 18 \qquad \text{or} \qquad 3x < -18.$$

Solving $3x > 18$ we obtain $x > 18/3 = 6$. In interval notation this is $(6, \infty)$. Solving $3x < -18$ we obtain $x < -18/3 = -6$. In interval notation this is $(-\infty, -6)$. The solution set is $(-\infty, -6) \cup (6, \infty)$. The graph of the solution set is

39. The inequality $|2x - 7| \leq 1$ is equivalent to $-1 \leq 2x - 7 \leq 1$. Solving, we obtain

$$-1 + 7 \leq 2x - 7 + 7 \leq 1 + 7$$

$$6 \leq 2x \leq 8$$

$$\frac{1}{2}(6) \leq \frac{1}{2}(2x) \leq \frac{1}{2}(8)$$

$$3 \leq x \leq 4.$$

The solution set is $[3, 4]$ and the graph of the solution set is

42. The inequality $|6x + 4| > 4$ is equivalent to

$$6x + 4 > 4 \qquad \text{or} \qquad 6x + 4 < -4.$$

Solving $6x + 4 > 4$ we obtain $6x > 0$, so $x > 0$. In interval notation this is $(0, \infty)$. Solving $6x + 4 < -4$ we obtain $6x < -8$, so $x < -\frac{8}{6} = -\frac{4}{3}$. In interval notation this is $(-\infty, -\frac{4}{3})$.

The solution set is $(-\infty, -\frac{4}{3}) \cup (0, \infty)$. The graph of the solution set is

45. The inequality is equivalent to $-0.01 < x - 5 < 0.01$. Solving, we obtain

$$-0.01 + 5 < x - 5 + 5 < 0.01 + 5$$

$$4.99 < x < 5.01.$$

The solution set is $(4.99, 5.01)$ and the graph of the solution set is

48. Since the solution set is a finite open interval, the form of the inequality will be $|x - b| < a$, where b is the midpoint, m, of the interval and a is the distance from m to b. Thus,

$$m = \frac{1 + 2}{2} = \frac{3}{2}$$

and

$$d(m, b) = b - \frac{3}{2} = \frac{1}{2} d(1, 2) = \frac{1}{2}(2 - 1) = \frac{1}{2}.$$

Then $b = \frac{1}{2} + \frac{3}{2} = \frac{4}{2} = 2$ and the inequality is $|x - \frac{3}{2}| < \frac{1}{2}$.

51. We want all numbers x such that the distance from x to -3 is greater than 2. That is, we want $d(-3, x) > 2$ or $|x - (-3)| \geq 2$. In interval notation this is $(-\infty, -5] \cup [-1, \infty)$.

Miscellaneous Applications

54. If your midterm score is M, then the average of M and 72 is $\frac{1}{2}(M + 72)$. Thus, you want

$$80 \le \frac{1}{2}(M + 72) \le 89$$

$$80 \le \frac{1}{2}M + 36 \le 89$$

$$80 - 36 \le \frac{1}{2}M \le 89 - 36$$

$$44 \le \frac{1}{2}M \le 53$$

$$2(44) \le 2\left(\frac{1}{2}M\right) \le 2(53).$$

$$88 \le M \le 106.$$

Assuming that the highest score you can get on a test is 100 (100%), you will need to get a score from 88% to 100%, inclusive, on the midterm to have a mid-semester grade of B.

1.3   The Rectangular Coordinate System

3.

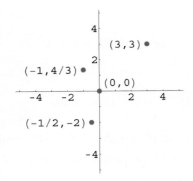

6. Since (a, b) is in the first quadrant, $a > 0$ and $b > 0$. Thus, $-b < 0$, and $(a, -b)$ is in the fourth quadrant.

9. Since (a, b) is in the first quadrant, $a > 0$ and $b > 0$. Thus, $-a < 0$ and $(-a, b)$ is in the second quadrant. Thus, $-b < 0$, and $(-b, a)$ is in the second quadrant.

12. Since (a, b) is in the first quadrant, $a > 0$ and $b > 0$. Thus, $-b < 0$ and $(b, -b)$ is in the fourth quadrant.

15. Since (a, b) is in the first quadrant, $a > 0$ and $b > 0$. Thus, $-a < 0$, and $(b, -a)$ is in the fourth quadrant.

18. $A : (-5, 0); \; B : (-3, -1); \; C : (0, -3); \; D : (-2, -2); \; E : (3, 3); \; F : (3, \frac{1}{2}); \; G : (6, -1)$

21. Since $xy = 0$ when $x = 0$ or when $y = 0$, the points that satisfy the condition constitute the coordinate axes.

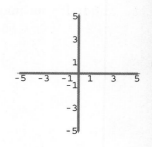

24. Since $x \leq 2$ and $y \geq -1$, the region lies to the left of the line $x = 2$ and above the line $y = -1$. The lines $x = 2$ and $y = -1$ are included, and are thus shown as solid lines in the figure.

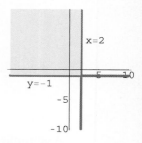

27. $d(A, B) = \sqrt{(-3 - 1)^2 + (4 - 2)^2} = \sqrt{(-4)^2 + 2^2} = \sqrt{16 + 4} = \sqrt{20} = 2\sqrt{5}$

30. $d(A, B) = \sqrt{(-5 - (-12))^2 + (-7 - (-3))^2} = \sqrt{(-5 + 12)^2 + (-7 + 3)^2} = \sqrt{7^2 + (-4)^2}$
$= \sqrt{49 + 16} = \sqrt{65}$

33. We find the distance between the vertices:

$$d(A, B) = \sqrt{(-3 - 8)^2 + (-1 - 1)^2} = \sqrt{(-11)^2 + (-2)^2} = \sqrt{121 + 4} = \sqrt{125}$$

$$d(B, C) = \sqrt{(10 - (-3))^2 + (5 - (-1))^2} = \sqrt{13^2 + 6^2} = \sqrt{169 + 36} = \sqrt{205}$$

$$d(C, A) = \sqrt{(8 - 10)^2 + (1 - 5)^2} = \sqrt{(-2)^2 + (-4)^2} = \sqrt{4 + 16} = \sqrt{20}.$$

The sum of the squares of the shorter sides is $(\sqrt{125})^2 + (\sqrt{20})^2 = 125 + 20 = 145$, which is not $(\sqrt{205})^2 = 205$, the square of the third side. Thus, the three points do not satisfy the Pythagorean Theorem and the triangle is not a right triangle.

36. We find the distances between the vertices:

$$d(A, B) = \sqrt{(1 - 4)^2 + (1 - 0)^2} = \sqrt{(-3)^2 + 1^2} = \sqrt{9 + 1} = \sqrt{10}$$

$$d(B, C) = \sqrt{(2 - 1)^2 + (3 - 1)^2} = \sqrt{1^2 + 2^2} = \sqrt{1 + 4} = \sqrt{5}$$

$$d(C, A) = \sqrt{(4 - 2)^2 + (0 - 3)^2} = \sqrt{2^2 + (-3)^2} = \sqrt{4 + 9} = \sqrt{13}.$$

The sum of the squares of the shorter sides is $(\sqrt{10})^2 + (\sqrt{5})^2 = 10 + 5 = 15$, which is not $(\sqrt{13})^2 = 13$, the square of the third side. Thus, the three points do not satisfy the Pythagorean Theorem and the triangle is not a right triangle.

39. (a)   If P is equidistant from A and B, then $d(A, P) = d(B, P)$ or

$$\sqrt{(x - (-1))^2 + (y - 2)^2} = \sqrt{(x - 3)^2 + (y - 4)^2}$$

Squaring both side and simplifying, we have

$$(x + 1)^2 + (y - 2)^2 = (x - 3)^2 + (y - 4)^2$$
$$x^2 + 2x + 1 + y^2 - 4y + 4 = x^2 - 6x + 9 + y^2 - 8y + 16$$
$$2x + 1 - 4y + 4 = -6x + 9 - 8y + 16$$
$$2x - 4y + 5 = -6x - 8y + 25$$
$$8x + 4y = 20$$
$$2x + y = 5.$$

Be careful when squaring both sides of an equation – you may introduce *extraneous solutions*. (See **Extraneous Roots** in **Part I, Topics in Algebra** of this manual.) In this case, no extraneous solutions were introduced because the expressions under the square root signs were both nonnegative.

(b)   The set of points in the plane that are equidistant from two given distinct points is the perpendicular bisector of the line segment joining A and B.

42. Let A be the point $(1/\sqrt{2}, 1/\sqrt{2})$, B be the point $(0.25, 0.97)$, and O be the origin. Then

$$d(O, A) = \sqrt{(1/\sqrt{2} - 0)^2 + (1/\sqrt{2} - 0)^2} = \sqrt{\tfrac{1}{2} + \tfrac{1}{2}} = \sqrt{1} = 1$$
$$d(O, B) = \sqrt{(0.25 - 0)^2 + (0.97 - 0)^2} = \sqrt{0.0625 + 0.9409} = \sqrt{1.0034} \approx 1.0017.$$

We see that $(1/\sqrt{2}, 1/\sqrt{2})$ is closer to the origin than $(0.25, 0.97)$.

45. Using the midpoint formula, we find that the coordinates of the midpoint are given by

$$\left(\frac{-1 - 8}{2}, \frac{0 + 5}{2}\right) \qquad \text{or} \qquad \left(-\frac{9}{2}, \frac{5}{2}\right).$$

48. Using the midpoint formula, we find that the coordinates of the midpoint are given by

$$\left(\frac{x + (-x)}{2}, \frac{x + (x + 2)}{2}\right) \qquad \text{or} \qquad \left(0, \frac{2x + 2}{2}\right) \qquad \text{or} \quad (0, x + 1).$$

51. If the coordinates of B are (x, y), then

$$-1 = \frac{1}{2}(5 + x) \qquad \text{or} \qquad x = -2 - 5 = -7$$

and

$$-1 = \frac{1}{2}(8 + y) \qquad \text{or} \qquad y = -2 - 8 = -10.$$

Thus, B is $(-7, -10)$.

54. Using the midpoint formula, we find that the coordinates of the midpoint are given by

$$\left(\frac{5+(-5)}{2}, \frac{2+(-6)}{2}\right) \qquad \text{or} \qquad (0,-2).$$

Since any point on the x-axis has y-coordinate 0, let $(x,0)$ be a point on the x-axis that is 3 units from $(0,-2)$. By the distance formula,

$$d\big((0,-2),(x,0)\big) = \sqrt{(x-0)^2 + (0-(-2))^2} = \sqrt{x^2+4} = 3.$$

Squaring both sides, we have $x^2 + 4 = 9$, $x^2 = 5$, and $x = \pm\sqrt{5}$. The points on the x-axis are $(-\sqrt{5},0)$, and $(\sqrt{5},0)$.

57. We first let $P_2(x_2,y_2)$ be the midpoint of the line segment joining $A(3,6)$ and $B(5,8)$. Then, by the midpoint formula, $x_2 = \frac{1}{2}(3+5) = \frac{1}{2}(8) = 4$ and $y_2 = \frac{1}{2}(6+8) = \frac{1}{2}(14) = 7$. Thus, P_2 is $(4,7)$. Now, P_1 is the midpoint of $A(3,6)$ and $P_2(4,7)$, so $x_1 = \frac{1}{2}(3+4) = \frac{7}{2}$ and $y_1 = \frac{1}{2}(6+7) = \frac{13}{2}$. Thus, P_1 is $\left(\frac{7}{2}, \frac{13}{2}\right)$. Finally, $P_3(x_3,y_3)$ is the midpoint of $P_2(4,7)$ and $B(5,8)$, so $x_3 = \frac{1}{2}(4+5) = \frac{9}{2}$ and $y_3 = \frac{1}{2}(7+8) = \frac{15}{2}$. Thus, P_3 is $\left(\frac{9}{2}, \frac{15}{2}\right)$.

1.4 Circles and Graphs

3. Writing the equation in the form

$$x^2 + (y-3)^2 = 7^2$$

we see that the center is $(0,3)$ and the radius is 7.

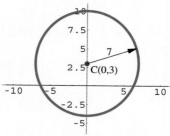

6. Writing the equation in the form

$$(x-(-3))^2 + (y-5)^2 = 5^2$$

we see that the center is $(-3,5)$ and the radius is 5.

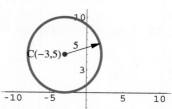

9. In order to find the center and radius, we want to write the equation in standard form. To do this, we rearrange the terms and complete the square:

$$x^2 + y^2 + 2x - 4y - 4 = 0$$

$$(x^2 + 2x \qquad\quad) + (y^2 - 4y \qquad\quad) = 4$$

$$\left[x^2 + 2x + \left(\frac{2}{2}\right)^2\right] + \left[y^2 - 4y + \left(\frac{-4}{2}\right)^2\right] = 4 + \left(\frac{2}{2}\right)^2 + \left(\frac{-4}{2}\right)^2$$

$$(x^2 + 2x + 1) + (y^2 - 4y + 4) = 4 + 1 + 4 = 9$$

$$(x+1)^2 + (y-2)^2 = 3^2.$$

From the last equation, we see that the circle is centered at $(-1, 2)$ and has radius 3.

12. In order to find the center and radius, we want to write the equation in standard form. To do this, we rearrange the terms and complete the square:

$$x^2 + y^2 + 3x - 16y + 63 = 0$$

$$(x^2 + 3x \qquad) + (y^2 - 16y \qquad) = -63$$

$$\left[x^2 + 3x + \left(\frac{3}{2}\right)^2\right] + \left[y^2 - 16y + \left(\frac{-16}{2}\right)^2\right] = -63 + \left(\frac{3}{2}\right)^2 + \left(\frac{-16}{2}\right)^2$$

$$\left(x^2 + 3x + \frac{9}{4}\right) + (y^2 - 16y + 64) = -63 + \frac{9}{4} + 64 = \frac{13}{4}$$

$$\left(x + \frac{3}{2}\right)^2 + (y - 8)^2 = \left(\frac{\sqrt{13}}{2}\right)^2.$$

From the last equation, we see that the circle is centered at $(-\frac{3}{2}, 8)$ and has radius $\sqrt{13}/2$.

15. Substituting $h = 0$, $k = 0$, and $r = 1$ in equation (2) in the text, we obtain

$$(x - 0)^2 + (y - 0)^2 = 1^2 \qquad \text{or} \qquad x^2 + y^2 = 1.$$

18. Substituting $h = -9$, $k = -4$, and $r = \frac{3}{2}$ in equation (2) in the text, we obtain

$$\left(x - (-9)\right)^2 + \left(y - (-4)\right)^2 = \left(\frac{3}{2}\right)^2 \qquad \text{or} \qquad (x + 9)^2 + (y + 4)^2 = \frac{9}{4}.$$

21. We identify $h = 0$ and $k = 0$. Since the radius is the distance from the center to a point on the circle, we have

$$\sqrt{(-1 - 0)^2 + (-2 - 0)^2} = \sqrt{1 + 4} = \sqrt{5}.$$

Thus, the equation of the circle is

$$x^2 + y^2 = \left(\sqrt{5}\right)^2 \qquad \text{or} \qquad x^2 + y^2 = 5.$$

24. Since the graph is tangent to the y-axis, the radius of the circle is 4, as seen in the figure. Thus, the equation of the circle is

$$\left(x - (-4)\right)^2 + \left(y - 3\right)^2 = 4^2$$

$$(x + 4)^2 + (y - 3)^2 = 16.$$

or

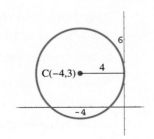

27. Squaring both sides of the equation and writing in the standard form of the equation of a circle, we get

$$x^2 = 1 - (y-1)^2 \qquad \text{or} \qquad x^2 + (y-1)^2 = 1 = 1^2.$$

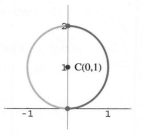

The graph of this equation is a circle with center $(0,1)$ and radius 1. Since the semicircle is defined by $x = \sqrt{1 - (y-1)^2}$, we see that $x \geq 0$. The graph is thus the right half of the circle.

30. To find the lower half of the circle, we solve for y and use the negative square root of the result:

$$(y-1)^2 = 9 - (x-5)^2$$
$$y - 1 = -\sqrt{9 - (x-5)^2}$$
$$y = 1 - \sqrt{9 - (x-5)^2}.$$

To find the left half of the circle, we solve for x and use the negative square root of the result:

$$(x-5)^2 = 9 - (y-1)^2$$
$$x - 5 = -\sqrt{9 - (y-1)^2}$$
$$x = 5 - \sqrt{9 - (y-1)^2}.$$

33. The inequality represents the region between the circles

$$x^2 + y^2 = 1 \qquad \text{and} \qquad x^2 + y^2 = 4.$$

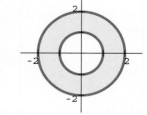

Because the inequalities both involve "$\leq$" (as opposed to "$<$"), the two circles are part of the region.

36. To find the x-intercepts, we set $y = 0$ in the equation. This gives $x^2 + 5x = 0$ or $x(x+5) - 0$. Thus, $x = 0$ and $x = -5$, so the x-intercepts are $(0,0)$ and $(-5,0)$. To find the y-intercepts we set $x = 0$ in the equation. This gives $y^2 - 6y = 0$ or $y(y-6) = 0$. Thus, $y = 0$ and $y = 6$, so the y-intercepts are $(0,0)$ and $(0,6)$. [Note that whenever the origin is an intercept, it is both an x- and a y-intercept.]

39. *Intercepts:* Setting $y = 0$ in $-x + 2y = 1$ we get $-x = 1$ or $x = -1$, so the x-intercept is $(-1,0)$. Setting $x = 0$ we have $2y = 1$ or $y = \frac{1}{2}$, so the y-intercept is $(0, \frac{1}{2})$.

Symmetry with respect to the x-axis: Replacing y by $-y$ in the equation $-x + 2y = 1$ we have $-x + 2(-y) = 1$ or $-x - 2y = 1$. This is not equivalent to the original equation, so the graph is not symmetric with respect to the x-axis.

Symmetry with respect to the y-axis: Replacing x by $-x$ in the equation $-x + 2y = 1$ we have $-(-x) + 2y = 1$ or $x + 2y = 1$. This is not equivalent to the original equation, so the graph is not symmetric with respect to the y-axis.

Symmetry with respect to the origin: Replacing x by $-x$ and y by $-y$ in the equation $-x+2y = 1$ we have $-(-x) + 2(-y) = 1$ or $x - 2y = 1$. This is not equivalent to the original equation, so the graph is not symmetric with respect to the origin.

42. *Intercepts*: Setting $y = 0$ in $y = x^3$ we get $0 = x^3$ or $x = 0$, so the x-intercept is $(0,0)$. This is also a y-intercept. Setting $x = 0$ we get $y = 0$, so $(0,0)$ is the only y-intercept.

Symmetry with respect to the x-axis: Replacing y by $-y$ in the equation $y = x^3$ we get $-y = x^3$ or $y = -x^3$. This is not equivalent to the original equation, so the graph is not symmetric with respect to the x-axis.

Symmetry with respect to the y-axis: Replacing x by $-x$ in the equation $y = x^3$ we get $y = (-x)^3$ or $y = -x^3$. This is not equivalent to the original equation, so the graph is not symmetric with respect to the y-axis.

Symmetry with respect to the origin: Replacing x by $-x$ and y by $-y$ in the equation $y = x^3$ we get $-y = (-x)^3$, $y = -(-x^3)$, so $y = x^3$. This is equivalent to the original equation, so the graph is symmetric with respect to the origin.

45. *Intercepts*: Setting $y = 0$ in $y = x^2 - 2x - 2$ we get $0 = x^2 - 2x - 2$. Using the quadratic formula, we find

$$x = \frac{-(-2) \pm \sqrt{(-2)^2 - 4(1)(-2)}}{2(1)} = \frac{2 \pm \sqrt{4+8}}{2} = \frac{2 \pm \sqrt{12}}{2} = \frac{2 \pm 2\sqrt{3}}{2} = 1 \pm \sqrt{3}.$$

The x-intercepts are $(1 - \sqrt{3}, 0)$ and $(1 + \sqrt{3}, 0)$. Setting $x = 0$ we get $y = 0^2 - 2(0) - 2 = -2$, so the y-intercept is $(0, -2)$.

Symmetry with respect to the x-axis: Replacing y by $-y$ in the equation $y = x^2 - 2x - 2$ we get $-y = x^2 - 2x - 2$ or $y = -x^2 + 2x + 2$. This is not equivalent to the original equation, so the graph is not symmetric with respect to the x-axis.

Symmetry with respect to the y-axis: Replacing x by $-x$ in the equation $y = x^2 - 2x - 2$ we get $y = (-x)^2 - 2(-x) - 2$ or $y = x^2 + 2x + 2$. This is not equivalent to the original equation, so the graph is not symmetric with respect to the y-axis.

Symmetry with respect to the origin: Replacing x by $-x$ and y by $-y$ in the equation $y = x^2 - 2x - 2$ we get $-y = (-x)^2 - 2(-x) - 2$ or $y = -x^2 - 2x + 2$. This is not equivalent to the original equation, so the graph is not symmetric with respect to the origin.

48. *Intercepts*: Setting $y = 0$ in $y = (x-2)^2(x+2)^2$ we get $0 = (x-2)^2(x+2)^2$, so the x-intercepts are $(-2, 0)$ and $(2, 0)$. Setting $x = 0$ we get $y = (0-2)^2(0+2)^2 = 4(4) = 16$, so the y-intercept is $(0, 16)$.

Symmetry with respect to the x-axis: Replacing y by $-y$ in the equation $y = (x - 2)^2(x + 2)^2$ we get $-y = (x - 2)^2(x + 2)^2$ or $y = -(x - 2)^2(x + 2)^2$. This is not equivalent to the original equation, so the graph is not symmetric with respect to the x-axis.

Symmetry with respect to the y-axis: Replacing x by $-x$ in the equation $y = (x-2)^2(x+2)^2$ we get $y = (-x-2)^2(-x+2)^2 = [-(x+2)]^2[-(x-2)]^2 = (x+2)^2(x-2)^2$. This is equivalent to the original equation, so the graph is symmetric with respect to the y-axis.

Symmetry with respect to the origin: Replacing x by $-x$ and y by $-y$ in the equation $y = (x-2)^2(x+2)^2$ we get $-y = (-x-2)^2(-x+2)^2 = [-(x+2)]^2[-(x-2)]^2 = (x+2)^2(x-2)^2$ or $y = -(x-2)^2(x+2)^2$. This is not equivalent to the original equation, so the graph is not symmetric with respect to the origin.

51. *Intercepts*: Setting $y = 0$ in $4y^2 - x^2 = 36$ we get $4(0)^2 - x^2 = 36$ or $x^2 = -36$. Since the square of a number must be nonnegative, the equation has no solution and there are no x-intercepts. Setting $x = 0$ we get $4y^2 - 0^2 = 36$, $y^2 = 9$, and $y = \pm 3$, so the y-intercepts are $(0, -3)$ and $(0, 3)$.

Symmetry with respect to the x-axis: Replacing y by $-y$ in the equation $4y^2 - x^2 = 36$ we get $4(-y)^2 - x^2 = 36$ or $4y^2 - x^2 = 36$. This is equivalent to the original equation, so the graph is symmetric with respect to the x-axis.

Symmetry with respect to the y-axis: Replacing x by $-x$ in the equation $4y^2 - x^2 = 36$ we get $4y^2 - (-x)^2 = 36$ or $4y^2 - x^2 = 36$. This is equivalent to the original equation, so the graph is symmetric with respect to the y-axis.

Symmetry with respect to the origin: Replacing x by $-x$ and y by $-y$ in the equation $4y^2 - x^2 = 36$ we get $4(-y)^2 - (-x)^2 = 36$ or $4y^2 - x^2 = 36$. This is equivalent to the original equation, so the graph is symmetric with respect to the origin.

54. *Intercepts*: Setting $y = 0$ in $y = (x^2 - 10)/(x^2 + 10)$ we get $0 = (x^2 - 10)/(x^2 + 10)$ so $x^2 - 10 = 0$ and $x = \pm\sqrt{10}$. The x-intercepts are $(-\sqrt{10}, 0)$ and $(\sqrt{10}, 0)$. Setting $x = 0$ we get $y = (0^2 - 10)/(0^2 + 10) = -10/10 = -1$, so the y-intercept is $(0, -1)$.

Symmetry with respect to the x-axis: Replacing y by $-y$ in the equation $y = (x^2 - 10)/(x^2 + 10)$ we get $-y = (x^2 - 10)/(x^2 + 10)$ or $y = -(x^2 - 10)/(x^2 + 10)$. This is not equivalent to the original equation, so the graph is not symmetric with respect to the x-axis.

Symmetry with respect to the y-axis: Replacing x by $-x$ in the equation $y = (x^2 - 10)/(x^2 + 10)$ we get $y = [(-x)^2 - 10]/[(-x)^2 + 10] = (x^2 - 10)/(x^2 + 10)$. This is equivalent to the original equation, so the graph is symmetric with respect to the y-axis.

Symmetry with respect to the origin: Replacing x by $-x$ and y by $-y$ in the equation $y = (x^2 - 10)/(x^2 + 10)$ we get $-y = [(-x)^2 - 10]/[(-x)^2 + 10] = (x^2 - 10)/(x^2 + 10)$, so $y = -(x^2 - 10)/(x^2 + 10)$. This is not equivalent to the original equation, so the graph is not symmetric with respect to the origin.

57. *Intercepts*: Setting $y = 0$ in $y = \sqrt{x} - 3$ we get $0 = \sqrt{x} - 3$ so $3 = \sqrt{x}$ and $x = 9$. Since we squared both sides of the equation to obtain this solution, we must check that $x = 9$ actually is a solution. It is, so the x-intercept is $(9, 0)$. Setting $x = 0$ we get $y = \sqrt{0} - 3 = -3$, so the y-intercept is $(0, -3)$.

Symmetry with respect to the x-axis: Replacing y by $-y$ in the equation $y = \sqrt{x} - 3$ we get $-y = \sqrt{x} - 3$ or $y = -(\sqrt{x} - 3)$. This is not equivalent to the original equation, so the graph is not symmetric with respect to the x-axis.

Symmetry with respect to the y-axis: Since the equation involves $\sqrt{x}$, x must be nonnegative. Hence, the graph of the equation cannot be symmetric with respect to the y-axis.

Symmetry with respect to the origin: Again, since the equation involves $\sqrt{x}$, x must be nonnegative. Hence, the graph of the equation cannot be symmetric with respect to the origin.

60. *Intercepts*: Setting $y = 0$ in $x = |y| - 4$ we get $x = |0| - 4 = -4$ so the x-intercept is $(-4, 0)$. Setting $x = 0$ we get $0 = |y| - 4$, $|y| = 4$, and $y = \pm 4$. Thus, the y-intercepts are $(0, -4)$ and $(0, 4)$.

Symmetry with respect to the x-axis: Replacing y by $-y$ in the equation $x = |y| - 4$ we get $x = |-y| - 4$ or $x = |y| - 4$. This is equivalent to the original equation, so the graph is symmetric with respect to the x-axis.

Symmetry with respect to the y-axis: Replacing x by $-x$ in the equation $x = |y| - 4$ we get $-x = |y| - 4$, so $x = -(|y| - 4)$. This is not equivalent to the original equation, so the graph is not symmetric with respect to the y-axis.

Symmetry with respect to the origin: Replacing x by $-x$ and y by $-y$ in the equation $x = |y| - 4$ we get $-x = |-y| - 4$, $-x = |y| - 4$ and $x = -(|y| - 4)$. This is not equivalent to the original equation, so the graph is not symmetric with respect to the origin.

63. We see that the graph is symmetric with respect to the x-axis, the y-axis, and the origin.

66. We see that the graph is symmetric with respect to the x-axis, the y-axis, and the origin.

In Problem 69 the given portion of the graph is shown in blue, while the reflected portion is shown in red.

69. Since the graph is to be symmetric with respect to the origin, each point in the first quadrant is reflected through the origin to a corresponding point on the graph in the third quadrant.

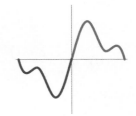

1.5 ∫ **Calculus** PREVIEW **Algebra and Limits**

3. (a) $\dfrac{x^2 - 7x + 6}{x - 1} = \dfrac{(x - 6)(x - 1)}{x - 1} = x - 6, \quad x \neq 1$

(b) $\displaystyle\lim_{x \to 1} \dfrac{x^2 - 7x + 6}{x - 1} = \lim_{x \to 1}(x - 6) = 1 - 6 = -5$

6. (a) $\dfrac{x^2-8x}{x^2-6x-16} = \dfrac{x(x-8)}{(x+2)(x-8)} = \dfrac{x}{x+2}, \quad x \neq 8$

(b) $\displaystyle\lim_{x\to 8}\dfrac{x^2-8x}{x^2-6x-16} = \lim_{x\to 8}\dfrac{x}{x+2} = \dfrac{8}{8+2} = \dfrac{8}{10} = \dfrac{4}{5}$

9. (a) $\dfrac{x^3-1}{x^2+3x-4} = \dfrac{(x-1)(x^2+x+1)}{(x-1)(x+4)} = \dfrac{x^2+x+1}{x+4}, \quad x \neq 1$

(b) $\displaystyle\lim_{x\to 1}\dfrac{x^3-1}{x^2+3x-4} = \lim_{x\to 1}\dfrac{x^2+x+1}{x+4} = \dfrac{1^2+1+1}{5} = \dfrac{3}{5}$

12. (a) $\dfrac{x^4-5x^3+4x-20}{x^4-5x^3+x-5} = \dfrac{x^3(x-5)+4(x-5)}{x^3(x-5)+(x-5)} = \dfrac{(x^3+4)(x-5)}{(x^3+1)(x-5)} = \dfrac{x^3+4}{x^3+1}, \quad x \neq 5$

(While x^3+1 factors with integer coefficients, x^3+4 does not, so there is no need to further factor the denominator. The fraction $(x^3+4)/(x^3+1)$ will not further simplify.)

(b) $\displaystyle\lim_{x\to 5}\dfrac{x^4-5x^3+4x-20}{x^4-5x^3+x-5} = \lim_{x\to 5}\dfrac{x^3+4}{x^3+1} = \dfrac{5^3+4}{5^3+1} = \dfrac{129}{126} = \dfrac{43}{42}$

15. (a) $\dfrac{(2x+1)^3-9}{x-1} = \dfrac{4x^2+4x+1-9}{x-1} = \dfrac{4x^2+4x-8}{x-1} = \dfrac{4(x^2+x-2)}{x-1} = \dfrac{4(x+2)(x-1)}{(x-1)}$

$= 4(x+2), \quad x \neq 1$

(b) $\displaystyle\lim_{x\to 1}\dfrac{(2x+1)^3-9}{x-1} = \lim_{x\to 1} 4(x+2) = 4(1+2) = 12$

18. (a) $\dfrac{(x+1)^3+(x-1)^3}{x} = \dfrac{(x^3+3x^2+3x+1)+(x^3-3x^2+3x-1)}{x} = \dfrac{2x^3+6x}{x}$

$= \dfrac{2x(x^2+3)}{x} = 2(x^3+3), \quad x \neq 0$

(b) $\displaystyle\lim_{x\to 0}\dfrac{(x+1)^3+(x-1)^3}{x} = \lim_{x\to 0} 2(x^3+3) = 2(0^3+3) = 6$

21. (a) $\dfrac{1}{x-2} - \dfrac{6}{x^2+2x-8} = \dfrac{1}{x-2} - \dfrac{6}{(x-2)(x+4)} = \dfrac{(x+4)-6}{(x-2)(x+4)} = \dfrac{x-2}{(x-2)(x+4)}$

$= \dfrac{1}{x+4}, \quad x \neq 2$

(b) $\displaystyle\lim_{x\to 2}\left[\dfrac{1}{x-2} - \dfrac{6}{x^2+2x-8}\right] = \lim_{x\to 2}\dfrac{1}{x+4} = \dfrac{1}{2+4} = \dfrac{1}{6}$

24. (a) $\dfrac{1}{x}\left[\dfrac{1}{9} - \dfrac{1}{x+9}\right] = \dfrac{1}{x}\left[\dfrac{(x+9)-9}{9(x+9)}\right] = \dfrac{x}{x(9)(x+9)} = \dfrac{1}{9(x+9)}, \quad x \neq 0$

(b) $\displaystyle\lim_{x\to 0} \dfrac{1}{x}\left[\dfrac{1}{9} - \dfrac{1}{x+9}\right] = \lim_{x\to 0} \dfrac{1}{9(x+9)} = \dfrac{1}{9(0+9)} = \dfrac{1}{81}$

27. (a) $\dfrac{\sqrt{x}-3}{x-9} = \dfrac{\sqrt{x}-3}{x-9}\dfrac{\sqrt{x}+3}{\sqrt{x}+3} = \dfrac{x-9}{(x-9)(\sqrt{x}+3)} = \dfrac{1}{\sqrt{x}+3}, \quad x \neq 9$

(b) $\displaystyle\lim_{x\to 9}\dfrac{\sqrt{x}-3}{x-9} = \lim_{x\to 9}\dfrac{1}{\sqrt{x}+3} = \dfrac{1}{\sqrt{9}+3} = \dfrac{1}{3+3} = \dfrac{1}{6}$

30. (a) $\dfrac{\sqrt{u+4}-3}{u-5} = \dfrac{\sqrt{u+4}-3}{u-5}\dfrac{\sqrt{u+4}+3}{\sqrt{u+4}+3} = \dfrac{u+4-9}{(u-5)(\sqrt{u+4}+3)} = \dfrac{u-5}{(u-5)(\sqrt{u+4}+3)}$

$$= \dfrac{1}{\sqrt{u+4}+3}, \quad u \neq 5$$

(b) $\displaystyle\lim_{u\to 5}\dfrac{\sqrt{u+4}-3}{u-5} = \lim_{u\to 5}\dfrac{1}{\sqrt{u+4}+3} = \dfrac{1}{\sqrt{5+4}+3} = \dfrac{1}{\sqrt{9}+3} = \dfrac{1}{3+3} = \dfrac{1}{6}$

33. (a) $\dfrac{4y^2}{\sqrt{y^2+y+1}-\sqrt{y+1}} = \dfrac{4y^2}{\sqrt{y^2+y+1}-\sqrt{y+1}}\dfrac{\sqrt{y^2+y+1}+\sqrt{y+1}}{\sqrt{y^2+y+1}+\sqrt{y+1}}$

$$= \dfrac{4y^2(\sqrt{y^2+y+1}+\sqrt{y+1})}{y^2+y+1-(y+1)} = \dfrac{4y^2(\sqrt{y^2+y+1}+\sqrt{y+1})}{y^2}$$

$$= 4(\sqrt{y^2+y+1}+\sqrt{y+1}), \quad y \neq 0$$

(b) $\displaystyle\lim_{y\to 0}\dfrac{4y^2}{\sqrt{y^2+y+1}-\sqrt{y+1}} = \lim_{y\to 0} 4(\sqrt{y^2+y+1}+\sqrt{y+1}) = 4(\sqrt{0^2+0+1}+\sqrt{0+1})$

$$= 4(\sqrt{1}+\sqrt{1}) = 8$$

36. $\dfrac{\dfrac{3}{(x+1)^2} - \dfrac{3}{(a+1)^2}}{x-a} = \dfrac{\dfrac{3(a+1)^2 - 3(x+1)^2}{(x+1)^2(a+1)^2}}{x-a} = \dfrac{3(a^2+2a+1) - 3(x^2+2x+1)}{(x-a)(x+1)^2(a+1)^2}$

$$= \dfrac{3a^2+6a-3x^2-6x}{(x-a)(x+1)^2(a+1)^2} = \dfrac{(3a^2-3x^2)+6(a-x)}{(x-a)(x+1)^2(a+1)^2}$$

$$= \dfrac{3(a-x)(a+x)+6(a-x)}{(x-a)(x+1)^2(a+1)^2} = \dfrac{3(a-x)[(a+x)+2]}{(x-a)(x+1)^2(a+1)^2}$$

$$= \dfrac{-3(x-a)(a+x+2)}{(x-a)(x+1)^2(a+1)^2} = \dfrac{-3(a+x+2)}{(x+1)^2(a+1)^2}, \quad x \neq a$$

39.
$$\frac{2x(-4x+6)^{1/2} - x^2(\tfrac{1}{2})(-4x+6)^{-1/2}(-4)}{[(-4x+6)^{1/2}]^2} = \frac{2x\left[(-4x+6)^{1/2} + x(-4x+6)^{-1/2}\right]}{-4x+6}$$

$$= \frac{2x\left[(-4x+6)^{1/2} + x(-4x+6)^{-1/2}\right]}{-4x+6} \frac{(-4x+6)^{1/2}}{(-4x+6)^{1/2}} = \frac{2x(-4x+6+x)}{(-4x+6)^{3/2}}$$

$$= \frac{-6x(x-2)}{(-4x+6)^{3/2}} = \frac{6x(2-x)}{(-4x+6)^{3/2}}$$

42. Multiply out the right-hand side and solve for y':

$$y' = 2(x-y) - 2(x-y)y'$$

$$2(x-y)y' + y' = 2x - 2y$$

$$[2(x-y) + 1]y' = 2x - 2y$$

$$y' = \frac{2x - 2y}{2x - 2y + 1}.$$

45. Multiply both sides of the equation by $(x-y)^2$ and solve for y':

$$(x-y)(1+y') - (x+y)(1-y') = (x-y)^2$$

$$(x-y) + (x-y)y' - (x+y) + (x+y)y' = (x-y)^2$$

$$(x-y+x+y)y' = (x-y)^2 - (x-y) + (x+y)$$

$$2xy' = x^2 - 2xy + y^2 + 2y$$

$$y' = \frac{x^2 - 2xy + y^2 + 2y}{2x}.$$

Chapter 1 Review Exercises

A. Fill in the Blanks

3. If (a,b) is in quadrant IV, then $a > 0$ and $b < 0$. Thus, (b,a) has a negative first coordinate and positive second coordinate, so it is in the second quadrant.

6. If a graph is symmetric with respect to the origin, then whenever (x,y) is on the graph, so is $(-x,-y)$. Thus, if $(-1,6)$ is on a graph which is symmetric with respect to the origin, then $(1,-6)$ is also on the graph.

9. Using the midpoint formula, we find that the coordinates of the midpoint are given by

$$\left(\frac{4+(-2)}{2}, \frac{-6+0}{2}\right) \qquad \text{or} \qquad (1,-3).$$

Using the distance formula, we find that the distance from $(0,0)$ to this midpoint is

$$d(0, M) = \sqrt{(1-0)^2 + (-3-0)^2} = \sqrt{1+9} = \sqrt{10}.$$

12. We complete the square to obtain the equation of the circle in standard form:

$$x^2 - 16y + y^2 = 0$$
$$(x^2 - 16y \quad) + (y-0)^2 = 0$$
$$(x^2 - 16y + 8^2) + (y-0)^2 = 8^2$$
$$(x-4)^2 + (y-0)^2 = 64.$$

The circle is centered at $(8,0)$ and is thus symmetric with respect to the x-axis.

15. Letting $x = -3$ in $x^2 + y^2 = 25$ we have

$$(-3)^2 + y^2 = 25$$
$$9 + y^2 = 25$$
$$y^2 = 25 - 9 = 16,$$

so $y = \pm 4$, The points on the circle are $(-3, -4)$ and $(-3, 4)$.

18. Letting $x = a$ and $y = a + \sqrt{3}$, we have

$$a + \sqrt{3} = 2a \qquad \text{or} \qquad a = \sqrt{3}.$$

B. True/False

3. True; $-3 \leq -1$.

6. False; if $a = -1$, then $-a = -(-1) = 1 > -1 = a$.

9. True

12. False; points on the coordinates axes do not lie in any quadrant. (See Figure 1.3.3 and Example 1 in Section 1.3 of the text.)

15. True; y-intercepts occur on the y-axis, which is the line $x = 0$.

18. True; see the figure.

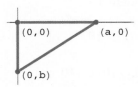

C. Review Exercises

3. Since b is positive, $a < a + b$.

6. Adding 2 to both sides of the inequality, we find $x \leq 9$.

9. First, add -1 to all three parts. This gives

$$-1 - 2 \leq -1 + 1 - x \leq -1 + 5$$
$$-3 \leq -x \leq 4 \qquad \leftarrow \text{ multiply by } -1$$
$$3 \geq x \geq -4.$$

Thus, $-4 \leq x \leq 3$.

12. We equate the two distances:

$$\sqrt{(x - 0)^2 + (y - 5)^2} = \sqrt{(x - x)^2 + (-5 - y)^2}$$
$$x^2 + (y - 5)^2 = 0^2 + (-5 - y)^2$$
$$x^2 + y^2 - 10y + 25 = 25 + 10y + y^2$$
$$x^2 = 20y.$$

The set of points is $y = \frac{1}{20}x^2$.

15. (a) $x \geq -4$ or $-4 \leq x < \infty$ \qquad (b) $[-4, \ \infty)$

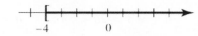

18. Using the properties of inequalities, we have

$$\frac{1}{4}x - 3 < \frac{1}{2}x + 1$$
$$-\frac{1}{4}x + \left(\frac{1}{4}x - 3\right) - 1 < -\frac{1}{4}x + \left(\frac{1}{2}x + 1\right) - 1$$
$$-4 < \frac{1}{4}x$$
$$-16 < x$$
$$x > -16.$$

In interval notation this is $(-16, \infty)$.

21. The inequality $|x| > 10$ is equivalent to

$$x > 10 \qquad \text{or} \qquad x < -10.$$

In interval notation this is $(-\infty, -10) \cup (10, \infty)$.

24. The inequality $|5-2x| \geq 7$ is equivalent to

$$5-2x \geq 7 \qquad \text{or} \qquad 5-2x \leq -7.$$

Solving $5-2x \geq 7$ we obtain $-2x \geq 2$, so $x \leq -1$. In interval notation this is $(-\infty, -1]$. Solving $5-2x \leq -7$ we obtain $-2x \leq -12$, so $x \geq 6$. In interval notation this is $[6, \infty)$. The solution set is $(-\infty, 1] \cup [6, \infty)$.

27. Rewriting the inequality and factoring, we have

$$\begin{aligned} x^3 &> x \\ x^3 - x &> 0 \\ x(x^2-1) &> 0 \\ x(x-1)(x+1) &> 0. \end{aligned}$$

Placing $x=-1$, $x=0$, and $x=1$ on the number line determines four intervals. The sign chart below, in turn, determines the solution set.

$x+1$	$-$	0	$+$	$+$	$+$	$+$	$+$
$x-1$	$-$	$-$	$-$	$-$	$-$	0	$+$
x	$-$	$-$	$-$	0	$+$	$+$	$+$
$x(x-1)(x+1)$	$-$	0	$+$	0	$-$	0	$+$

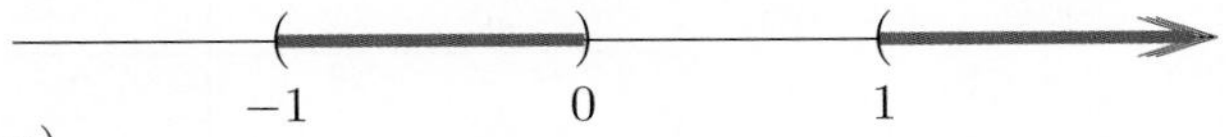

The solution set is $(-1, 0) \cup (1, \infty)$.

30. First, rewrite the inequality as a simple fraction:

$$\begin{aligned} \frac{2x-6}{x-1} &> 1 \\ \frac{2x-6}{x-1} - 1 &> 0 \\ \frac{2x-6-(x-1)}{x-1} &> 0 \\ \frac{x-5}{x-1} &> 0. \end{aligned}$$

Placing $x=1$ and $x=5$ on the number line determines three intervals. The sign chart below, in turn, determines the solution set.

$x-1$	$-$	0	$+$	$+$	$+$
$x-5$	$-$	$-$	$-$	0	$+$
$(x-5)/(x-1)$	$+$	undefined	$-$	0	$+$

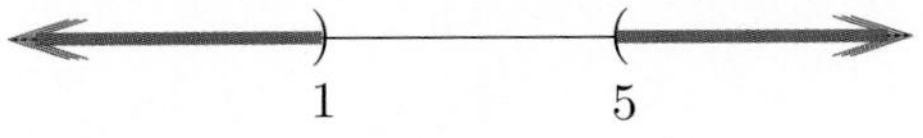

The solution set is $(-\infty, 1) \cup (5, \infty)$.

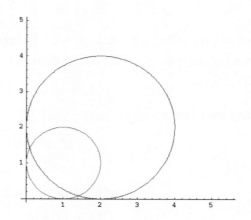

33. Choose the center to be any point on the negative y-axis and the radius to be the distance from the point to the origin. Circles satisfying the conditions are

$$x^2 + (y+1)^2 = 1$$
$$x^2 + (y+2)^2 = 4$$

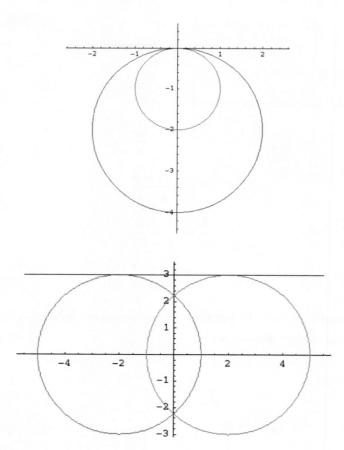

36. To solve $|x - 3| + |x + 1| < 10$ we subdivide the real number line as shown in the figure.

For $x \geq 3$ and $x \geq -1 \Rightarrow x \geq 3$ and so $x - 3 + x + 1 < 10 \Rightarrow 2x < 12$ $\Rightarrow x < 6$. The inequality is true for $3 \leq x < 6$.

For $x < 3$ and $x \geq -1 \Rightarrow -1 \leq x < 3 \Rightarrow -x + 3 + x + 1 < 10 \Rightarrow 4 < 10$ which is always true, so we have $-1 \leq x < 3$.

Finally for $x < -1 \Rightarrow x < 3 \Rightarrow -x + 3 - x - 1 < 10 \Rightarrow -2x < 8 \Rightarrow x > -4$, that is, $-4 < x < -1$.

The solution set is the union of intervals $(-4, -1) \cup [-1, 3) \cup [3, 6) = (-4, 6)$.

39. (a) Rationalizing the denominator, we have

$$\frac{x^2 - 16}{\sqrt{x} - 2} = \frac{x^2 - 16}{\sqrt{x} - 2} \frac{\sqrt{x} + 2}{\sqrt{x} + 2} = \frac{(x - 4)(x + 4)(\sqrt{x} + 2)}{x - 4} = (x + 4)(\sqrt{x} + 2), \quad x \neq 4.$$

(b) $\lim\limits_{x \to 4} \dfrac{x^2 - 16}{\sqrt{x} - 2} = \lim\limits_{x \to 4} (x + 4)(\sqrt{x} + 2) = (4 + 4)(\sqrt{4} + 2) = 8(4) = 32.$

Chapter 2

Functions

2.1 | **Functions and Graphs**

3. $f(-1) = \sqrt{-1+1} = \sqrt{0} = 0$

$f(0) = \sqrt{0+1} = \sqrt{1} = 1$

$f(3) = \sqrt{3+1} = \sqrt{4} = 2$

$f(5) = \sqrt{5+1} = \sqrt{6}$

6. $f(-\sqrt{2}) = \dfrac{(-\sqrt{2})^2}{(-\sqrt{2})^3 - 2} = \dfrac{2}{-2\sqrt{2} - 2} = \dfrac{1}{-\sqrt{2} - 1} = -\dfrac{1}{\sqrt{2} + 1}$

$f(-1) = \dfrac{(-1)^2}{(-1)^3 - 2} = \dfrac{1}{-1 - 2} = \dfrac{1}{-3} = -\dfrac{1}{3}$

$f(0) = \dfrac{0^2}{0^3 - 2} = \dfrac{0}{-2} = 0$

$f(\tfrac{1}{2}) = \dfrac{(\tfrac{1}{2})^2}{(\tfrac{1}{2})^3 - 2} = \dfrac{\tfrac{1}{4}}{\tfrac{1}{8} - 2} = \dfrac{\tfrac{1}{4}}{\tfrac{1}{8} - 2}\left(\dfrac{8}{8}\right) = \dfrac{2}{1 - 16} = \dfrac{2}{-15} = -\dfrac{2}{15}$

9. Setting $f(x) = 23$ and solving for x, we find

$$6x^2 - 1 = 23$$
$$6x^2 = 24$$
$$x^2 = 4$$
$$x = \pm 2.$$

When we compute $f(-2)$ and $f(2)$ we obtain 23 in both cases, so $x = \pm 2$ is the answer.

12. The domain of $f(x) = \sqrt{15 - 5x}$ is the set of all x for which $15 - 5x \geq 0$. This is equivalent to

$$15 \geq 5x$$
$$3 \geq x$$
$$x \leq 3.$$

The domain of $f(x)$ is $(-\infty, 3]$.

15. The domain of $f(x) = (2x - 5)/x(x - 3)$ is the set of all x for which $x(x - 3) \neq 0$. Since $x(x - 3) = 0$ when $x = 0$ or $x = 3$, the domain of $f(x)$ is $\{x | x \neq 0, \ x \neq 3\}$.

18. The domain of $f(x) = (x + 1)/(x^2 - 4x - 12)$ is the set of all x for which $x^2 - 4x - 12 \neq 0$. Since $x^2 - 4x - 12 = (x + 4)(x - 6) = 0$ when $x = -4$ or $x = 6$, the domain of $f(x)$ is $\{x | x \neq -4, \ x \neq 6\}$.

21. The domain of $f(x) = \sqrt{25 - x^2}$ is the set of all x for which $25 - x^2 \geq 0$. We solve this inequality by writing $25 - x^2 = (5 + x)(5 - x) = 0$ and placing $x = -5$ and $x = 5$ on the number line. This determines three intervals, and the sign chart below then determines the solution set.

$$
\begin{array}{lccccc}
5 + x & - & 0 & + & + & + \\
5 - x & + & + & + & 0 & - \\
(5 + x)(5 - x) & - & 0 & + & 0 & - \\
\end{array}
$$

The domain of the function is $[-5, 5]$.

24. The domain of $f(x) = \sqrt{x^2 - 3x - 10}$ is the set of all x for which $x^2 - 3x - 10 \geq 0$. We solve this inequality by writing $x^2 - 3x - 10 = (x + 2)(x - 5) = 0$ and placing $x = -2$ and $x = 5$ on the number line. This determines three intervals, and the sign chart below then determines the solution set.

$$
\begin{array}{lccccc}
x + 2 & - & 0 & + & + & + \\
x - 5 & - & - & - & 0 & + \\
(x + 2)(x - 5) & + & 0 & - & 0 & + \\
\end{array}
$$

The domain of the function is $(-\infty, -2] \cup [5, \infty)$.

27. Since the y-axis (a vertical line) intersects the graph in more than one point (three points in this case), the graph is not that of a function.

30. Since the y-axis (a vertical line) intersects the graph in more than one point (three points in this case), the graph is not that of a function.

33. Horizontally, the graph extends between $x = 1$ and $x = 9$ and terminates at both ends, as indicated by the solid dots. Thus, the domain is $[1, 9]$. Vertically, the graph extends between $y = 1$ and $y = 6$, so the range is $[1, 6]$.

36. We solve $f(x) = -2x + 9 = 0$:

$$-2x + 9 = 0$$
$$-2x = -9$$
$$x = \frac{-9}{-2} = \frac{9}{2}.$$

Thus, the zero of the function is $x = \frac{9}{2}$.

39. Setting $f(x) = x(3x - 1)(x + 9) = 0$ we see that the zeros of the function occur where $x = 0$, $3x - 1 = 0$ or $x = \frac{1}{3}$, and $x = -9$.

42. We solve $f(x) = 2 - \sqrt{4 - x^2} = 0$:

$$2 - \sqrt{4 - x^2} = 0$$
$$2 = \sqrt{4 - x^2} \quad \leftarrow \quad \text{square both sides of the equality}$$
$$4 = 4 - x^2$$
$$-4 + 4 = -4 + 4 - x^2$$
$$0 = -x^2$$
$$x^2 = 0$$
$$x = 0.$$

This result is easily checked, verifying that $x = 0$ is a solution, so the only zero of f is 0.

45. *x-intercepts*: We solve $f(x) = 4(x - 2)^2 - 1 = 0$:

$$4(x - 2)^2 - 1 = 0$$
$$4(x - 2)^2 = 1$$
$$(x - 2)^2 = \frac{1}{4}$$
$$x - 2 = \pm\sqrt{\frac{1}{4}} = \pm\frac{1}{2}$$
$$x = 2 \pm \frac{1}{2}.$$

Both of these check with the original equation, so the x-intercepts are $\left(\frac{3}{2}, 0\right)$ and $\left(\frac{5}{2}, 0\right)$.

y-intercept: Since $f(0) = 4(0 - 2)^2 - 1 = 4(4) - 1 = 15$, the y-intercept is $(0, 15)$.

48. *x-intercepts*: We solve $f(x) = x(x + 1)(x - 6)/(x + 8) = 0$:

$$\frac{x(x + 1)(x - 6)}{x + 8} = 0$$
$$x(x + 1)(x - 6) = 0, \quad x \neq -8.$$

Thus, $x = 0$, $x = -1$, and $x = 6$. The x-intercepts are $(0, 0)$, $(-1, 0)$, and $(6, 0)$.

y-intercept: Since $f(0) = 0$, the y-intercept is $(0,0)$.

Note that whenever the origin is an x-intercept of a function, it is also the y-intercept.

51. Solving $x = y^2 - 5$ for y, we obtain

$$x = y^2 - 5$$
$$x + 5 = y^2$$
$$y^2 = x + 5$$
$$y = \pm\sqrt{x + 5}.$$

The two functions are $f_1(x) = -\sqrt{x+5}$ and $f_2(x) = \sqrt{x+5}$. The domains are both $[-5, \infty)$.

54. $f(x) = -2x^2 + kx$

$f(2) = -3 \Rightarrow f(2) = -8 + 2k = -3 \Rightarrow k = \frac{5}{2} \Rightarrow f(x) = -2x^2 + \frac{5}{2}x$

57. $f(x) = \dfrac{2x - k}{x}$

$f(2) = -3 \Rightarrow f(2) = \dfrac{4 - k}{2} = -3 \Rightarrow k = 10 \Rightarrow f(x) = \dfrac{2x - 10}{x}$

60. The function values are the directed distances from the x-axis at the given value of x.

$$f(-3) \approx 1; \quad f(-2) \approx 1.4; \quad f(-1) \approx 1.8; \quad f(1) \approx -0.8; \quad f(2) \approx -1.3; \quad f(3) \approx 0.8.$$

Miscellaneous Calculus-Related Problems

63. (a) $f(2) = 2! = 2 \cdot 1 = 2$

$f(3) = 3! = 3 \cdot 2 \cdot 1 = 6$

$f(5) = 5! = 5 \cdot 4 \cdot 3 \cdot 2 \cdot 1 = 120$

$f(7) = 7! = 7 \cdot 6 \cdot 5 \cdot 4 \cdot 3 \cdot 2 \cdot 1 = 5040$

Note that we could have simplified the computation of 7! in this case by writing

$$7! = 7 \cdot 6 \cdot 5! = 7 \cdot 6 \cdot 120 = 5040.$$

(b) $f(n+1) = (n+1)! = (n+1)n! = n!(n+1) = f(n)(n+1)$

(c) $\dfrac{f(n+2)}{f(n)} = \dfrac{(n+2)!}{n!} = \dfrac{(n+2)(n+1)n!}{n!} = (n+2)(n+1)$

2.2 | Symmetry and Transformations

3. Since

$$f(-x) = (-x)^3 - (-x) + 4 = -x^3 + x + 4$$

is neither $f(x)$ nor $-f(x)$, the function is neither even nor odd.

6. Since

$$f(-x) = \frac{-x}{(-x)^2 + 1} = -\frac{x}{x^2 + 1} = -f(x),$$

the function is odd. Because the only function that is both even and odd is $f(x) = 0$, and f is odd in this case, it cannot also be even. See **Even and Odd Functions** in **Part I, Topics in Algebra**, of this manual.

9. Since

$$f(-x) = |(-x)^3| = |-x^3| = x^3 = f(x),$$

the function is even. Because the only function that is both even and odd is $f(x) = 0$, and f is even in this case, it cannot also be odd. See **Even and Odd Functions** in **Part I, Topics in Algebra**, of this manual.

12. The function is odd because we can see from the graph that $f(-x) = -f(x)$.

15. (a) An even function is symmetric with respect to the y-axis.

(b) An odd function is symmetric with respect to the origin.

18. (a) An even function is symmetric with respect to the y-axis.

(b) An odd function is symmetric with respect to the origin.

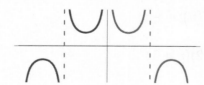

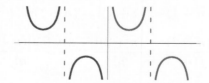

21. Since g is an odd function, $g(-x) = -g(x)$. Thus, taking $x = -1$, we have $g(-(-1)) = -g(-1) = -(-5) = 5$, and taking $x = 4$, we have $g(-4) = -g(4) = -8$.

24. Shifting $(2, 1)$ down 5 units we get $(2, 1 - 5)$ or $(2, -4)$. Shifting $(3, -4)$ down 5 units we get $(3, -4 - 5)$ or $(3, -9)$.

27. Shifting $(-2, 1)$ up 1 unit and left 4 units we get $(-2 - 4, 1 + 1)$ or $(-6, 2)$. Shifting $(3, -4)$ up 1 unit and left 4 units we get $(3 - 4, -4 + 1)$ or $(-1, -3)$.

30. Reflecting $(-2, 1)$ in the x-axis changes the y-coordinate from 1 to -1. The reflected point is then $(-2, -1)$. Reflecting $(3, -4)$ in the x-axis changes the y-coordinate from -4 to 4. The reflected point is then $(3, 4)$.

33. In **(a)** the graph is shifted up 2 units; in **(b)** it is shifted down 2 units; in **(c)** it is shifted left 2 units; in **(d)** it is shifted right 3 units; in **(e)** it is reflected in the x-axis; and in **(f)** it is reflected in the y-axis.

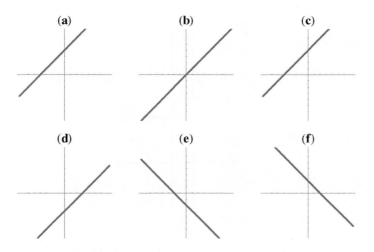

36. In **(a)** the graph is shifted up 2 units; in **(b)** it is shifted down 2 units; in **(c)** it is shifted left 2 units; in **(d)** it is shifted right 3 units; in **(e)** it is reflected in the x-axis; and in **(f)** it is reflected in the y-axis.

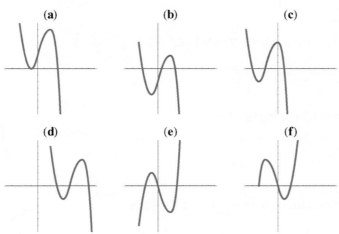

39. If $f(x)$ is shifted up 5 units and right 1 unit, the new function is $y = f(x - 1) + 5$. Since $f(x) = x^3$, this becomes $y = (x - 1)^3 + 5$.

42. If $f(x)$ is reflected in the y-axis, then shifted left 5 units and down 10 units, the new function is $y = -f(x+5) - 10$. Since $f(x) = 1/x$, this becomes $y = -1/(x+5) - 10$.

45. The graph of $1/x$ is shifted 1 unit to the right to obtain $y = 1/(x-1)$. A reflection in the x-axis gives $y = -1/(x-1)$. A vertical stretch by a factor of 2 then gives $y = -2/(x-1)$. Finally, by shifting the last graph up 2 units we get $y = 2 - 2/(x-1)$.

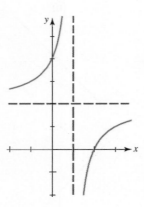

2.3   Linear Functions

3. The slope of the line through $(5, 2)$ and $(4, -3)$ is

$$m = \frac{-3 - 2}{4 - 5} = \frac{-5}{-1} = 5.$$

The graph is shown to the right.

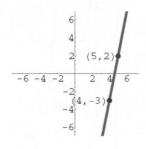

6. The slope of the line through $(8, -\frac{1}{2})$ and $(2, \frac{5}{2})$ is

$$m = \frac{\frac{5}{2} - \left(-\frac{1}{2}\right)}{2 - 8} = \frac{3}{-6} = -\frac{1}{2}.$$

The graph is shown to the right.

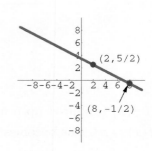

9. To find the x-intercept we set $y = 0$. This gives $3x + 12 = 0$ or $x = -4$. The x-intercept is $(-4, 0)$. Now, write the equation in slope-intercept form by solving for y:

$$3x - 4y + 12 = 0$$
$$-4y = -3x - 12$$
$$y = \frac{3}{4}x + 3.$$

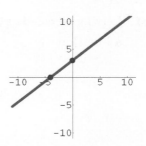

The slope of the line is $m = \frac{3}{4}$ and the y-intercept is $(0, 3)$. The graph of the line is shown to the right.

12. To find the x-intercept we set $y = 0$. This gives $-4x + 6 = 0$ or $x = \frac{3}{2}$. The x-intercept is $(\frac{3}{2}, 0)$. Now, write the equation in slope-intercept form by solving for y:

$$-4x - 2y + 6 = 0$$
$$-2y = 4x - 6$$
$$y = -2x + 3.$$

The slope of the line is $m = -2$ and the y-intercept is $(0, 3)$. The graph of the line is shown to the right.

15. To find the x-intercept we set $y = 0$. This gives $\frac{2}{3}x = 1$ or $x = \frac{3}{2}$. The x-intercept is $(\frac{3}{2}, 0)$. Solving for y, we have

$$y + \frac{2}{3}x = 1 \qquad \text{or} \qquad y = -\frac{2}{3}x + 1.$$

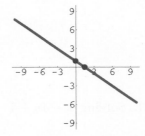

The slope of the line is $m = -\frac{2}{3}$ and the y-intercept is $(0, 1)$. The graph of the line is shown to the right.

18. Letting $m = \frac{1}{10}$, $x_1 = 1$, and $y_1 = 2$, we obtain from the point-slope form of the equation of a line that

$$y - 2 = \frac{1}{10}(x - 1)$$
$$y - 2 = \frac{1}{10}x - \frac{1}{10}$$
$$y = \frac{1}{10}x + \frac{19}{10}.$$

21. Letting $m = -1$, $x_1 = 1$, and $y_1 = 2$, we obtain from the point-slope form of the equation of a line that

$$y - 2 = -1(x - 1)$$
$$y - 2 = -x + 1$$
$$y = -x + 3.$$

24. The slope of the line is

$$m = \frac{0 - (-6)}{4 - 5} = -6.$$

The form of the equation of the line is $y = -6x + b$. Letting $x = 4$ and $y = 0$, we have $0 = -6(4) + b$, so $b = 24$. The equation of the line is $y = -6x + 24$.

27. Since the x-coordinates of the two points are the same, the points determine the vertical line $x = -2$.

30. Solving $2x - 5y + 4 = 0$ for y, we obtain

$$-5y = -2x - 4 \qquad \text{or} \qquad y = \frac{2}{5}x + \frac{4}{5}.$$

Thus, the slope of the parallel line is $\frac{2}{5}$. Now, using the point-slope form of the equation of a line, we identify $m = \frac{2}{5}$ $x_1 = 1$, and $y_1 = -3$, so that

$$y - (-3) = \frac{2}{5}(x - 1)$$
$$y + 3 = \frac{2}{5}x - \frac{2}{5}$$
$$y = \frac{2}{5}x - \frac{17}{5}.$$

33. Solving $x - 4y + 1 = 0$ for y, we obtain

$$-4y = -x - 1 \qquad \text{and} \qquad y = \frac{1}{4}x + \frac{1}{4}.$$

Thus, the slope of a line perpendicular to this one is $m = -1/(1/4) = -4$. We identify $x_1 = 2$ and $y_1 = 3$ and use the point-slope form of the equation of a line:

$$y - 3 = -4(x - 2)$$
$$y - 3 = -4x + 8$$
$$y = -4x + 11.$$

36. A line perpendicular to another line with slope 2 has slope $m = -\frac{1}{2}$. Since the line passes through the origin, the y-intercept is 0. Using the slope-intercept form of the equation of a line, we then have $y = -\frac{1}{2}x + 0$ or $y = -\frac{1}{2}x$.

39. We solve each equation for y to find its slope:

(a) $y = \dfrac{3}{5}x + \dfrac{9}{5}$; $m = \dfrac{3}{5}$

(b) $y = -\dfrac{5}{3}x$; $m = -\dfrac{5}{3}$

(c) $y = \dfrac{3}{5}x + \dfrac{2}{5}$; $m = \dfrac{3}{5}$

(d) $y = -\dfrac{3}{5}x - \dfrac{4}{5}$; $m = -\dfrac{3}{5}$

(e) $y = -\dfrac{5}{3}x + \dfrac{8}{3}$; $m = -\dfrac{5}{3}$

(f) $y = \dfrac{5}{3}x + \dfrac{2}{3}$; $m = \dfrac{5}{3}$

Since parallel lines have equal slopes and the slopes of perpendicular lines satisfy $m_1 m_2 = -1$, we see that **(a)** and **(c)**, and **(b)** and **(e)** are parallel; while **(a)** and **(b)**, **(a)** and **(e)**, **(b)** and **(c)**, **(c)** and **(e)**, and **(d)** and **(f)** are perpendicular.

42. We solve each equation for y to find its slope:

(a) $y = -5$;   line is horizontal

(b) $x = 7$;   line is vertical

(c) $y = -\dfrac{2}{3}x + \dfrac{1}{2}$; $m = -\dfrac{2}{3}$

(d) $y = \dfrac{4}{3}x + \dfrac{7}{9}$; $m = \dfrac{4}{3}$

(e) $y = \dfrac{2}{3}x - \dfrac{2}{3}$; $m = \dfrac{2}{3}$

(f) $y = -\dfrac{3}{4}x + \dfrac{11}{4}$; $m = -\dfrac{3}{4}$

Since parallel lines have equal slopes and the slopes of perpendicular lines satisfy $m_1 m_2 = -1$, we see that none of the lines are parallel, while **(a)** and **(b)**, and **(d)** and **(f)** are perpendicular.

45. We want to solve the simultaneous equations

$$y = -2x + 1$$
$$y = 4x + 6.$$

This implies that $-2x + 1 = 4x + 6$ or $-5 = 6x$. Thus, $x = -\dfrac{5}{6}$. Substituting into the first of the simultaneous equations, we find $y = -2(-\dfrac{5}{6}) + 1 = \dfrac{5}{3} + 1 = \dfrac{8}{3}$. Thus, the point of intersection is $(-\dfrac{5}{6}, \dfrac{8}{3})$.

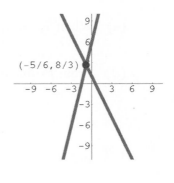

48. We want to solve the simultaneous equations

$$y = 2x - 10$$
$$y = -3x.$$

This implies that $2x - 10 = -3x$ or $5x = 10$. Thus, $x = 2$, and from the second equation, we see that $y = -3(2) = -6$. The point of intersection is $(2, -6)$.

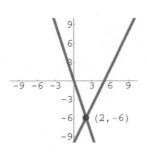

51. When $x = -1$, the corresponding point on the blue curve has y-coordinate $y = (-1)^2 + 1 = 2$. When $x = 2$, the corresponding point on the blue curve has y-coordinate $y = 2^2 + 1 = 5$. The slope of the line through $(-1, 2)$ and $(2, 5)$ is

$$m = \frac{5 - 2}{2 - (-1)} = \frac{3}{3} = 1.$$

The form of the equation of the line is then $y = 1x + b = x + b$. Using $x = -1$ and $y = 2$ we have $2 = -1 + b$ or $b = 3$. Thus, the equation of the line is $y = x + 3$.

Miscellaneous Applications

54. We are looking for a and b where $T_K = aT_C + b$.

    (a) Setting $T_C = 0$ with $T_K = 273$ and $T_C = 27$ with $T_K = 300$, we have

$$273 = a(0) + b = b$$
$$300 = a(27) + b,$$

    so $b = 273$ and $27a = 300 - b = 300 - 273 = 27$. Thus, $a = 1$ and $T_K = T_C + 273$.

    (b) When $T_C = 100$, $T_K = 100 + 273 = 373$. Thus, the boiling point in degrees Kelvin is 373 K.

    (c) When $T_K = 0$, we have $0 = T_C + 273$, so $T_C = -273$. Thus, absolute zero is $-273°$ C.

    (d) From Problem 53 we have $T_F = \frac{9}{5}T_C + 32$, so

$$T_C = \frac{5}{9}T_F - \frac{160}{9}$$

    and

$$T_K = T_C + 273 = \frac{5}{9}T_F - \frac{160}{9} + 273 = \frac{5}{9}T_F + \frac{2297}{9}.$$

    (e) When $T_K = 0$, $T_F = -\frac{9}{5}\left(\frac{2297}{9}\right) = -\frac{2297}{5} = -459.4°$ F.

2.4 | Quadratic Functions

We obtain the graphs in Problems 3 and 6 by a combination of stretches, translations, and reflections.

3. The graph of $f(x) = 2x^2 - 2$ is a parabola that opens upward because the coefficient of x^2 is positive. Identifying $a = 2$ and $b = 0$, we see that the vertex is at $x = -b/2a = 0$. Thus, the vertex is $(0, f(0)) = (0, -2)$. To find the x-intercepts we solve $2x^2 - 2 = 0$:

$$x^2 - 1 = 0$$
$$x^2 = 1$$
$$x = \pm 1.$$

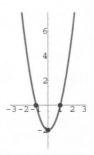

Thus, $(-1, 0)$ and $(1, 0)$ are on the graph. Plotting the three points, $(-1, 0)$, $(0, -2)$, and $(1, 0)$, we obtain the graph of the parabola.

6. The graph of $f(x) = -2x^2 - 3$ is a parabola that opens downward because the coefficient of x^2 is negative. Identifying $a = -2$ and $b = 0$ we see that the vertex is at $x = -b/2a = 0$. Thus, the vertex is $(0, f(0)) = (0, -3)$. Since $-2x^2 - 3 = 0$ is equivalent to $x^2 = -\frac{3}{2}$, the equation has no solutions and there are no x-intercepts. In this case, the only intercept is the vertex. To graph the function we need another point on the graph. To obtain such a point simply choose any x other than $x = 0$ and find $f(x)$. For example, if we let $x = 1$, we obtain $f(1) = -2(1)^2 - 3 = -2(1) - 3 = -5$. Thus, the point $(1, -5)$ is on the graph. Since f is an even function, $(-1, -5)$ is also on the graph and we plot the parabola through $(-1, -5)$, $(0, -3)$, and $(1, -5)$.

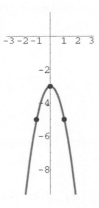

9. (a) *x-intercepts*: Solving $(3 - x)(x + 1) = 0$ we obtain $x = 3$ and $x = -1$, so the x-intercepts are $(-1, 0)$ and $(3, 0)$.

   y-intercept: Since $f(0) = (3 - 0)(0 + 1) = 3$, the y-intercept is $(0, 3)$.

(b) We expand the expression and complete the square:

$$
\begin{aligned}
f(x) &= (3 - x)(x + 1) \\
&= -x^2 + 2x + 3 \\
&= -(x^2 - 2x \quad) + 3 \\
&= -(x^2 - 2x + 1) + 3 + 1 \\
&= -(x - 1)^2 + 4.
\end{aligned}
$$

(c) Identifying $h = 1$ and $k = 4$ we see that the vertex is $(1, 4)$. The axis of symmetry is $x = 1$.

(d) Since the coefficient of x^2 is negative, the graph opens downward. We use the vertex and intercepts to draw the graph.

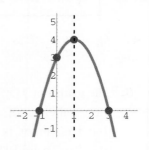

12. (a) *x-intercepts*: Factoring, we obtain

$$f(x) = -x^2 + 6x - 5 = -(x^2 - 6x + 5) = -(x - 1)(x - 5).$$

Solving $-(x - 1)(x - 5) = 0$ we obtain $x = 1$ and $x = 5$, so the x-intercepts are $(1, 0)$ and $(5, 0)$.

y-intercept: Since $f(0) = -5$, the y-intercept is $(0, -5)$.

(b) To complete the square and obtain the standard form, we start by factoring -1 from the two x-terms:

$$\begin{aligned}
f(x) &= -x^2 + 6x - 5 \\
&= -(x^2 - 6x \quad) - 5 \\
&= -(x^2 - 6x + 9) - 5 + 9 \\
&= -(x - 3)^2 + 4.
\end{aligned}$$

(c) Identifying $h = 3$ and $k = 4$ we see that the vertex is $(3, 4)$. The axis of symmetry is $x = 3$.

(d) Since the coefficient of x^2 is negative, the graph opens downward. We use the vertex and intercepts to draw the graph.

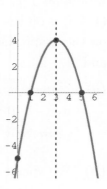

15. (a) *x-intercepts*: Factoring, we obtain

$$f(x) = -\frac{1}{2}x^2 + x + 1 = -\frac{1}{2}(x^2 - 2x - 2).$$

Using the quadratic formula we find that the roots of $x^2 - 2x - 2$ are $(2 \pm \sqrt{4 + 8})/2 = 1 \pm \sqrt{3}$, so the x-intercepts are $(1 - \sqrt{3}, 0)$ and $(1 + \sqrt{3}, 0)$.

y-intercept: Since $f(0) = 1$, the y-intercept is $(0, 1)$.

(b) To complete the square and obtain the standard form we start by factoring $-\frac{1}{2}$ from the two x-terms:

$$f(x) = -\frac{1}{2}x^2 + x + 1$$
$$= -\frac{1}{2}(x^2 - 2x \quad) + 1$$
$$= -\frac{1}{2}(x^2 - 2x + 1) + 1 + \frac{1}{2}(1)$$
$$= -\frac{1}{2}(x - 1)^2 + \frac{3}{2}.$$

(c) Identifying $h = 1$ and $k = \frac{3}{2}$ we see that the vertex is $(1, \frac{3}{2})$. The axis of symmetry is $x = 1$.

(d) Since the coefficient of x^2 is negative, the parabola opens down.

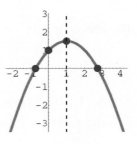

18. (a) *x-intercepts*: Factoring, we obtain

$$f(x) = -x^2 + 6x - 9 = -(x^2 - 6x + 9) = -(x - 3)^2.$$

Setting $f(x) = -(x - 3)^2 = 0$ we obtain $x = 3$, so the x-intercept is $(3, 0)$.

y-intercept: Since $f(0) = -9$, the y-intercept is $(0, -9)$.

(b) The function in standard form is $f(x) = -(x - 3)^2$.

(c) Identifying $h = 3$ and $k = 0$ we see that the vertex is $(3, 0)$. The axis of symetry is $x = 3$.

(d) Since the coefficient of x^2 is negative, the graph opens down.

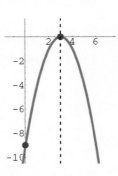

21. The function is in standard form, so we identify $h = 0$. Since the coefficient of x^2 is positive, the parabola opens upward and the function is decreasing on $(-\infty, 0]$ and increasing on $[0, \infty)$.

24. To put the function in standard form we complete the square:

$$f(x) = x^2 + 8x - 1$$
$$= (x^2 + 8x \qquad) - 1$$
$$= (x^2 + 8x + 16) - 1 - 16$$
$$= (x+4)^2 - 17.$$

We identify $h = -4$. Since the coefficient of x^2 is positive, the parabola opens upward and the function is decreasing on $(-\infty, -4]$ and increasing on $[-4, \infty)$.

27. The vertex of the graph of $f(x) = -\frac{1}{3}(x+4)^2 + 9$ is $(-4, 9)$, so the graph of $f(x)$ is the graph of $y = x^2$ reflected in the x-axis, compressed by a factor of $\frac{1}{3}$, shifted to the left by 4 units, and shifted up 9 units.

30. Since $(a-b)^2 = (b-a)^2$ for any choice of a and b, the vertex of the graph of

$$f(x) = -(1-x)^2 + 1 = -(x-1)^2 + 1$$

is $(1, -1)$. Thus, the graph of $f(x)$ is the graph of $y = x^2$ reflected in the x-axis, shifted to the right by 1 unit, and shifted up 1 unit.

33. The graph of $y = x^2$ has been reflected in he x-axis and shifted down by 1 unit. Thus, its equation is $y = -x^2 - 1$.

36. The graph of $y = x^2$ has been reflected in the x-axis and shifted up by 3 units. Thus, its equation is $y = -x^2 + 3$.

39. Since the vertex of f is $(1, 2)$, we identify $h = 1$ and $k = 2$. Then, the equation is $f(x) = a(x-1) + 2$. Since

$$f(2) = a(2-1) + 2 = a + 2 = 6,$$

we have $a = 4$ and $f(x) = 4(x-1)^2 + 2$.

42. To find the points of intersection of the two graphs, we solve $2x-2 = 1 - x^2$, since both sides of this equation are equal to y:

$$2x - 2 = 1 - x^2$$
$$x^2 + 2x - 3 = 0$$
$$(x+3)(x-1) = 0.$$

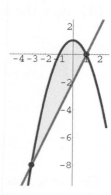

When $x = -3$, $y = 2(-3) - 2 = -8$ and when $x = 1$, $y = 2(1) - 2 = 0$, so the points of intersection are $(-3, -8)$ and $(1, 0)$. The region between the graphs is shown in the figure.

45. In $f(x) = -x + 6\sqrt{x} + 10$ we let $t = \sqrt{x} \Rightarrow t^2 = x \Rightarrow f(t) = -t^2 + 6t + 10 = -(t-3)^2 + 19 \Rightarrow$ $f(3) = 19$ is the maximum value for $f(t)$. But $t = 3 \Rightarrow x = 9 \Rightarrow f(9) = 19$ is the maximum value of $f(x)$.

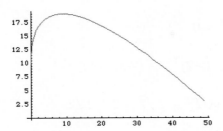

48. The parabola passes through $(0, 6)$ and $(200, 0)$, so we have the equations

$$a \cdot 0^2 + 0 + c = 6 \qquad\qquad c = 6$$
$$\text{or}$$
$$a \cdot 200^2 + 200 + c = 0 \qquad 40{,}000a + c = -200.$$

Since $c = 6$, $40{,}000a + 6 = -200$, and $a = -\dfrac{206}{40{,}000} = -\dfrac{103}{20{,}000}$. Thus, the equation of the parabola is

$$y = -\frac{103}{20{,}000}x^2 + x + 6$$

$$= -\frac{103}{20{,}000}\left[x^2 - \frac{20{,}000}{103}x + \left(-\frac{10{,}000}{103}\right)^2\right] + \frac{103}{20{,}000}\left(\frac{10{,}000}{103}\right)^2 + 6$$

$$= -\frac{103}{20{,}000}\left(x - \frac{10{,}000}{103}\right)^2 + \frac{5{,}000}{103} + 6$$

$$= -\frac{103}{20{,}000}\left(x - \frac{10{,}000}{103}\right)^2 + \frac{5{,}618}{103}.$$

The maximum height attained by the arrow is $\dfrac{5{,}618}{103} \approx 54.54$ feet.

51. (a) As discussed in the text, $s(t) = \frac{1}{2}gt^2 + v_0 t + s_0$ and $v(t) = gt + v_0$. We identify $g = -9.8$, $s_0 = 122.5$, and, since the ball is simply dropped, $v_0 = 0$. Thus, $s(t) = -4.9t^2 + 122.5$ and $v(t) = -9.8t$. When $t = 1$, $s(1) = -4.9 + 122.5 = 117.6$ m and $v(1) = -9.8$ m/s.

(b) We find when the ball hits the ground by solving $s(t) = 0$:

$$-4.9t^2 + 122.5 = 0$$

$$t^2 = \frac{122.5}{4.9} = 25.$$

Thus, the ball hits the ground at $t = \sqrt{25} = 5$ s.

(c) When $t = 5$, the velocity is $v(5) = -9.8(5) = -49$ m/s.

2.5   Piecewise-Defined Functions

3. Since $1 \geq 1$, $f(1) = 1^2 + 2(1) = 1 + 2 = 3$.

Since $0 < 1$, $f(0) = -0^3 = 0$.

Since $-2 < 1$, $f(-2) = -(-2)^3 = -(-8) = 8$.

Since $\sqrt{2} \geq 1$, $f(\sqrt{2}) = (\sqrt{2})^2 + 2\sqrt{2} = 2 + 2\sqrt{2}$.

6. The y-intercept occurs where $x = 0$. In this case, since 0 is a rational number, $f(0) = 1$, and the y-intercept is $(0, 1)$.

9. The y-intercept is at $f(0) = -0 = 0$, so the y-intercept is $(0, 0)$. To find the x-intercepts, we set each part of the function equal to 0, solve for x, and then check to see if the value of x is in the interval determined by the appropriate part of the function:

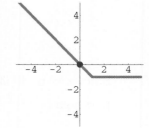

$$-x = 0 \qquad \text{so} \qquad x = 0.$$

Since $0 \leq 1$, $(0, 0)$ is an x-intercept. (We could also have determined that it is an x-intercept because whenever the origin is the y-intercept, it must also be an x-intercept.) Because -1 is never 0, there is no x-intercept for $x > 0$. Since the graph has no holes, the function is continuous.

12. The y-intercept is at $f(0) = 0$, so the y-intercept is $(0, 0)$. To find the x-intercepts, we set each part of the function equal to 0, solve for x, and then check to see if the value of x is in the interval determined by the appropriate part of the function:

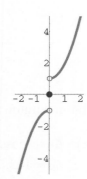

$$-x^2 - 1 = 0 \qquad \text{so} \qquad x^2 = -1.$$

Since x^2 must be nonnegative, there is no x-intercept for $x < 0$. Since $(0, 0)$ is the y-intercept, it must also be an x-intercept. Setting $x^2 + 1 = 0$, we have $x^2 = -1$, so there is no x-intercept for $x > 0$. The function is discontinuous at every integer value of x.

15. The y-intercept is at $f(0) = 0$, so the y-intercept is $(0, 0)$. To find the x-intercepts, we solve $-[[x]] = 0$. This is equivalent to $[[x]] = 0$, and we see from Figure 2.5.3 in the text that $[[x]] = 0$ for all x in the interval $[0, 1)$. Thus, the x-intercepts are $[a, 0)$ for $0 \leq a < 1$. The function is discontinuous at every integer value of x.

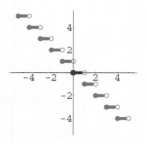

18. The y-intercept is at $f(0) = -|0 - 4| = -4$, so the y-intercept is $(0, -4)$. To find the x-intercepts, we solve $-|x - 4| = 0$. This is equivalent to $x = 4$, so the x-intercept is $(4, 0)$. Since the graph has no holes, the function is continuous.

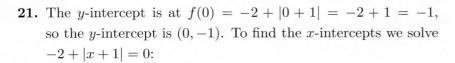

21. The y-intercept is at $f(0) = -2 + |0 + 1| = -2 + 1 = -1$, so the y-intercept is $(0, -1)$. To find the x-intercepts we solve $-2 + |x + 1| = 0$:

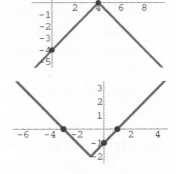

$$-2 + |x + 1| = 0$$
$$|x + 1| = 2$$
$$x + 1 = \pm 2$$
$$x = -1 \pm 2.$$

Thus, $x = -1 - 2 = -3$ and $x = -1 + 2 = 1$. The x-intercepts are $(-3, 0)$ and $(1, 0)$. Since the graph has no holes, the function is continuous.

24. The y-intercept is at $f(0) = |2(0) - 5| = 5$, so the y-intercept is $(0, 5)$. To find the x-intercepts we solve

$$|2x - 5| = 0 \quad \text{or} \quad 2x - 5 = 0.$$

Thus, $x = \frac{5}{2}$ and the x-intercept is $(\frac{5}{2}, 0)$. Since the graph has no holes, the function is continuous.

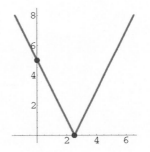

27. The y-intercept is at $f(0) = |0^2 - 2(0)| = 0$, so the y-intercept is $(0, 0)$. Since the origin is a y-intercept, it is also an x-intercept. To find any other x-intercepts we set $|x^2 - 2x| = 0$. This is equivalent to $x^2 - 2x = x(x - 2) = 0$. Thus a second x-intercept occurs at $x = 2$ and the x-intercepts are $(0, 0)$ and $(2, 0)$. Since the graph has no holes, the function is continuous.

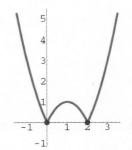

30. The y-intercept is at $f(0) = |\sqrt{0} - 2| = 2$, so the y-intercept is $(0, 2)$. To find the x-intercepts we solve $|\sqrt{x} - 2| = 0$:

$$|\sqrt{x} - 2| = 0$$
$$\sqrt{x} - 2 = 0$$
$$\sqrt{x} = 2$$
$$x = 4.$$

Thus, the x-intercept is $(4, 0)$. Since the graph has no holes, the function is continuous on $[0, \infty)$.

33. Since 0 is in the interval determined by $0 \le x \le 2$, the y-intercept is at $f(0) = |0 - 1| = 1$, so the y-intercept is $(0, 1)$. Since $y = 1$ for $x < 0$ and for $x > 2$, there can be no x-intercepts in these intervals. Setting $|x - 1| = 0$, we have $x - 1 = 0$, so the x-intercept is $(1, 0)$. Since the graph has no holes, the function is continuous.

36. The graphs are shown below with the graph of $y = 2[[x]]$ on the left and the graph of $y = [[2x]]$ on the right. The scales on the two graphs are the same so it is easy to compare them. The steps on the graph of $y = 2[[x]]$ are longer and further apart than those on the graph of $y = [[2x]]$.

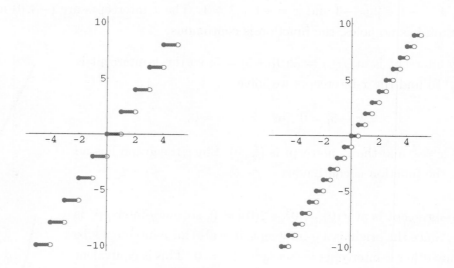

39. The line in the second quadrant is a portion of the graph of $y = -x$, while the graph of the line in the first quadrant is the graph of $y = x$. The semicircle is the graph of the upper half of the circle whose equation is $x^2 + y^2 = 9$, so this portion of the function is defined by $y = \sqrt{9 - x^2}$. The piecewise-defined function is

$$f(x) = \begin{cases} -x, & x < -3 \\ \sqrt{9 - x^2}, & -3 \le x < 3 \\ x, & x \ge 3, \end{cases}$$

where the use of $<$ versus $\le$ is based on the appearance of a circle versus a solid dot in the graph.

42. To graph the absolute value of the function determined by the given graph, simply reflect through the x-axis any portions of the graph that are below the x-axis.

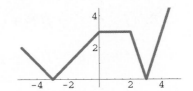

$$f(x) = \begin{cases} -x - 3, & x \le -3 \\ x + 3, & -3 < x \le 0 \\ 3, & 0 < x \le 2 \\ -3x + 9, & 2 < x \le 3 \\ 3x - 9, & 3 < x \end{cases}$$

45. Since $2 \le 2$, $f(2) = \frac{1}{2}(2) + 1 = 1 + 1 = 2$. We need to find k so that $k(2) = f(2) = 2$. This is equivalent to $2k = 2$ or $k = 1$.

48.

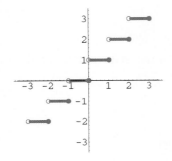

2.6 Combining Functions

3. $(f + g)(x) = f(x) + g(x) = x + \sqrt{x - 1}$

$(f - g)(x) = f(x) - g(x) = x - \sqrt{x - 1}$

$(fg)(x) = f(x)g(x) = x\sqrt{x - 1}$

$(f/g)(x) = \dfrac{f(x)}{g(x)} = \dfrac{x}{\sqrt{x - 1}}$

The domain of f is $(-\infty, \infty)$ and the domain of g is $\{x \mid x - 1 \ge 0\} = \{x \mid x \ge 1\}$ or $[1, \infty)$. The intersection of these sets is $(-\infty, \infty) \cap [1, \infty)$. Thus, the domains of $f + g$, $f - g$ and fg are all $[1, \infty)$.

The domain of g also includes the provision that $\sqrt{x - 1} \ne 0$. This is equivalent to $x - 1 \ne 0$ or $x \ne 1$. Thus, the domain of f/g is $(1, \infty)$.

6. $(f + g)(x) = f(x) + g(x) = \dfrac{4}{x-6} + \dfrac{x}{x-3} = \dfrac{4(x-3) + x(x-6)}{(x-6)(x-3)} = \dfrac{x^2 - 2x - 12}{x^2 - 9x + 18}$

$(f - g)(x) = f(x) - g(x) = x - \sqrt{x-1}$

$(fg)(x) = f(x)g(x) = \dfrac{4}{x-6}\left(\dfrac{x}{x-3}\right) = \dfrac{4x}{x^2 - 9x + 18}$

$(f/g)(x) = \dfrac{f(x)}{g(x)} = \dfrac{4}{x-6} \Big/ \dfrac{x}{x-3} = \dfrac{4(x-3)}{x(x-6)} = \dfrac{4x - 12}{x^2 - 6x}$

The domain of f is $\{x \mid x \neq 6\}$ and the domain of g is $\{x \mid x \neq 3\}$. The intersection of these sets is $(-\infty, 3) \cup (3, 6) \cup (6, \infty)$. Thus, the domains of $f + g, f - g$ and fg are all $(-\infty, 3) \cup (3, 6) \cup (6, \infty)$.

The domain of g also includes the provision that $x/(x-3) \neq 0$. This is equivalent to $x \neq 0$. Thus, the domain of f/g is $(-\infty, 0) \cup (0, 3) \cup (3, 6) \cup (6, \infty)$. [*Note:* The formula for f/g,

$$\left(\frac{f}{g}\right)(x) = \frac{4x - 12}{x^2 - 6x},$$

does not require that $x \neq 3$. Nevertheless, the value $x = 3$ must still be excluded from the domain of f/g because $g(x)$ is not defined for $x = 3$.]

9. To fill in the bottom row we compute

$$(f \circ g)(0) = f(g(0)) = f(2) = 10$$
$$(f \circ g)(1) = f(g(1)) = f(3) = 8$$
$$(f \circ g)(2) = f(g(2)) = f(0) = -1$$
$$(f \circ g)(3) = f(g(3)) = f(1) = 2$$
$$(f \circ g)(4) = f(g(4)) = f(4) = 0$$

x	0	1	2	3	4
$f(x)$	-1	2	10	8	0
$g(x)$	2	3	0	1	4
$(f \circ g)(x)$	10	8	-1	2	0

12. $(f \circ g)(x) = f(g(x)) = f(-x + 4) = (-x + 4)^2 - (-x + 4) + 5 = x^2 - 8x + 16 + x - 4 + 5$
$\qquad = x^2 - 7x + 17$

$(g \circ f)(x) = g(f(x)) = g(x^2 - x + 5) = -(x^2 - x + 5) + 4 = -x^2 + x - 1$

The domains of $f \circ g$ and $g \circ f$ are both $(-\infty, \infty)$.

15. $(f \circ g)(x) = f(g(x)) = f\left(\tfrac{1}{2}(x + 3)\right) = 2[\tfrac{1}{2}(x + 3)] - 3 = x + 3 - 3 = x$

$(g \circ f)(x) = g(f(x)) = g(2x - 3) = \tfrac{1}{2}[(2x - 3) + 3] = \tfrac{1}{2}(2x) = x$

18. $(f \circ g)(x) = f(g(x)) = f(x^2) = \sqrt{x^2 - 4}$

$(g \circ f)(x) = g(f(x)) = g(\sqrt{x-4}) = (\sqrt{x-4})^2 = x - 4, \quad x \geq 4$

[*Note:* Since $f(x) = \sqrt{x-4}$ is only defined for $x - 4 \geq 0$ or $x \geq 4$, the domain of $g \circ f$ must be restricted to $x \geq 4$, even though the final expression $x - 4$ is defined for all x.]

21. $(f \circ f)(x) = f(f(x)) = f(2x + 6) = 2(2x + 6) + 6 = 4x + 18$

$\left(f \circ \dfrac{1}{f}\right)(x) = f\left(\dfrac{1}{f}(x)\right) = f\left(\dfrac{1}{f(x)}\right) = f\left(\dfrac{1}{2x + 6}\right) = 2\left(\dfrac{1}{2x + 6}\right) + 6$

$= \dfrac{1}{x + 3} + 6 = \dfrac{1 + 6(x + 3)}{x + 3} = \dfrac{6x + 19}{x + 3}$

24. $(f \circ f)(x) = f(f(x)) = f\left(\dfrac{x + 4}{x}\right) = \dfrac{\dfrac{x+4}{x} + 4}{\dfrac{x+4}{x}} = \dfrac{\dfrac{x+4}{x} + 4}{\dfrac{x+4}{x}}\left(\dfrac{x}{x}\right) = \dfrac{x + 4 + 4}{x + 4} = \dfrac{x + 8}{x + 4}$

$\left(f \circ \dfrac{1}{f}\right)(x) = f\left(\dfrac{1}{f}(x)\right) = f\left(\dfrac{1}{f(x)}\right) = f\left(\dfrac{1}{\dfrac{x+4}{x}}\right) = f\left(\dfrac{x}{x + 4}\right)$

$= \dfrac{\dfrac{x}{x+4} + 4}{\dfrac{x}{x+4}} = \dfrac{\dfrac{x}{x+4} + 4}{\dfrac{x}{x+4}}\left(\dfrac{x+4}{x+4}\right) = \dfrac{x + 4(x + 4)}{x} = \dfrac{5x + 16}{x}$

27. $(f \circ g \circ g)(x) = f(g(g(x))) = f(g(3x^2)) = f(3(3x^2)^2)$

$= f(27x^4) = 2(27x^4) + 7 = 54x^4 + 7$

30. $(f \circ f \circ f)(x) = f(f(f(x))) = f(f(x^2 - 1)) = f((x^2 - 1)^2 - 1) = f(x^4 - 2x^2 + 1 - 1)$

$= f(x^4 - 2x^2) = (x^4 - 2x^2)^2 - 1 = x^8 - 4x^6 + 4x^4 - 1$

33. In $f \circ g$ the first function to operate on x is g. One way to determine a choice for g is to ask yourself "if I were evaluating the function $F(x) = (x - 3)^2 + 4\sqrt{x - 3}$ at a specific number, say 7, what would be the first thing I would do with the 7?" You could answer this question by saying that you would first subtract 3 from the 7. In this case, then, $g(x) = x - 3$. Then, what is the next thing you would do? This would be to square the $x - 3$ and add it to 4 times the square root of $x - 3$. Thus, $f(x) = x^2 + 4\sqrt{x}$. Note that x is used here, not $x - 3$.

36. First, we find formulas for $f \circ g$ and $g \circ f$:

$$(f \circ g)(x) = f(g(x)) = f(|x|) = [[|x| - 1]]$$
$$(g \circ f)(x) = g(f(x)) = g([[x - 1]]) = |[[x - 1]]|.$$

The graphs are shown below:

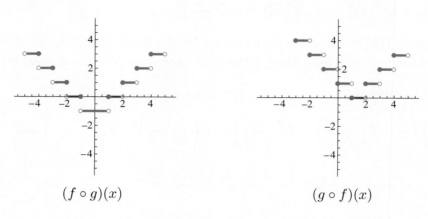

$$(f \circ g)(x) \qquad\qquad (g \circ f)(x)$$

One way to obtain these graphs is to plot some points for values of x like

$$-3, \quad -2.5, \quad -2, \quad -1.5, \quad -1, \quad -0.5, \quad 0, \quad 0.5, \quad 1.5, \quad \text{and} \quad 2.5.$$

39. The graphs of $f(x) = |x - 1|$ and $g(x) = |x|$ are shown together with the graph of $(f+g)(x)$, which is shown with a blue curve. In this case, it is helpful to plot points at values of x like 0, 0.5, 1, 0.75, 2, and 3. Note also from the graphs of f and g, that the graph of $f+g$ is symmetric around the line $x = \frac{1}{2}$.

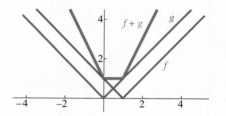

42. The graphs of $f(x) = x$ and $g(x) = [[x]]$ are shown together with the graph of $(fg)(x)$, which is shown with a blue curve. Here you could discern the graph of fg by plotting points at values of x like 0, 0.5, 1, 1.5, 1.9, 2, 2.9, 3, -1, -1.1, and -2.

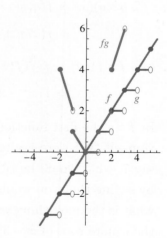

45. (a) Solving $x^2 - 1 = 1 - x$ we have

$$x^2 + x - 2 = 0$$
$$(x - 1)(x + 2) = 0$$

and thus,

$$x = 1, -2.$$

When $x = 1$, $y = 1 - 1 = 0$ (or $y = 1^2 - 1 = 0$), and when $x = -2$, $y = 1 - (-2) = 3$ (or $y = (-2)^2 - 1 = 3$). The points of intersection are $(1, 0)$ and $(-2, 3)$.

(b) To find d we subtract:

$$(1 - x) - (x^2 - 1) = -x^2 - x + 2.$$

Thus, $d(x) = -x^2 - x + 2$.

(c) The graph of $d(x)$ is a parabola opening downward (because the coefficient of x^2 is negative). Its vertex thus represents both the value at which the maximum value of $d(x)$ occurs as well as the actual maximum value of d. To find the vertex we express the quadratic function in standard form by completing the square:

$$\begin{aligned}
d &= -x^2 - x + 2 \\
&= -(x^2 + x \qquad) + 2 \\
&= -\left(x^2 + x + \frac{1}{4}\right) + 2 + \frac{1}{4} \\
&= -\left(x + \frac{1}{2}\right)^2 + \frac{9}{4}.
\end{aligned}$$

The vertex is $(-\frac{1}{2}, \frac{9}{4})$. We thus see that the maximum value of d occurs when $x = -\frac{1}{2}$ and the maximum value of d is $\frac{9}{4}$.

Miscellaneous Applications

48. (a) The area of a circle is given by $A = \pi r^2$, so

$$A(t) = \pi[r(t)]^2 = \pi\left(4 - \frac{4}{t^2 + 1}\right)^2 = 4^2\pi\left(1 - \frac{1}{t^2 + 1}\right)^2 = 16\pi\left(\frac{t^2 + 1 - 1}{t^2 + 1}\right)^2$$

$$= 16\pi\left(\frac{t^2}{t^2 + 1}\right)^2 = 16\pi\frac{t^4}{(t^2 + 1)^2}.$$

(b) The circumference C of a circle is given by $C = 2\pi r$, so

$$C(t) = 2\pi r(t) = 2\pi\left(4 - \frac{4}{t^2 + 1}\right) = 8\pi\left(1 - \frac{1}{t^2 + 1}\right)$$

$$= 8\pi\left(\frac{t^2 + 1 - 1}{t^2 + 1}\right) = 8\pi\left(\frac{t^2}{t^2 + 1}\right).$$

2.7 | Inverse Functions

3. Since there exists a horizontal line that intersects the graph of f in more than one point, f is not a one-to-one function.

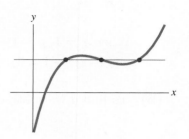

6. Since there exists a horizontal line that intersects the graph of f in more than one point, f is not a one-to-one function.

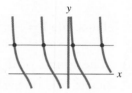

9. Since there exists a horizontal line that intersects the graph of f in more than one point, f is not a one-to-one function.

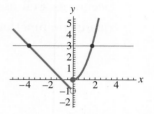

12. As in Example 2(b) in this section of the text, we need to find two values of x that have the same outputs. In this case we see that $f(2) = 0$ and $f(-1) = 0$, which shows that f is not one-to-one.

15. As in Example 2(b) in this section of the text, we assume that $f(x_1) = f(x_2)$. Then

$$\frac{2}{5x_1 + 8} = \frac{2}{5x_2 + 8} \qquad \leftarrow \text{take reciprocals}$$

$$\frac{5x_1 + 8}{2} = \frac{5x_2 + 8}{2} \qquad \leftarrow \text{multiply by 2}$$

$$5x_1 + 8 = 5x_2 + 8 \qquad \leftarrow \text{subtract 8}$$

$$5x_1 = 5x_2 \qquad \leftarrow \text{divide by 5}$$

$$x_1 = x_2$$

Since $f(x_1) = f(x_2)$ implies $x_1 = x_2$, we conclude that f is one-to-one.

18. As in Example 2(b) in this section of the text, we assume that $f(x_1) = f(x_2)$. Then

$$\frac{1}{x_1^3 + 1} = \frac{1}{x_2^3 + 1} \qquad \leftarrow \text{take reciprocals}$$

$$x_1^3 + 1 = x_2^3 + 1 \qquad \leftarrow \text{subtract 1}$$

$$x_1^3 = x_2^3 \qquad \leftarrow \text{take cube roots}$$

$$x_1 = x_2$$

Since $f(x_1) = f(x_2)$ implies $x_1 = x_2$, we conclude that f is one-to-one.

21. $f(g(x)) = \dfrac{1}{(1/\sqrt[3]{x})^3} = \dfrac{1}{1/x} = x$ and $g(f(x)) = g(1/x^3) = \dfrac{1}{\sqrt[3]{1/x^3}} = \dfrac{1}{1/x} = x$

24. $f(g(x)) = f\left(\dfrac{x+3}{1-x}\right) = \dfrac{\dfrac{x+3}{1-x} - 3}{\dfrac{x+3}{1-x} + 1} = \dfrac{\dfrac{4x}{1-x}}{\dfrac{4}{1-x}} = x$ and

$g(f(x)) = g\left(\dfrac{x-3}{x+1}\right) = \dfrac{\dfrac{x-3}{x+1} + 3}{1 - \dfrac{x-3}{x+1}} = \dfrac{\dfrac{4x}{x+1}}{\dfrac{4}{x+1}} = x$

27. We write the function as $y = 2/\sqrt{x}$. Solving for x, we have

$$y = \frac{2}{\sqrt{x}}$$

$$\sqrt{x} = \frac{2}{y}$$

$$x = \frac{4}{y^2}.$$

Relabeling the variables, we obtain $y = 4/x^2$ Thus $f^{-1}(x) = 4/x^2$. The domain of f^{-1} is the range of f or $\{x \mid x > 0\}$ and the range of f^{-1} is the domain of f or $\{y \mid y > 0\}$.

30. We solve $y = -2x + 1$ for x:

$$y = -2x + 1$$

$$2x = 1 - y$$

$$x = \frac{1}{2}(1 - y).$$

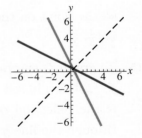

Relabeling the variables, we have $f^{-1}(x) = \frac{1}{2}(1 - x)$. The graph of f is shown in blue and the graph of f^{-1} in red.

33. We solve $y = 2 - \sqrt{x}$ for x:

$$y = 2 - \sqrt{x}$$

$$\sqrt{x} = 2 - y$$

$$x = (2 - y)^2.$$

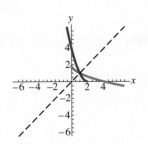

Relabeling the variables, we have $f^{-1}(x) = (2 - x)^2 = (x - 2)^2$. Since $y = (x - 2)^2$ is a parabola, it is not one-to-one, and we must restrict the values of x. To do this, we can note from the graph of f^{-1} (shown in red) that $x \leq 2$. Alternatively, the range of f is $\{y \mid y \leq 2\}$ so the domain of f^{-1} must be $\{x \mid x \leq 2\}$. Hence, $f^{-1}(x) = (x - 2)^2, \quad x \leq 2$.

36. The domain of f is $\{x \mid 5x + 8 \neq 0\} = \{x \mid x \neq \frac{8}{5}\}$. Then the range of f^{-1} is $\{y \mid y \neq \frac{8}{5}\}$. To find the domain of f^{-1} we find a formula for f^{-1} by solving $y = 2/(5x + 8)$ for x:

$$y = \frac{2}{5x + 8}$$

$$5x + 8 = \frac{2}{y}$$

$$5x = \frac{2}{y} - 8 = \frac{2 - 8y}{y}$$

$$x = \frac{2 - 8y}{5y} \, .$$

Thus,

$$f^{-1}(x) = \frac{2 - 8x}{5x}$$

and the domain of f^{-1} is $\{x \mid x \neq 0\}$. Hence, the range of f is $\{y \mid y \neq 0\}$.

39. When $x = 2$, $f(2) = 2(2)^3 + 2(2) = 2(8) + 4 = 20$, so the point on the graph of f is $(2, 20)$. Then, the corresponding point on the graph of f^{-1} is $(20, 2)$.

42. When $x = \frac{1}{2}$,

$$f\left(\frac{1}{2}\right) = \frac{4(1/2)}{1/2 + 1} = \frac{2}{3/2} = \frac{4}{3},$$

so the point on the graph of f is $(\frac{1}{2}, \frac{4}{3})$. Then, the corresponding point on the graph of f^{-1} is $(\frac{4}{3}, \frac{1}{2})$.

45. To graph f, shown in blue, we use the fact that f is the inverse of f^{-1}, shown in red, and that the graph of the inverse of a function is the reflection of the graph of the original function reflected through the line $y = x$.

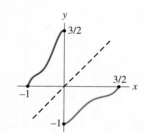

48. Solving for $y = (3 - 2x)^2$ for x we have

$$y = (3 - 2x)^2$$

$$-\sqrt{y} = 3 - 2x$$

$$2x = 3 + \sqrt{y}$$

$$x = \frac{1}{2}\left(3 + \sqrt{y}\right),$$

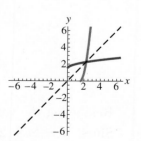

where we use the negative square root of y because, for $x \geq \frac{3}{2}$, $3 - 2x \leq 0$ and $\sqrt{y} \geq 0$ for all y. Relabeling the variables, we have $f^{-1}(x) = \frac{1}{2}\left(3 + \sqrt{x}\right)$, which has domain $\{x \mid x \geq 0\}$ or $[0, \infty)$. The graph of f is shown in blue, and the graph of f^{-1} is shown in red.

51. Since $f(x) = x^3$ we have $f^{-1}(x) = x^{1/3}$; and since $g(x) = 4x + 5$ we have $g^{-1}(x) = \frac{1}{4}x - \frac{5}{4}$. Then

$$(f \circ g)(x) = f(g(x)) = f(4x + 5) = (4x + 5)^{1/3}.$$

To find $(f \circ g)^{-1}(x)$ we solve $y = (f \circ g)(x) = (4x + 5)^{1/3}$ for x:

$$y^3 = 4x + 5$$
$$4x = y^3 - 5$$
$$x = \frac{1}{4}y^3 - \frac{5}{4}.$$

Thus $(f \circ g)^{-1}(x) = \frac{1}{4}x^3 - \frac{5}{4}$. But

$$(g^{-1} \circ f^{-1})(x) = g^{-1}(f^{-1}(x)) = g^{-1}(x^{1/3}) = \frac{1}{4}x^{1/3} - \frac{5}{4},$$

so $(f \circ g)^{-1} = g^{-1} \circ f^{-1}$.

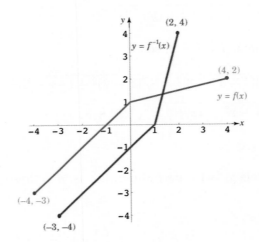

2.8   Building a Function from Words

3. Let x and y be the nonnegative numbers. Then $x + y = 1$. Now, the sum of the square of x and twice the square of y is $x^2 + 2y^2$. From $x + y = 1$ we have $y = 1 - x$, so the function in this case is

$$s(x) = x^2 + 2(1 - x)^2 = x^2 + 2(1 - 2x + x^2) = 3x^2 - 4x + 2.$$

Since x and y are both nonnegative, we must have $0 \le x \le 1$ and $0 \le y \le 1$. (If, say $y > 1$, then we would have $x < 0$.) Thus

$$s(x) = 3x^2 - 4x + 2, \quad 0 \le x \le 1.$$

Alternatively, if we choose the independent variable of s to be y, we have $x = 1 - y$ and

$$s(y) = (1 - y)^2 + 2y^2 = 1 - 2y + y^2 + 2y^2 = 3y^2 - 2y + 1, \quad 0 \le y \le 1.$$

6. Let the sides of the rectangle be x and y as shown in the figure. Then $A = xy = 400$, $x, y > 0$. The perimeter of the rectangle is $P = 2x + 2y$. To express P in terms of just x, we use $y = 400/x$. Then

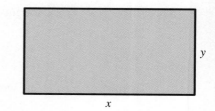

$$P(x) = 2x + 2\left(\frac{400}{x}\right) = 2x + \frac{800}{x} = \frac{2x^2 + 800}{x}, \quad x > 0.$$

Alternatively, from $xy = 400$ we have $x = 400/y$ and

$$P(x) = 2y + 2\left(\frac{400}{y}\right) = 2y + \frac{800}{y} = \frac{2y^2 + 800}{y}, \quad y > 0.$$

9. The distance between (x, y) and $(2, 3)$ is given by

$$d = \sqrt{(x-2)^2 + (y-3)^2}.$$

Since $x + y = 1$, we have $y = 1 - x$, so

$$d = \sqrt{(x-2)^2 + [(1-x)-3]^2} = \sqrt{(x-2)^2 + (-x-2)^2}$$
$$= \sqrt{x^2 - 4x + 4 + x^2 + 4x + 4} = \sqrt{2x^2 + 8}.$$

The domain of d is $(-\infty, \infty)$.

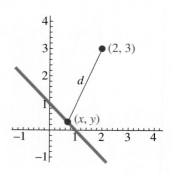

12. The diameter of a circle is twice its radius, that is, $d = 2r$ so $r = \frac{1}{2}d$. The area of a circle is then given by

$$A = \pi r^2 = \pi \left(\frac{1}{2}d\right)^2 = \frac{1}{4}\pi d^2.$$

15. Let the sides of the equilateral triangle each be of length x. Then, refering to the figure and using the Pythagorean theorem, we have

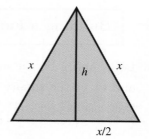

$$\left(\frac{x}{2}\right)^2 + h^2 = x^2$$

$$h^2 = x^2 - \frac{x^2}{4} = \frac{3}{4}x^2$$

$$x^2 = \frac{4}{3}h^2$$

$$x = \frac{2}{\sqrt{3}}h.$$

The area of the triangle is

$$A = \frac{1}{2}xh = \frac{1}{2}\left(\frac{2}{\sqrt{3}}h\right)h = \frac{1}{\sqrt{3}}h^2 = \frac{\sqrt{3}}{3}h^2.$$

(Whether or not you rationalize $1/\sqrt{3}$ as above is up to your instructor.) In this problem, $h > 0$ and the domain of the function is $(0, \infty)$.

18. As shown in the figure, we will bend the portion of the wire of length x into a square and the portion of length $x - L$ into a circle. Then the sides of the square are each $x/4$ and the circumference of the circle is $C = x - L$. Since the circumference of a circle is related to its radius by $C = 2\pi r$, we have $r = C/2\pi = (x - L)/2\pi$. The sum of the areas is

$$A = \text{area of square} + \text{area of circle}$$
$$= \left(\frac{x}{4}\right)^2 + \pi \left(\frac{x - L}{2\pi}\right)^2 = \frac{x^2}{16} + \pi \left(\frac{(x - L)^2}{4\pi^2}\right)$$
$$= \frac{x^2}{16} + \frac{(x - L)^2}{4\pi}.$$

In this problem, $0 < x < L$.

21. Let w be the width of the box, and h the height of the box. Then the length of the box is $3w$ and the volume of the box is

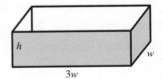

$$V = (3w)(w)(h) = 3w^2 h = 450.$$

Since the box is open, its surface area is

$$S = 2wh + 2(3w)h + 3w(w) = 8wh + 3w^2.$$

From $3w^2 h = 450$ we have $h = 150/w^2$, so

$$S = 8w\left(\frac{150}{w^2}\right) + 3w^2 = \frac{1200}{w} + 3w^2 = \frac{1200 + 3w^3}{w}.$$

In this problem, $w > 0$.

24. As shown in the figure, suppose the lower of the two airliners is traveling at 550 mi/h and the higher airliner is traveling at 500 mi/h. Then, after time t, the lower plane has traveled $550t$ miles and the higher plane has traveled $500t$ miles. From the figure, we see, using the Pythagorean Theorem, that the distance between the two planes is

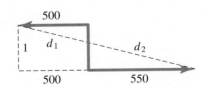

$$d = d_1 + d_2 = \sqrt{(500t + 550t)^2 + 1^2} = \sqrt{(1050t)^2 + 1} = \sqrt{1{,}102{,}500t^2 + 1}.$$

In this problem, $t > 0$.

27. We want the maximum value of the difference of two numbers, x and x^2. The objective function is $D(x) = x - x^2$, where the domain of D is all real numbers.

30. We want to maximize the area as a function of the lengths of the sides of the plot of land. Let x be the side of the plot parallel to the two interior fences and let y be the length of the other side of each of the three subplots. See the figure.

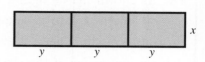

The total length of fencing used is $4x + 6y = 8000$. The area function is $A = x(3y) = 3xy$. Solving $4x + 6y = 8000$ for y, we have $6y = 8000 - 4x$ or $y = 4000/x - 2x/3$. Thus, the objective function is

$$A(x) = 3xy = 3x \left(\frac{4000}{3} - \frac{2}{3} x \right) = 4000x - 2x^2.$$

In this problem, $x > 0$.

33. We want to maximize area. The perimeter of the rectangle is $p = 2x + 2y$ and the area is $A = xy$. Solving $p = 2x + 2y$ for y, we obtain $2y = p - 2x$ or $y = \frac{1}{2} p - x$. Then, the objective function is

$$A(x) = xy = x \left(\frac{1}{2} p - x \right) = \frac{1}{2} px - x^2.$$

In this problem, since $A(x)$ must be positive, $0 \le x \le \frac{1}{2} p$.

36. The figure for Problem 35 in the text applies, except that the top of the box is closed. In this case we want to maximize volume. The total cost to construct each box is

$$2x^2 + xy + xy + xy + xy = 2x^2 + 4xy = 36.$$

The volume is $V = x^2 y$. Solving the cost equation for y, we obtain

$$2x^2 + 4xy = 36$$
$$4xy = 36 - 2x^2$$
$$y = \frac{36 - 2x^2}{4x} = \frac{18 - x^2}{2x}.$$

The objective function is

$$V(x) = x^2 y = x^2 \left(\frac{18 - x^2}{2x} \right) = 9x - \frac{1}{2} x^3.$$

In this problem, $x > 0$.

39. We want to minimize the area of the entire page, so let the width and height of the page be x and y, respectively, as shown in the figure. The area of the printed portion is $P = (x-4)(y-2) = 32$. The area of the entire page is $A = xy$. Solving $(x-4)(y-2) = 32$ for y, we obtain

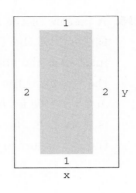

$$(x-4)(y-2) = 32$$

$$y - 2 = \frac{32}{x-4}$$

and thus,

$$y = \frac{32}{x-4} + 2 = \frac{32 + 2x - 8}{x-4} = \frac{24 + 2x}{x-4}.$$

The objective function is

$$A(x) = xy = x\left(\frac{24 + 2x}{x-4}\right) = \frac{24x + 2x^2}{x-4}.$$

In this problem, $x > 4$, rather than $x > 0$, to allow for the side margins.

An alternative way to solve the problem is to let the dimensions of the printed portion be x and y, where x is the height of the printed portion. This leads to $A(x) = 40 + 4x + 64/x$.

42. Figures for this problem are shown in the text. We want to minimize the surface area of the can. The volume is $V = \pi r^2 h = 32$. The surface area is

$$S = \text{(area of bottom)} + \text{(area of top)} + \text{(area of lateral side)}$$
$$= \pi r^2 + \pi r^2 + 2\pi rh = 2\pi r^2 + 2\pi rh.$$

Solving $\pi r^2 h = 32$ for h, we have $h = 32/\pi r^2$. The objective function is

$$S(r) = 2\pi r^2 + 2\pi rh = 2\pi r^2 + 2\pi r \left(\frac{32}{\pi r^2}\right) = 2\pi r^2 + \frac{64\pi}{r}$$
$$= \frac{2\pi r^3 + 64\pi}{r}.$$

In this problem, we must have $r > 0$.

45. A picture of the trough is shown in the text and a figure for the end of the trough is shown to the right. We want to maximize the volume of the trough. The area of the end of the trough is the area of an isosceles triangle with sides 4, 4, and x. If we let h be the height of the triangle, then the area is $A = \frac{1}{2}xh$. To represent this strictly in terms of x, we use the Pythagorean theorem and solve for h:

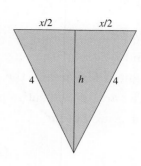

$$\left(\frac{x}{2}\right)^2 + h^2 = 4^2$$

$$\frac{x^2}{4} + h^2 = 16$$

$$x^2 + 4h^2 = 64$$

$$4h^2 = 64 - x^2$$

$$h^2 = \frac{64 - x^2}{4}$$

$$h = \frac{1}{2}\sqrt{64 - x^2}.$$

Thus, the area of the triangle is

$$A = \frac{1}{2}x\left(\frac{1}{2}\sqrt{64 - x^2}\right) = \frac{1}{4}x\sqrt{64 - x^2},$$

and the objective function is

$$S(x) = (\text{area of base}) \times (\text{height of trough}) = \left(\frac{1}{4}x\sqrt{64 - x^2}\right)(20) = 5x\sqrt{64 - x^2}.$$

In this problem, we must have $0 < x < 8$. (If $x \geq 8$, the three line segments will not form a triangle.)

48. Let x represent the distance from the left flagpole to the point on the ground. The objective function, or total length of wire used, is given by

$$L(x) = \sqrt{x^2 + 400} + \sqrt{(30 - x)^2 + 100}$$

2.9 $\int$ **Calculus** PREVIEW   **The Tangent Line Problem**

3. (a) We first compute $f(1 + h)$:

$$f(1 + h) = (1 + h)^2 - 3(1 + h)$$
$$= 1 + 2h + h^2 - 3 - 3h$$
$$= h^2 - h - 2.$$

Next, since $f(1) = 1^2 - 3(1) = 1 - 3 = -2$, we have

$$f(1 + h) - f(1) = h^2 - h - 2 - (-2) = h^2 - h = h(h - 1).$$

Finally,

$$\frac{f(1 + h) - f(1)}{h} = \frac{h(h - 1)}{h} = h - 1.$$

(b) $m_{\tan} = \lim_{h \to 0} \dfrac{f(1 + h) - f(1)}{h} = \lim_{h \to 0} (h - 1) = -1$

(c) The y-coordinate of the point of tangency is $f(1) = -2$, so using the point-slope form for the equation of a line we see that the tangent line is

$$y - (-2) = -1(x - 1) \qquad \text{or} \qquad y = -x - 1.$$

6. (a) We first compute $f(\frac{1}{2} + h)$:

$$
\begin{aligned}
f\left(\frac{1}{2} + h\right) &= 8\left(\frac{1}{2} + h\right)^3 - 4 \\
&= 8\left(\frac{1}{8} + \frac{3}{4}h + \frac{3}{2}h^2 + h^3\right) - 4 \\
&= 1 + 6h + 12h^2 + 8h^3 - 4 \\
&= 8h^3 + 12h^2 + 6h - 3.
\end{aligned}
$$

Next, since $f(\frac{1}{2}) = 8(\frac{1}{2})^3 - 4 = 8(\frac{1}{8}) - 4 = 1 - 4 = -3$, we have

$$f\left(\frac{1}{2} + h\right) - f\left(\frac{1}{2}\right) = 8h^3 + 12h^2 + 6h - 3 - (-3) = h(8h^2 + 12h + 6).$$

Finally,

$$\frac{f(\frac{1}{2} + h) - f(\frac{1}{2})}{h} = \frac{h(h^2 + 12h + 6)}{h} = h^2 + 12h + 6.$$

(b) $m_{\tan} = \lim_{h \to 0} \dfrac{f(\frac{1}{2} + h) - f(\frac{1}{2})}{h} = \lim_{h \to 0} (8h^2 + 12h + 6) = 6$

(c) The y-coordinate of the point of tangency is $f(\frac{1}{2}) = -3$, so, using the point-slope form for the equation of a line, we see that the tangent line is

$$y - (-3) = 6\left(x - \frac{1}{2}\right) \qquad \text{or} \qquad y = 6x - 6.$$

9. (a) We first compute $f(4+h)$:

$$f(4+h) = \sqrt{4+h}.$$

Next, since $f(4) = \sqrt{4} = 2$, we have

$$f(4+h) - f(4) = \sqrt{4+h} - 2.$$

Finally,

$$\frac{f(4+h) - f(4)}{h} = \frac{\sqrt{4+h} - 2}{h} = \frac{\sqrt{4+h} - 2}{h}\left(\frac{\sqrt{4+h} + 2}{\sqrt{4+h} + 2}\right)$$

$$= \frac{4+h-4}{h(\sqrt{4+h} + 2)} = \frac{h}{h(\sqrt{4+h} + 2)}$$

$$= \frac{1}{\sqrt{4+h} + 2}.$$

(b) $m_{\tan} = \lim\limits_{h \to 0} \dfrac{f(4+h) - f(4)}{h} = \lim\limits_{h \to 0} \dfrac{1}{\sqrt{4+h} + 2} = \dfrac{1}{2+2} = \dfrac{1}{4}$

(c) The y-coordinate of the point of tangency is $f(4) = 2$ so, using the point-slope form for the equation of a line, we see that the tangent line is

$$y - 2 = \frac{1}{4}(x - 4) \qquad \text{or} \qquad y = \frac{1}{4}x + 1.$$

12. (a) We first find $f(x+h)$:

$$f(x+h) = -3(x+h) + 8 = -3x - 3h + 8.$$

Then

$$f(x+h) - f(x) = (-3x - 3h + 8) - (-3x + 8) = -3h,$$

and so

$$\frac{f(x+h) - f(x)}{h} = \frac{-3h}{h} = -3.$$

(b) The derivative of f is

$$f'(x) = \lim\limits_{h \to 0} \frac{f(x+h) - f(x)}{h} = \lim\limits_{h \to 0}(-3) = -3.$$

15. (a) We first find $f(x+h)$:

$$f(x+h) = 3(x+h)^2 - (x+h) + 7 = 3x^2 + 6xh + 3h^2 - x - h + 7.$$

Then

$$f(x+h) - f(x) = (3x^2 + 6xh + 3h^2 - x - h + 7) - (3x^2 - x + 7)$$

$$= 6xh + 3h^2 - h$$

$$= h(6x + 3h - 1),$$

and so

$$\frac{f(x+h)-f(x)}{h} = \frac{h(6x+3h-1)}{h} = 6x+3h-1.$$

(b)   The derivative of f is

$$f'(x) = \lim_{h\to 0}\frac{f(x+h)-f(x)}{h} = \lim_{h\to 0}(6x+3h-1) = 6x-1.$$

18. (a)   We first find $f(x+h)$:

$$f(x+h) = 2(x+h)^3 + (x+h)^2$$

$$= 2(x^3 + 3x^2h + 3xh^2 + h^3) + (x^2 + 2xh + h^2)$$

$$= 2x^3 + 6x^2h + 6xh^2 + 2h^3 + x^2 + 2xh + h^2.$$

Then

$$f(x+h) - f(x) = (2x^3 + 6x^2h + 6xh^2 + 2h^3 + x^2 + 2xh + h^2) - (2x^3 + x^2)$$

$$= 6x^2h + 6xh^2 + 2h^3 + 2xh + h^2$$

$$= h(6x^2 + 6xh + 2h^2 + 2x + h)$$

and so

$$\frac{f(x+h)-f(x)}{h} = \frac{h(6x^2 + 6xh + 2h^2 + 2x + h)}{h}$$

$$= 6x^2 + 6xh + 2h^2 + 2x + h.$$

(b)   The derivative of f is

$$f'(x) = \lim_{h\to 0}\frac{f(x+h)-f(x)}{h} = \lim_{h\to 0}(6x^2 + 6xh + 2h^2 + 2x + h) = 6x^2 + 2x.$$

21. (a)   We first find $f(x+h)$:

$$f(x+h) = \frac{x+h}{x+h-1}.$$

Then

$$f(x+h) - f(x) = \frac{x+h}{x+h-1} - \frac{x}{x-1}$$

$$= \frac{(x+h)(x-1) - x(x+h-1)}{(x+h-1)(x-1)}$$

$$= \frac{x^2 + xh - x - h - x^2 - xh + x}{(x+h-1)(x-1)}$$

$$= -\frac{h}{(x+h-1)(x-1)}$$

and so

$$\frac{f(x+h) - f(x)}{h} = \frac{-\dfrac{h}{(x+h-1)(x-1)}}{h} = -\frac{h}{h(x+h-1)(x-1)}$$

$$= -\frac{1}{(x+h-1)(x-1)}.$$

(b) The derivative of f is

$$f'(x) = \lim_{h \to 0} \frac{f(x+h) - f(x)}{h} = \lim_{h \to 0} \left(-\frac{1}{(x+h-1)(x-1)} \right)$$

$$= -\frac{1}{(x-1)(x-1)} = -\frac{1}{(x-1)^2}.$$

24. (a) We first find $f(x+h)$:

$$f(x+h) = \frac{1}{(x+h)^2}.$$

Then

$$f(x+h) - f(x) = \frac{1}{(x+h)^2} - \frac{1}{x^2} = \frac{x^2 - (x+h)^2}{(x+h)^2 x^2}$$

$$= \frac{x^2 - x^2 - 2xh - h^2}{(x+h)^2 x^2} = -\frac{2xh - h^2}{(x+h)^2 x^2},$$

and so

$$\frac{f(x+h) - f(x)}{h} = \frac{-\dfrac{2xh - h^2}{(x+h)^2 x^2}}{h} = -\frac{2xh - h^2}{h(x+h)^2 x^2}$$

$$= -\frac{2x - h}{(x+h)^2 x^2}.$$

(b) The derivative of f is

$$f'(x) = \lim_{h \to 0} \frac{f(x+h) - f(x)}{h} = \lim_{h \to 0} \left(-\frac{2x - h}{(x+h)^2 x^2} \right)$$

$$= -\frac{2x}{x^2 x^2} = -\frac{2}{x^3}.$$

27. From Problem 15, $f'(x) = 6x - 1$, so when $x = 2$ the slope of the tangent line is

$$m_{\tan} = f'(2) = 6(2) - 1 = 11.$$

The y-coordinate of the point on the graph of $y = f(x)$, when $x = 2$, is

$$f(2) = 3(2)^2 - 2 + 7 = 17,$$

so the point of tangency is $(2, 17)$. Using the point-slope form of the equation of a line, we then see that the tangent line is

$$y - 17 = 11(x - 2) \qquad \text{or} \qquad y = 11x - 5.$$

30. From Problem 18, $f'(x) = 6x^2 + 2x$, so at $x = -\frac{1}{2}$ the slope of the tangent line is

$$m_{\tan} = f'\left(-\frac{1}{2}\right) = 6\left(-\frac{1}{2}\right)^2 + 2\left(-\frac{1}{2}\right) = \frac{3}{2} - 1 = \frac{1}{2}.$$

The y-coordinate of the point on the graph of $y = f(x)$ when $x = -\frac{1}{2}$ is

$$f\left(-\frac{1}{2}\right) = 2\left(-\frac{1}{2}\right)^3 + \left(-\frac{1}{2}\right)^2 = 2\left(-\frac{1}{8}\right) + \frac{1}{4} = 0.$$

Using the point-slope form of the equation of a line, we then see that the tangent line is

$$y - 0 = \frac{1}{2}\left[x - \left(-\frac{1}{2}\right)\right] \qquad \text{or} \qquad y = \frac{1}{2}x + \frac{1}{4}.$$

33. (a) We have $f(a) = 3a^2 + 1$ and

$$f(x) - f(a) = (3x^2 + 1) - (3a^2 + 1) = 3(x^2 - a^2) = 3(x + a)(x - a).$$

Then

$$\frac{f(x) - f(a)}{x - a} = \frac{3(x + a)(x - a)}{x - a} = 3(x + a).$$

(b) The derivative at $x = a$ is

$$f'(a) = \lim_{x \to a} \frac{f(x) - f(a)}{x - a} = \lim_{x \to a} 3(x + a) = 3(2a) = 6a.$$

36. (a) We have $f(a) = a^4$ and

$$f(x) - f(a) = x^4 - a^4 = (x - a)(x^3 + ax^2 + a^2x + a^3).$$

This result can be obtained by using synthetic division to factor $x - a$ out of $x^4 - a^4$. Now,

$$\frac{f(x) - f(a)}{x - a} = \frac{(x - a)(x^3 + ax^2 + a^2x + a^3)}{x - a} = x^3 + ax^2 + a^2x + a^3.$$

(b) The derivative at $x = a$ is

$$f'(a) = \lim_{x \to a} \frac{f(x) - f(a)}{x - a} = \lim_{x \to a}(x^3 + ax^2 + a^2x + a^3) = a^3 + a^3 + a^3 + a^3 = 4a^3.$$

39. (a) We have $f(a) = \sqrt{7a}$ and

$$f(x) - f(a) = \sqrt{7x} - \sqrt{7a} = \sqrt{7}\left(\sqrt{x} - \sqrt{a}\right).$$

Then

$$\frac{f(x) - f(a)}{x - a} = \frac{\sqrt{7}\left(\sqrt{x} - \sqrt{a}\right)}{x - a} = \frac{\sqrt{7}\left(\sqrt{x} - \sqrt{a}\right)}{x - a}\left(\frac{\sqrt{x} + \sqrt{a}}{\sqrt{x} + \sqrt{a}}\right)$$

$$= \frac{\sqrt{7}(x - a)}{(x - a)\left(\sqrt{x} + \sqrt{a}\right)} = \frac{\sqrt{7}}{\sqrt{x} + \sqrt{a}}.$$

(b) The derivative at $x = a$ is

$$f'(a) = \lim_{x \to a} \frac{f(x) - f(a)}{x - a} = \lim_{x \to a} \frac{\sqrt{7}}{\sqrt{x} + \sqrt{a}} = \frac{\sqrt{7}}{\sqrt{a} + \sqrt{a}} = \frac{\sqrt{7}}{2\sqrt{a}}.$$

Chapter 2 Review Exercises

A. Fill in the Blanks

3. We need $5 - x > 0$ (not $= 0$ because $\sqrt{5 - x}$ is in the denominator), so $x < 5$. The domain is $(-\infty, 5)$.

6. Symmetry with respect to the y-axis means $f(-x) = f(x)$.

9. The slope of the line through $(-2, 0)$ and $(0, -3)$ is

$$m = \frac{-3 - 0}{0 - (-2)} = -\frac{3}{2}.$$

12. The graph is a parabola opening downward, so we put the function in standard form to find the vertex:

$$\begin{aligned} f(x) &= -x^2 + 6x - 21 \\ &= -(x^2 - 6x \quad) - 21 \\ &= -(x^2 - 6x + 9) - 21 + 9 \\ &= -(x - 3)^2 - 12. \end{aligned}$$

The y-coordinate of the vertex is $(3, -12)$, so the range of the function is $(-\infty, -12]$.

15. The graph of $y = -5(x - 10)^2 + 2$ is the graph of $f(x) = x^2$ shifted 10 units to the right and 2 units up. Therefore, the vertex of the parabola $y = -5(x - 10)^2 + 2$ is $(10, 2)$.

18. To find the inverse function, we set $y = (x - 5)/(2x + 1)$ and solve for x:

$$y = \frac{x - 5}{2x + 1}$$
$$(2x + 1)y = x - 5$$
$$2xy + y = x - 5$$
$$2xy - x = -y - 5$$
$$x(2y - 1) = -y - 5$$
$$x = -\frac{y + 5}{2y - 1}.$$

Interchanging x and y, we see that the inverse function is

$$f^{-1}(x) = -\frac{x + 5}{2x - 1}.$$

21. If the entire graph of a one-to-one function f lies in the fourth quadrant, then graph of f^{-1} lies in the second quadrant.

24. The range of f is $[-0.6,\ 3]$.

27. f is increasing on the intervals $[2,\ 4]$ and $[6,\ 7]$.

30. $f(1) = $ approximately 1.

B. True/False

3. False. When $f(x) = x^2$, $f(-1) = f(1)$.

6. False. Since 0 is not in the domain of f, it cannot be in the domain of g/f.

9. True

12. True. The graph of $f(x) = 2x^2 + 16x - 2$ is a parabola opening up, with vertex at $x = -16/2(2) = -4$. Since $[-7, -5]$ is to the left of the line $x = -4$, the function is decreasing on the interval.

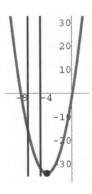

15. False. Let $f(x) = x$. Then $f^{-1}(x) = x \neq 1/x = 1/f(x)$.

18. False. Suppose $f(x) = x^2$. Then

$$(f \circ (g + h))(x) = f((g + h)(x)) = f(g(x) + h(x)) = (g(x) + h(x))^2$$

and

$$(f \circ g + f \circ h)(x) = (f \circ g)(x) + (f \circ h)(x) = f(g(x)) + f(h(x))$$
$$= (g(x))^2 + (h(x))^2.$$

Since, in general $(g(x)+h(x))^2 \neq (g(x))^2+(h(x))^2$, it is not generally the case that $f \circ (g+h) = f \circ g + f \circ h$.

21. False. The graph of $y = (x+2)^2$ is the graph of $f(x) = x^2$ shifted horizontally to the *left*.

C. Review Exercises

3. To shift the graph of a function to the left a units, we add a to each occurrence of the independent variable x; to shift to the right, we subtract a from x; to shift the graph up by b units, we add b to function output; to shift down, we subtract b. To reflect the graph of $f(x)$ in the y-axis, we change each x to $-x$; to reflect in the x-axis, we negate the output of f. To vertically stretch the graph by a factor of a, the output of the function is multiplied by a.

(a) $f(x+3) = (x+3)^3 - 2$

(b) $f(x) - 5 = x^3 - 2 - 5 = x^3 - 7$

(c) $f(x-1) + 2 = (x-1)^3 - 2 + 2 = (x-1)^3$

(d) $-f(x) = -(x^3 - 2) = -x^3 + 2$

(e) $f(-x) = (-x)^3 - 2 = -x^3 - 2$

(f) $3f(x) = 3(x^3 - 2) = 3x^3 - 6$

6. The graph of f^{-1} is obtained by reflecting the graph of f through the line $y = x$. The graph of f is shown in blue, and the graph of f^{-1} is shown in red.

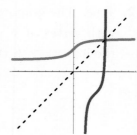

9. To graph f, we write

$$f(x) = x^2 - 6x + 10 = x^2 - 6x + 9 + 1 = (x-3)^2 + 1,$$

and note that the vertex is at $(3, 1)$. Since the graph of f is always above the x-axis, $f(x) = x^2 - 6x + 10$ is always positive and the domain of $g(x) = \sqrt{f(x)}$ is $(-\infty, \infty)$.

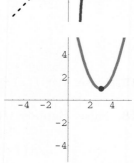

12. The domain of a composition $g \circ f$ is the set of all x in the domain of f such that $f(x)$ is in the domain of g. For $f(x) = \dfrac{1}{x+1}$, $g(x) = \dfrac{5}{x-2}$ we can exclude $x = -1$ from the domain of the composition $g \circ f$ since f is not defined at -1. Similarly, we can exclude x in the domain of f for which $f(x) = 2$ since g is not defined at 2. Solving

$$f(x) = 2 \Rightarrow \frac{1}{x+1} = 2 \Rightarrow 1 = 2x + 2 \Rightarrow x = -\tfrac{1}{2}$$

the domain of $f \circ g$ is $\left\{ x \mid x \neq -1, \, x \neq -\tfrac{1}{2} \right\}$.

15. See Figure 2.R.3 in the text. The area of the shaded region is the area of the square with opposite corners at $(0,0)$ and (h,h) minus the area of the quarter circle centered at (h,h) with radius h. That is, the area of the shaded region is

$$A = h^2 - \frac{1}{4}\pi h^2 = \left(1 - \frac{\pi}{4}\right) h^2.$$

18. See Figure 2.R.6 in the text. Using the fact that the radius r at the top of the cylinder is one leg of the right triangle whose other leg has length $h/2$ and whose hypotenuse has length 1, we have

$$r^2 + \left(\frac{h}{2}\right)^2 = 1^2 \qquad \text{or} \qquad r^2 = 1 - \frac{h^2}{4}.$$

The volume of the cylinder is

$$V = \pi r^2 h = \pi \left(1 - \frac{h^2}{4}\right) h = \pi \left(h - \frac{1}{4}h^3\right).$$

21. If x is the number, the objective function is the sum

$$S(x) = 20x + 5\left(\frac{1}{x}\right) = 20x + \frac{5}{x}, \quad x > 0,$$

where $x > 0$ because the number is positive.

24. See Figure 2.R.10 in the text. Since the pipe is $4 - x$ miles long on the land and $\sqrt{x^2 + 4^2} = \sqrt{x^2 + 16}$ miles long over the swamp, the cost of construction is

$$C(x) = 20{,}000(4 - x) + 25{,}000\sqrt{x^2 + 16}, \quad 0 \le x \le 4.$$

This is the objective function.

27. To find the tangent line, we first find the derivative of $f(x)$:

$$f(x+h) = \frac{-1}{2(x+h)^2} = -\frac{1}{2x^2 + 4xh + 2h^2}$$

$$f(x+h) - f(x) = -\frac{1}{2x^2 + 4xh + 2h^2} - \frac{-1}{2x^2} = \frac{-2x^2 + 2x^2 + 4xh + 2h^2}{2x^2(2x^2 + 4xh + 2h^2)}$$

$$= \frac{4xh + 2h^2}{2x^2(2x^2 + 4xh + 2h^2)} = \frac{2xh + h^2}{x^2(2x^2 + 4xh + 2h^2)}$$

$$\frac{f(x+h) - f(x)}{h} = \frac{\dfrac{2xh + h^2}{x^2(2x^2 + 4xh + 2h^2)}}{h} = \frac{2xh + h^2}{hx^2(2x^2 + 4xh + 2h^2)}$$

$$= \frac{2x + h}{x^2(2h^2 + 4xh + 2h^2)}$$

$$f'(x) = \lim_{h \to 0} \frac{f(x+h) - f(x)}{h} = \lim_{h \to 0} \frac{2x + h}{x^2(2x^2 + 4xh + 2h^2)}$$

$$= \frac{2x}{x^2(2x^2)} = \frac{x}{x^4} = \frac{1}{x^3}.$$

The slope of tangent line when $x = \frac{1}{2}$ is

$$m_{\text{tan}} = f'\left(\frac{1}{2}\right) = \frac{1}{(1/2)^3} = 8,$$

and the y-coordinate of the function when $x = \frac{1}{2}$ is

$$f\left(\frac{1}{2}\right) = \frac{-1}{2(1/2)^2} = \frac{-1}{2(1/4)} = \frac{-1}{1/2} = -2.$$

Using the point-slope form of the equation of a line, we see that the tangent line is

$$y - (-2) = 8\left(x - \frac{1}{2}\right) \qquad \text{or} \qquad y = 8x - 6.$$

30. The derivative of the function $f(x) = x - 4\sqrt{x}$ in Problem 28 is $f'(x) = 1 - \dfrac{2}{\sqrt{x}} = \dfrac{\sqrt{x} - 2}{\sqrt{x}}$.
At point where the tangent is horizontal $f'(x) = 0 \Rightarrow \sqrt{x} - 2 = 0 \Rightarrow x = 4$. So the only point on the graph is $(4,\ f(4)) = (4,\ -4)$.

Chapter 3

Polynomial and Rational Functions

3.1 | **Polynomial Functions**

3. The graph of $y = (x - 2)^3 + 2$, shown in blue, is the graph of $y = x^3$, shown in red, shifted 2 units to the right and then shifted upward 2 units.

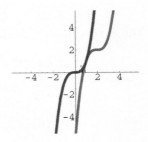

6. The graph of $y = x^4 - 1$, shown in blue, is the graph of $y = x^4$, shown in red, shifted downward 1 unit.

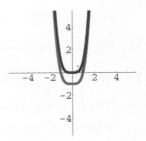

9. The function $f(x) = -2x^3 + 4x$ is odd because all powers of x are odd.

12. Since

$$f(x) = x^3(x + 2)(x - 2) = x^3(x^2 - 4) = x^5 - 4x^3,$$

we see that the function is odd because all powers of x are odd.

To solve Problems 13–18, first note that all of the functions have zeros at $x = 0$ and $x = 1$. Next, note whether the exponents of x and $x - 1$ are 1, even, or odd and greater than 1. Finally, use the facts that when the exponent of $x - a$ is

1:   *the graph passes directly through the x-axis at $x = a$;*

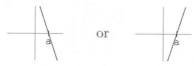

even: *the graph is tangent to, but does not pass through the x-axis at $x = a$;*

odd and greater than 1:   *the graph is tangent to, and passes through the x-axis.*

The forms of the functions in (a)–(f) are

(a) $x^{\text{even}}(x - 1)^{\text{even}}$

(b) $x^{\text{odd}}(x - 1)^1$

(c) $x^{\text{odd}}(x - 1)^{\text{odd}}$

(d) $x^1(x - 1)^{\text{odd}}$

(e) $x^{\text{even}}(x - 1)^1$

(f) $x^{\text{odd}}(x - 1)^{\text{even}}$

where "odd" means "odd and greater than 1." Because each of these is distinct, it is not necessary in this set of problems to consider whether the lead coefficient is positive or negative.

15. The form of the function must be $x^{\text{even}}(x - 1)^1$, so this graph corresponds to (e).

18. The form of the function must be $x^1(x - 1)^{\text{odd}}$, so this graph corresponds to (d).

21. The function $f(x) = -7x(x - 1)(x + 3)^2$ has

- four real zeros,
- 1 is a simple zero,
- -3 is a zero of multiplicity 2,
- and its end behavior is like $y = -7x^4$.

24. *End Behavior:* For large x, the graph is like that of $y = -x^3$.

Symmetry: Since the powers of x are all odd, the graph is symmetric with respect to the origin.

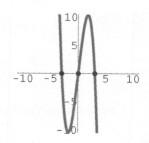

Intercepts: Since $f(0) = 0$, the y-intercept is $(0, 0)$. Solving

$$f(x) = 9x - x^3 = x(3 + x)(3 - x) = 0$$

we see that the x-intercepts are $(-3, 0), (0, 0)$, and $(3, 0)$.

The Graph: From $f(x) = -x^1(x + 3)^1(x - 3)^1$ we see that $0, -3$, and 3 are all simple zeros.

27. *End Behavior:* For large x, the graph is like that of $y = x^3$.

Symmetry: Since the powers of x in

$$f(x) = (x + 1)(x - 2)(x - 4) = x^3 - 5x^2 + 2x + 8$$

are both even and odd, the graph has no symmetry with respect to the origin or y-axis.

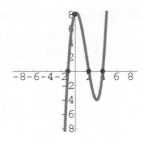

Intercepts: Since $f(0) = 8$, the y-intercept is $(0, 8)$. Solving

$$f(x) = (x + 1)(x - 2)(x - 4) = 0$$

we see that the x-intercepts are $(-1, 0), (2, 0)$, and $(4, 0)$.

The Graph: From $f(x) = (x + 1)^1(x - 2)^1(x - 4)^1$ we see that -1, 2, and 4 are all simple zeros.

30. *End Behavior:* For large x, the graph is like that of $y = x^4$.

Symmetry: Since the powers of x in

$$f(x) = x^2(x - 2)^2 = x^4 - 4x^3 + 4x^2$$

are both even and odd, the graph has no symmetry with respect to the origin or y-axis.

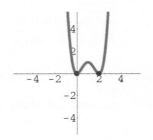

Intercepts: Since $f(0) = 0$, the y-intercept is $(0, 0)$. Solving

$$f(x) = x^2(x - 2)^2 = 0$$

we see that the x-intercepts are $(0, 0)$ and $(2, 0)$.

The Graph: From $f(x) = x^2(x - 2)^2$ we see that the graph is tangent to the x-axis at $x = 0$ and $x = 2$.

33. *End Behavior:* For large x, the graph is like that of $y = x^4$.

Symmetry: Since the powers of x in

$$f(x) = (x^2 - 1)(x^2 + 9) = x^4 + 8x^2 - 9$$

are all even, the graph is symmetric with respect to the y-axis.

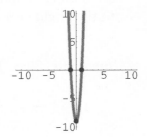

Intercepts: Since $f(0) = -9$, the y-intercept is $(0, -9)$. Solving

$$f(x) = (x + 1)(x - 1)(x^2 + 9) = 0$$

we see that the x-intercepts are $(-1, 0)$ and $(1, 0)$.

The Graph: From $f(x) = (x + 1)^1(x - 1)^1(x^2 + 9)$ we see that -1 and 1 are simple zeros.

36. *End Behavior:* For large x, the graph is like that of $y = x^4$.

Symmetry: Since the powers of x are all even, the graph is symmetric with respect to the y-axis.

Intercepts: Since $f(0) = 9$, the y-intercept is $(0, 9)$. Solving

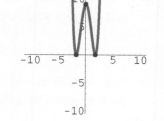

$$f(x) = x^4 - 6x^2 + 9 = (x^2 - 3)^2 = \left(x + \sqrt{3}\right)^2 \left(x - \sqrt{3}\right)^2 = 0$$

we see that the x-intercepts are $\left(-\sqrt{3}, 0\right)$ and $\left(\sqrt{3}, 0\right)$.

The Graph: From $f(x) = \left(x + \sqrt{3}\right)^2 \left(x - \sqrt{3}\right)^2$ we see that the graph is tangent to the x-axis at $x = -\sqrt{3}$ and $x = \sqrt{3}$.

39. *End Behavior:* For large x, the graph is like that of $y = x^5$.

Symmetry: Since the powers of x are all odd, the graph is symmetric with respect to the origin.

Intercepts: Since $f(0) = 0$, the y-intercept is $(0, 0)$. Solving

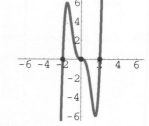

$$f(x) = x^5 - 4x^3 = x^3(x + 2)(x - 2) = 0$$

we see that the x-intercepts are $(0, 0), (-2, 0)$, and $(2, 0)$.

The Graph: From $f(x) = x^3(x + 2)^1(x - 2)^1$ we see that -2 and 2 are simple zeros and the graph is tangent to, but passes through, the x-axis.

42. *End Behavior:* For large x, the graph is like that of $y = x^5$.

Symmetry: Since the powers of x in

$$f(x) = (x + 1)^2(x - 1)^3 = x^5 - x^4 - 2x^3 + 2x^2 + x - 1$$

are both even and odd, the graph has no symmetry with respect to the origin or y-axis.

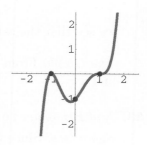

Intercepts: Since $f(0) = -1$, the y-intercept is $(0, -1)$. Solving

$$f(x) = (x + 1)^2(x - 1)^3 = 0$$

we see that the x-intercepts are $(-1, 0)$ and $(1, 0)$.

The Graph: From $f(x) = (x + 1)^2(x - 1)^3$ we see that there are no simple zeros and the graph is tangent to the x-axis at $x = -1$. It is also tangent to, but passes through, the x-axis at $x = 1$.

45. (a) The graph of $f(x)$ is shown in blue, and the graph of $g(x) = f(x) + 2$ is shown in red.

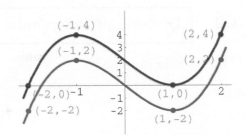

(b) We see from the graph in part **(a)** that the x-intercepts of $g(x)$ are at $(-2, 0)$ and $(1, 0)$. Also the graph of $g(x)$ is tangent to the x-axis at $x = 1$. Thus, the form of $g(x)$ is

$$g(x) = a(x + 2)(x - 1)^2.$$

Since $g(2) = 4$, we have

$$a(2 + 2)(2 - 1)^2 = 4a = 4,$$

so $a = 1$ and

$$g(x) = (x + 2)(x - 1)^2 = x^3 - 3x + 2.$$

48. Since $f(-1) = k_1(-1)^4 - k_2(-1)^3 - 1 - 4 = 0$ and $f(1) = k_1 - k_2 + 1 - 4 = 0$, we have the system of equations

$$k_1 + k_2 = 5$$
$$k_1 - k_2 = 3.$$

Adding these equations, we find $2k_1 = 8$, so $k_1 = 4$. Then, from the first equation, $k_2 = 5 - k_1 = 5 - 4 = 1$.

51. The graph of $f(x) = (x - 1)^{n+2}(x + 1)$ will cross the x-axis at $(1, 0)$ when $n + 2$ is an odd positive integer. This will occur when n is an odd integer greater than or equal to -1. But, since it is given that n is positive, n must be a positive odd integer.

Miscellaneous Calculus-Related Problems

54. In this case, the area of the base is $(30 - 2x)(40 - 4x)$ and the height is x, so the volume is

$$V(x) = x(30 - 2x)(40 - 4x).$$

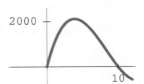

Since $40 - 4x = 0$ when $x = 10$, the domain is $[0, 10]$.

3.2 | Division of Polynomial Functions

3. By long division,

$$
\begin{array}{r}
5x - 12 \\
x^2 + x - 1 \,\overline{\big)\, 5x^3 - 7x^2 + 4x + 1} \\
\underline{5x^3 + 5x^2 - 5x} \\
-12x^2 + 9x + 1 \\
\underline{-12x^2 - 12x + 12} \\
21x - 11
\end{array}
$$

The quotient is $q(x) = 5x - 12$, the remainder is $r(x) = 21x - 1$, and

$$5x^3 - 7x^2 + 4x + 1 = (x^2 + x - 1)(5x - 12) + (21x - 11).$$

6. Using $(2x+1)^2 = 4x^2 + 4x + 1$ and long division, we have

$$
\begin{array}{r}
\tfrac{1}{4}\,x \\
4x^2 + 4x + 1 \,\overline{\big)\, x^3 + x^2 + x + 1} \\
\underline{x^3 + x^2 + \tfrac{1}{4}\,x} \\
\tfrac{3}{4}\,x + 1
\end{array}
$$

The quotient is $q(x) = \frac{1}{4}\,x$, the remainder is $r(x) = \frac{3}{4}\,x + 1$, and

$$x^3 + x^2 + x + 1 = (4x^2 + 4x + 1)\left(\frac{1}{4}\,x\right) + \left(\frac{3}{4}\,x + 1\right).$$

9. By long division,

$$
\begin{array}{r}
6x^2 + 4x + 1 \\
x^3 - 2 \,\overline{\big)\, 6x^5 + 4x^4 + x^3 + 0x^2 + 0x + 0} \\
\underline{6x^5 + 0x^4 + 0x^3 - 12x^2} \\
4x^4 + x^3 + 12x^2 + 0x + 0 \\
\underline{4x^4 + 0x^3 + 0x^2 - 8x} \\
x^3 + 12x^2 + 8x + 0 \\
\underline{x^3 + 0x^2 + 0x - 2} \\
12x^2 + 8x + 2
\end{array}
$$

The quotient is $q(x) = 6x^2 + 4x + 1$, the remainder is $r(x) = 12x^2 + 8x + 2$, and

$$6x^5 + 4x^4 + x^3 = (x^3 - 2)(6x^2 + 4x + 1) + (12x^2 + 8x + 2).$$

12. From the Remainder Theorem, the remainder r, when $f(x) = 3x^2 + 7x - 1$ is divided by $x + 3$, is $f(-3)$. Thus,

$$r = f(-3) = 3(-3)^2 + 7(-3) - 1 = 5.$$

15. From the Remainder Theorem, the remainder r, when $f(x) = x^4 - x^3 + 2x^2 + 3x - 5$ is divided by $x - 3$, is $f(3)$. Thus,

$$r = f(3) = 3^4 - 3^3 + 2(3)^2 + 3(3) - 5 = 76.$$

18. To find $f\left(\frac{1}{4}\right)$ we use the Remainder Theorem with dividend $f(x) = 6x^2 + 4x - 2$ and divisor $d(x) = x - \frac{1}{4}$:

$$
\begin{array}{r}
6x + \quad \frac{11}{2} \\
x - \frac{1}{4} \overline{\smash{\big)}\ 6x^2 + \ 4x - \ 2} \\
\underline{6x^2 - \ \frac{3}{2}x} \\
\frac{11}{2}x - \ 2 \\
\underline{\frac{11}{2}x - \frac{11}{8}} \\
-\frac{5}{8}
\end{array}
$$

The remainder is $r = -\frac{5}{8}$, so $f\left(\frac{1}{4}\right) = r = -\frac{5}{8}$.

21. To find $f\left(\frac{1}{2}\right)$ we use the Remainder Theorem with dividend $f(x) = 3x^4 - 5x^2 + 20$ and divisor $d(x) = x - \frac{1}{2}$:

$$
\begin{array}{r}
3x^3 + \frac{3}{2}x^2 - \frac{17}{4}x - \quad \frac{17}{8} \\
x - \frac{1}{2} \overline{\smash{\big)}\ 3x^4 + \ 0x^3 - \ 5x^2 + \ 0x + \ 20} \\
\underline{3x^4 - \frac{3}{2}x^3} \\
\frac{3}{2}x^3 - \ 5x^2 + \ 0x + \ 20 \\
\underline{\frac{3}{2}x^3 - \frac{3}{4}x^2} \\
-\frac{17}{4}x^2 + \ 0x + \ 20 \\
\underline{-\frac{17}{4}x^2 + \frac{17}{8}x} \\
-\frac{17}{8}x + \ 20 \\
\underline{-\frac{17}{8}x + \frac{17}{16}} \\
\frac{303}{16}
\end{array}
$$

The remainder is $r = \frac{303}{16}$, so $f\left(\frac{1}{2}\right) = r = \frac{303}{16}$.

24. Using synthetic division to divide $f(x) = 4x^2 - 8x + 6$ by $x - \frac{1}{2}$, we have

$$
\begin{array}{r|rrr}
\frac{1}{2} & 4 & -8 & 6 \\
  &   & 2 & -3 \\
\hline
  & 4 & -6 & \boxed{3} = r
\end{array}
$$

The quotient is $q(x) = 4x - 6$ and the remainder is $r = 3$.

27. Using synthetic division to divide $f(x) = x^4 + 16$ by $x - 2$, we have

$$
\begin{array}{r|rrrrr}
2 & 1 & 0 & 0 & 0 & 16 \\
  &   & 2 & 4 & 8 & 16 \\
\hline
  & 1 & 2 & 4 & 8 & \boxed{32 = r} \\
\end{array}
$$

The quotient is $q(x) = x^3 + 2x^2 + 4x + 8$ and the remainder is $r = 32$.

30. Using synthetic division to divide $f(x) = 2x^6 + 3x^3 - 4x^2 - 1$ by $x + 1$, we have

$$
\begin{array}{r|rrrrrrr}
-1 & 2 & 0 & 0 & 3 & -4 & 0 & -1 \\
   &   & -2 & 2 & -2 & -1 & 5 & -5 \\
\hline
   & 2 & -2 & 2 & 1 & -5 & 5 & \boxed{-6 = r} \\
\end{array}
$$

The quotient is $q(x) = 2x^5 - 2x^4 + 2x^3 + x^2 - 5x + 5$ and the remainder is $r = -6$.

33. From synthetic division

$$
\begin{array}{r|rrr}
-3 & 4 & -2 & 9 \\
   &   & -12 & 42 \\
\hline
   & 4 & -14 & \boxed{51 = r = f(-3)} \\
\end{array}
$$

we see that $f(-3) = r = 51$.

36. We express $f(x)$ using all the coefficients, including 0. Then, by synthetic division, we have

$$
\begin{array}{r|rrrrrr}
-2 & 3 & 0 & 0 & 1 & 0 & -16 \\
   &   & -6 & 12 & -24 & 46 & -92 \\
\hline
   & 3 & -6 & 12 & -23 & 46 & \boxed{-108 = r = f(-2)} \\
\end{array}
$$

so $f(-2) = r = -108$.

39. We are looking for a value of k so that when $f(x) = x^4 + x^3 + 3x^2 + kx - 4$ is divided by $d(x) = x^2 - 1$, the remainder is 0. Using long division, we have

$$
\begin{array}{r}
x^2 + x + 4 \\
x^2 - 1 \overline{\smash{\big)}\, x^4 + x^3 + 3x^2 + kx - 4} \\
\underline{x^4 - x^2} \\
x^3 + 4x^2 + kx - 4 \\
\underline{x^3 - x} \\
4x^2 + (k+1)x - 4 \\
\underline{4x^2 - 4} \\
(k+1)x + 0
\end{array}
$$

The remainder is $r(x) = (k+1)x$, which we want to be 0, so choose $k = -1$. Then, when $k = -1$, $f(x) = x^4 + x^3 + 3x^2 - x - 4$ is divisible by $d(x) = x^2 - 1$.

42. We are looking for a value of k so that when $f(x) = x^3 + kx^2 - 2kx + 4$ is divided by $d(x) = x + 2$, the remainder is 0. Using synthetic division, we have

$$
\begin{array}{r|cccc}
-2 & 1 & k & -2k & 4 \\
& & -2 & -2k+4 & 8k-8 \\
\hline
& 1 & k-2 & -4k+4 & \boxed{8k-4} = r
\end{array}
$$

The remainder is $r = 8k - 4$, which we want to be 0, so choose $k = \frac{1}{2}$. Then, when $k = \frac{1}{2}$, $f(x) = x^3 + \frac{1}{2}x^2 - x + 4$ is divisible by $d(x) = x + 2$.

3.3 | Zeros and Factors of Polynomial Functions

In Problems 3–9 below we could determine if the given number is a zero of the polynomial function $f(x)$ by evaluating $f(x)$ at this number to see if the output is 0. Instead, we use synthetic division because, in addition to giving us the value of the function in the form of the remainder, we also obtain the quotient which is a polynomial of degree one less than the degree of $f(x)$. As such, it is more likely to be useful in finding any remaining zeros.

3. By synthetic division

$$
\begin{array}{r|cccc}
5 & 1 & -6 & 6 & 5 \\
& & 5 & -5 & 5 \\
\hline
& 1 & -1 & 1 & \boxed{10} = r
\end{array}
$$

so $f(5) = 10$. Since $f(5) \neq 0$, we conclude that 5 is not a zero of $f(x)$.

6. By synthetic division

$$
\begin{array}{r|cccc}
-2 & 1 & -4 & -2 & 20 \\
& & -2 & 12 & -20 \\
\hline
& 1 & -6 & 10 & \boxed{0} = r
\end{array}
$$

we see that $f(-2) = 0$. Thus, $x + 2$ is a factor of $f(x)$ and $f(x) = (x+2)(x^2 - 6x + 10)$. We can find the zeros of $q(x) = x^2 - 6x + 10$ by the quadratic formula:

$$
x = \frac{-(-6) \pm \sqrt{(-6)^2 - 4(1)(10)}}{2(1)} = \frac{6 \pm \sqrt{-4}}{2} = 3 \pm i.
$$

Thus, the complete factorization of $f(x)$ is

$$
f(x) = (x + 2)(x - 3 + i)(x - 3 - i).
$$

9. By synthetic division

$$
\begin{array}{r|rrrrr}
1 & 9 & 69 & -29 & -41 & -8 \\
  &   & 9 & 78 & 49 & 8 \\
\hline
  & 9 & 78 & 49 & 8   & \boxed{0} = r
\end{array}
$$

so 1 is a zero of $f(x)$ and

$$f(x) = (x-1)q_1(x) = (x-1)(9x^3 + 78x^2 + 49x + 8).$$

Using synthetic division again, we confirm that $-\frac{1}{3}$ is a zero of $q_1(x)$:

$$
\begin{array}{r|rrrr}
-\frac{1}{3} & 9 & 78 & 49 & 8 \\
             &   & -3 & -25 & -8 \\
\hline
             & 9 & 75 & 24 & \boxed{0} = r
\end{array}
$$

Thus, $-\frac{1}{3}$ is also a zero of $f(x)$, and

$$f(x) = (x-1)q_1(x) = (x-1)\left(x + \frac{1}{3}\right)q_2(x) = (x-1)\left(x + \frac{1}{3}\right)(9x^2 + 75x + 24)$$

$$= 3(x-1)\left(x + \frac{1}{3}\right)(3x^2 + 25x + 8) = 3(x-1)\left(x + \frac{1}{3}\right)(3x+1)(x+8)$$

$$= 9(x-1)\left(x + \frac{1}{3}\right)^2 (x+8).$$

12. Since 0 is a zero, x must be a factor of f, that is, $f(x) = x(2x^4 - 17x^3 + 40x^2 - 16x)$. But by inspection x is still another factor $\Rightarrow f(x) = x^2(2x^3 - 17x^2 + 40x - 16)$. Then synthetic division of $2x^3 - 17x^2 + 40x - 16$ by $x - \frac{1}{2}$ yields $f(x) = x^2(x - \frac{1}{2})(2x^2 - 16x + 32)$.

The quadratic term in the preceding expression factors: $2x^2 - 16x + 32 = 2(x^2 - 8x + 16) = 2(x-4)^2$. Therefore

$$f(x) = x^2(x - \tfrac{1}{2})2(x-4)^2 = x^2(2x - 1)(x-4)^2.$$

15. Using synthetic division

$$
\begin{array}{r|rrrr}
1 & 1 & 0 & 1 & -2 \\
  &   & 1 & 1 & 2 \\
\hline
  & 1 & 1 & 2 & \boxed{0} = r
\end{array}
$$

we see that 1 is a zero of $f(x)$, and $f(x) = (x-1)(x^2 + x + 2)$. Since $x^2 + x + 2$, does not factor with integer coefficients, we find the remaining zeros of $f(x)$ using the quadratic formula:

$$x = \frac{-1 \pm \sqrt{1^2 - 4(1)(2)}}{2(1)} = \frac{-1 \pm \sqrt{7}\,i}{2}.$$

The complete factorization of $f(x)$ is

$$f(x) = (x-1)\left(x + \frac{1}{2} - \frac{\sqrt{7}}{2}i\right)\left(x + \frac{1}{2} + \frac{\sqrt{7}}{2}i\right)$$

18. Using synthetic division

$$
\begin{array}{r|rrrr}
2 & 1 & -6 & -16 & 48 \\
  &   & 2 & -8 & -48 \\
\hline
  & 1 & -4 & -24 & \boxed{\;0 = r\;}
\end{array}
$$

we see that 2 is a zero of $f(x)$, and $f(x) = (x-2)(x^2 - 4x - 24)$. Since $x^2 - 4x - 24$ does not factor with integer coefficients, we find the remaining zeros of $f(x)$ using the quadratic formula:

$$x = \frac{-(-4) \pm \sqrt{(-4)^2 - 4(1)(-24)}}{2(1)} = \frac{4 \pm \sqrt{112}}{2} = \frac{4 \pm 4\sqrt{7}}{2} = 2 \pm 2\sqrt{7}.$$

The complete factorization of $f(x)$ is

$$f(x) = (x-2)(x - 2 - 2\sqrt{7})(x - 2 + 2\sqrt{7}).$$

21. Since $(x-1)^2 = x^2 - 2x + 1$ is not linear, we use long division:

$$
\begin{array}{r}
2x^2 + 5x + 3 \\
x^2 - 2x + 1 \overline{\smash{\big)}\ 2x^4 + x^3 - 5x^2 - x + 3} \\
\underline{2x^4 - 4x^3 + 2x^2} \\
5x^3 - 7x^2 - x + 3 \\
\underline{5x^3 - 10x^2 + 5x} \\
3x^2 - 6x + 3 \\
\underline{3x^2 - 6x + 3} \\
0
\end{array}
$$

Since the remainder is 0, $(x-1)^2$ is a factor of $f(x) = 2x^4 + x^3 - 5x^2 - x + 3$, and $x = 1$ is a zero of multiplicity 2. The quotient is $q(x) = 2x^2 + 5x + 3 = (2x+3)(x+1)$, so the remaining zeros of $f(x)$ are $-\frac{3}{2}$ and -1. The complete factorization of $f(x)$ is

$$f(x) = (x-1)^2(2x - 3)(x + 1).$$

24. Using $i^2 = -1$ and $i^3 = -i$ we have

$$f\left(\frac{1}{2}i\right) = 12\left(\frac{1}{2}i\right)^3 + 8\left(\frac{1}{2}i\right)^2 + 3\left(\frac{1}{2}i\right) + 2 = 12\left(\frac{1}{8}i^3\right) + 8\left(\frac{1}{4}i^2\right) + \frac{3}{2}i + 2$$

$$= -\frac{3}{2}i - 2 + \frac{3}{2}i + 2 = 0.$$

Thus, $x = \frac{1}{2}i$ is a zero of $f(x)$. Since the coefficients of $f(x)$ are real numbers, another zero is $x = -\frac{1}{2}i$, and

$$\left(x - \frac{1}{2}i\right)\left(x + \frac{1}{2}i\right) = x^2 - \frac{1}{4}i^2 = x^2 + \frac{1}{4}$$

is a factor of $f(x)$. To find the remaining factors we use long division:

$$
\begin{array}{r}
12x + 8 \\
x^2 + \frac{1}{4} \overline{\smash{)}\ 12x^3 + 8x^2 + 3x + 2} \\
\underline{12x^3 + 0x^2 + 3x} \\
8x^2 + 0x + 2 \\
\underline{8x^2 + 0x + 2} \\
0
\end{array}
$$

Thus, the other factor of $f(x)$ is $12x + 8 = 4(3x + 2)$, so the zeros of $f(x)$ are $\frac{1}{2}i$, $-\frac{1}{2}i$, and $-\frac{2}{3}$, and the complete factorization of $f(x)$ is

$$f(x) = 4\left(x - \frac{1}{2}i\right)\left(x + \frac{1}{2}i\right)(3x + 2).$$

27. We use $(1 + 2i)^2 = -3 + 4i$, $(1 + 2i)^3 = 11 - 2i$, and $(1 + 2i)^4 = -7 - 24i$. Then

$$
\begin{aligned}
f(1 + 2i) &= (1 + 2i)^4 - 2(1 + 2i)^3 - 4(1 + 2i)^2 + 18(1 + 2i) - 45 \\
&= -7 - 24i - 2(11 - 2i) - 4(-3 + 4i) + 18(1 + 2i) - 45 \\
&= -7 - 24i + 22 + 4i + 12 - 16i + 18 + 36i - 45 = 0,
\end{aligned}
$$

and we see that $x = 1 + 2i$ is a zero of $f(x)$. Since the coefficients of $f(x)$ are real numbers, another zero is $x = 1 - 2i$, and

$$
\begin{aligned}
[x - (1 + 2i)][x - (1 - 2i)] &= [(x - 1) - 2i][(x - 1) + 2i] \\
&= (x - 1)^2 - 4i^2 = x^2 - 2x + 5
\end{aligned}
$$

is a factor of $f(x)$. To find the remaining factors we use long division:

$$
\begin{array}{r}
x^2 - 9 \\
x^2 - 2x + 5 \overline{\smash{)}\ x^4 - 2x^3 - 4x^2 + 18x - 45} \\
\underline{x^4 - 2x^3 + 5x^2} \\
-9x^2 + 18x - 45 \\
\underline{-9x^2 + 18x - 45} \\
0
\end{array}
$$

Since $x^2 - 9 = (x + 3)(x - 3)$, the zeros of $f(x)$ are $1 + 2i$, $1 - 2i$, 3, and -3. The complete factorization of $f(x)$ is

$$
\begin{aligned}
f(x) &= [x - (1 + 2i)][x - (1 - 2i)](x + 3)(x - 3) \\
&= (x - 1 - 2i)(x - 1 + 2i)(x + 3)(x - 3).
\end{aligned}
$$

30. Since $-4i$ is a zero, so is $4i$. Thus, a fifth degree polynomial with the given zeros is

$$f(x) = (x + 4i)(x - 4i)\left(x + \frac{1}{3}\right)\left(x - \frac{1}{2}\right)^2$$

$$= \left(x^2 - 16i^2\right)\left(x + \frac{1}{3}\right)\left(x^2 - x + \frac{1}{4}\right)$$

$$= (x^2 + 16)\left(x + \frac{1}{3}\right)\left(x^2 - x + \frac{1}{4}\right)$$

$$= x^5 - \frac{2}{3}x^4 + \frac{191}{12}x^3 - \frac{127}{12}x^2 - \frac{4}{3}x + \frac{4}{3}.$$

33. Since $1 - 6i$ is a zero, so is $1 + 6i$. Thus, a second degree polynomial with the given zeros is

$$f(x) = [x - (1 - 6i)][x - (1 + 6i)] = [(x - 1) + 6i][(x - 1) - 6i]$$

$$= \left[(x - 1)^2 - 36i^2\right] = x^2 - 2x + 37.$$

36. Factoring, we have

$$x^4 + 6x^3 + 9x^2 = x^2(x^2 + 6x + 9) = x^2(x + 3)^2.$$

Setting $x = 0$ and $x + 3 = 0$ we see that $x = 0$ and $x = -3$ are each zeros of multiplicity 2.

39. Since 3 is a zero of $f(x)$, we solve

$$f(3) = 2(3)^3 - 2(3)^2 + k = 54 - 18 + k = 36 + k = 0,$$

which gives $k = -36$. Thus $f(x) = 2x^3 - 2x^2 - 36$, and by synthetic division

$$
\begin{array}{r|rrrr}
3 & 2 & -2 & 0 & -36 \\
  &   & 6 & 12 & 36 \\
\hline
  & 2 & 4 & 12 & 0 = r
\end{array}
$$

so

$$f(x) = (x - 3)(2x^2 + 4x + 12) = 2(x - 3)(x^2 + 2x + 6).$$

By the quadratic formula, $q(x) = x^2 + 2x + 6$ has zeros $-1 \pm \sqrt{5}\,i$, and so the complete factorization of $f(x)$ is

$$f(x) = 2x^3 - 2x^2 - 36 = 2(x - 3)(x + 1 - \sqrt{5}\,i)(x + 1 + \sqrt{5}\,i).$$

42. Since the graph passes through the x-axis at $x = -2$ and $x = -1$ without being tangent to the x-axis at either point, $x = -2$ and $x = -1$ are simple zeros. Since the graph passes through the x-axis at $x = 1$ but is flattened at that point, $x = 1$ is a zero of odd multiplicity greater than 1. Since the degree of $f(x)$ is 5, $x = 1$ is a zero of multiplicity 3, and

$$f(x) = a(x - 1)^3(x + 2)(x + 1) = a(x^5 - 4x^3 + 2x^2 + 3x - 2).$$

When $x = 0$ we see from the graph that $y = f(0) = 2$. From the above, $f(0) = -2a$ so $a = -1$ and

$$f(x) = -x^5 + 4x^3 - 2x^2 - 3x + 2.$$

3.4 | Real Zeros of Polynomial Functions

3. The factors of $a_0 = -3$ and $a_3 = 1$ are

$$p : \pm 1, \ \pm 3 \qquad \text{and} \qquad s : \pm 1,$$

respectively, so possible rational zeros are

$$\frac{p}{s} : \ \pm 1, \ \pm 3.$$

It is easily seen by direct substitution into $f(x)$ that $x = -1$, $x = 1$, and $x = -3$ are not zeros of $f(x)$. We then use synthetic division to test $x = 3$:

$$
\begin{array}{r|rrrr}
3 & 1 & 0 & -8 & -3 \\
  &   & 3 & 9 & 3 \\
\hline
  & 1 & 3 & 1 & \boxed{0 = r}
\end{array}
$$

Thus, $x = 3$ is a zero of $f(x)$ and

$$f(x) = (x - 3)(x^2 + 3x - 1).$$

Since $x^2 + 3x - 1$ does not factor, the only rational zero of $f(x)$ is 3.

6. The factors of $a_0 = -2$ and $a_4 = 8$ are

$$p : \pm 1, \ \pm 2 \qquad \text{and} \qquad s : \pm 1, \ \pm 2, \ \pm 4, \ \pm 8,$$

respectively, so possible rational zeros are

$$\frac{p}{s} : \ \pm 1, \ \pm \frac{1}{2}, \ \pm \frac{1}{4}, \ \pm \frac{1}{8}, \ \pm 2.$$

There are a number of other possible ratios, such as $\frac{2}{8}$, but they can all be reduced to one of the forms listed – for example, $\frac{2}{8} = \frac{1}{4}$. As usual, it is just as easy to check -1 and 1 directly. In this case, $f(-1) = 27$ and $f(1) = 15$ so neither -1 nor 1 is a zero of $f(x)$. Using synthetic division, we find $f(-\frac{1}{2}) = \frac{9}{2}$, so $-\frac{1}{2}$ is not a zero. Next, we use synthetic division to compute $f(\frac{1}{2})$:

$$
\begin{array}{r|rrrrr}
\frac{1}{2} & 8 & -2 & 15 & -4 & -2 \\
  &   & 4 & 1 & 8 & 2 \\
\hline
  & 8 & 2 & 16 & 4 & \boxed{0 = r}
\end{array}
$$

Thus, $x = \frac{1}{2}$ is a zero of $f(x)$ and

$$f(x) = \left(x - \frac{1}{2} \right)(8x^3 + 2x^2 + 16x + 4).$$

Since all coefficients of

$$q(x) = 8x^3 + 2x^2 + 16x + 4$$

are positive, we need only look for negative zeros. We use synthetic division to compute $q(-\frac{1}{4})$:

$$
\begin{array}{r|rrrr}
-\frac{1}{4} & 8 & 2 & 16 & 4 \\
             &   & -2 & 0 & -4 \\
\hline
             & 8 & 0 & 16 & \boxed{0} = r
\end{array}
$$

Thus, $f(-\frac{1}{4}) = (\frac{1}{4} - \frac{1}{2})q(\frac{1}{4}) = 0$, $x = -\frac{1}{4}$ is a zero of $f(x)$, and

$$f(x) = \left(x - \frac{1}{2}\right)\left(x + \frac{1}{4}\right)(8x^2 + 16).$$

Since $8x^2 + 16$ has no real zeros, the rational zeros of $f(x)$ are $\frac{1}{2}$ and $-\frac{1}{4}$.

9. The factors of $a_0 = 3$ and $a_4 = 6$ are

$$p : \pm 1,\ \pm 3 \qquad \text{and} \qquad s : \pm 1,\ \pm 2,\ \pm 3,\ \pm 6,$$

respectively, so possible rational zeros are

$$\frac{p}{s} : \pm 1,\ \frac{1}{2},\ \pm\frac{1}{3},\ \pm\frac{1}{6},\ \pm 3,\ \pm\frac{3}{2}.$$

There are a number of other possible ratios, such as $\frac{3}{6}$, but they can all be reduced to one of the forms listed – for example, $\frac{3}{6} = \frac{1}{2}$. As usual, it is just as easy to check -1 and 1 directly. In this case, $f(-1) = 20$ and $f(1) = -6$, so neither -1 nor 1 is a zero of $f(x)$. Using synthetic division, we find $f(-\frac{1}{2}) = \frac{15}{2}$, $f(\frac{1}{2}) = -\frac{7}{4}$, and $f(-\frac{1}{3}) = \frac{154}{27}$, so $-\frac{1}{2}$, $\frac{1}{2}$, and $-\frac{1}{3}$ are not zeros. We use synthetic division to compute $f(\frac{1}{3})$:

$$
\begin{array}{r|rrrrr}
\frac{1}{3} & 6 & -5 & -2 & -8 & 3 \\
            &   & 2 & -1 & -1 & -3 \\
\hline
            & 6 & -3 & -3 & -9 & \boxed{0} = r
\end{array}
$$

Thus, $x = \frac{1}{3}$ is a zero of $f(x)$ and

$$f(x) = \left(x - \frac{1}{3}\right)(6x^3 - 3x^2 - 3x - 9) = 3\left(x - \frac{1}{3}\right)(2x^3 - x^2 - x - 3).$$

Possible rational zeros of $g(x) = 2x^3 - x^2 - x - 3$ are ± 1, $\pm\frac{1}{2}$, ± 3, and $\pm\frac{3}{2}$. Note that we have eliminated $\pm\frac{1}{6}$ as possible zeros. Using synthetic division, we find $f(-3) = 630$, $f(3) = 312$, and $f(-\frac{3}{2}) = \frac{231}{4}$, so -3, 3, and $-\frac{3}{2}$ are not zeros. Finally, we use synthetic division to compute $f(\frac{3}{2})$:

$$
\begin{array}{r|rrrr}
\frac{3}{2} & 2 & -1 & -1 & -3 \\
            &   & 3 & 3 & 3 \\
\hline
            & 2 & 2 & 2 & \boxed{0} = r
\end{array}
$$

Thus, $x = \frac{3}{2}$ is a zero of $f(x)$ and

$$f(x) = 3\left(x - \frac{1}{3}\right)\left(x - \frac{3}{2}\right)(2x^2 + 2x + 2)$$

$$= 6\left(x - \frac{1}{3}\right)\left(x - \frac{3}{2}\right)(x^2 + x + 1).$$

We use the quadratic formula to find the zeros of $h(x) = x^2 + x + 1$:

$$x = \frac{-1 \pm \sqrt{1^2 - 4(1)(1)}}{2(1)} = -\frac{1}{2} \pm \frac{1}{2}\sqrt{-3} = -\frac{1}{2} \pm \frac{1}{2}\sqrt{3}\, i.$$

Thus, the only rational zeros of $f(x)$ are $\frac{1}{3}$ and $\frac{3}{2}$.

12. We begin by noting that $x = 0$ is a zero of $f(x)$ and

$$f(x) = x(x^4 - 2x - 12).$$

Working with $q(x) = x^4 - 2x - 12$ we see that the possible rational zeros of $q(x)$ are ± 1, ± 2, ± 3, ± 4, ± 6, and ± 12. Using direct substitution or synthetic division, we find $q(-1) = -9$, $q(1) = -13$, $q(-2) = 8$, and $q(2) = 0$. Thus, 2 is a zero of $q(x)$ and hence of $f(x)$. We then find from synthetic division that

$$f(x) = x(x^4 - 2x - 12) = x(x - 2)(x^3 + 2x^2 + 4x + 6).$$

We now use synthetic division with $p(x) = x^3 + 2x^2 + 4x + 6$, which has possible rational zeros ± 1, ± 2, ± 3, and ± 6. We already know that -1, 1, and -2 cannot be zeros (since they would then also be zeros of $q(x)$). Using synthetic division, we find $p(2) = 30$, $p(-3) = -15$, $p(3) = 63$, $p(-6) = -162$, and $p(6) = 318$. Thus, the only rational zeros of $f(x)$ are 0 and 2.

15. Since $f(x)$ has noninteger coefficients, we begin by multiplying $f(x)$ by the least common denominator 4 of the coefficients. This gives

$$g(x) = 4f(x) = 4\left(\frac{1}{2}x^3 - \frac{9}{4}x^2 + \frac{17}{4}x - 3\right) = 2x^3 - 9x^2 + 17x - 12.$$

Since the zeros of $f(x)$ and $g(x)$ are the same, we consider the set of all possible zeros of $g(x)$:

$$\frac{p}{s}: \ \pm 1, \ \pm\frac{1}{2}, \ \pm 2, \ \pm 3, \ \pm\frac{3}{2}, \ \pm 4, \ \pm 6, \ \pm 12.$$

Testing these, we find $g(-1) = -40$, $g(1) = -2$, $g(-2) = -98$, $g(2) = 2$, $g(-3) = -198$, $g(3) = 12$, $g(-\frac{3}{2}) = -\frac{129}{2}$, and $g(\frac{3}{2}) = 0$. From synthetic division, we have

$$f(x) = \frac{1}{4}g(x) = \frac{1}{4}(2x^3 - 9x^2 + 17x - 12) = \frac{1}{4}\left(x - \frac{3}{2}\right)(2x^2 - 6x + 8)$$

$$= \frac{1}{2}\left(x - \frac{3}{2}\right)(x^2 - 3x + 4).$$

Since $x^2 - 3x + 4$ does not factor with integer coefficients, there are no more rational zeros of $f(x)$. Thus, the only rational zero is $\frac{3}{2}$.

18. Since $f(x)$ has noninteger coefficients, we begin by multiplying $f(x)$ by the least common denominator 12 of the coefficients. This gives

$$g(x) = 12f(x) = 12\left(\frac{3}{4}x^3 + \frac{9}{4}x^2 + \frac{5}{3}x + \frac{1}{3}\right) = 9x^3 + 27x^2 + 20x + 4.$$

Since the zeros of $f(x)$ and $g(x)$ are the same, we consider the set of all possible zeros of $g(x)$:

$$\frac{p}{s}: \quad -1, \ -\frac{1}{3}, \ -\frac{1}{9}, \ -2, \ -\frac{2}{3}, \ -\frac{2}{9}, \ -4, \ -\frac{4}{3}, \ -\frac{4}{9}.$$

Note that we need only consider negative possible zeros because all coefficients of $g(x)$ are positive. Testing the possible zeros using synthetic division, we find $g(-1) = 2$ and $g(-\frac{1}{3}) = 0$. Since $-\frac{1}{3}$ is a zero, we use the result of the synthetic division to write

$$f(x) = \frac{1}{12}g(x) = \frac{1}{12}(9x^3 + 27x^2 + 20x + 4) = \frac{1}{12}\left(x + \frac{1}{3}\right)(9x^2 + 24x + 12)$$

$$= \frac{3}{12}\left(x + \frac{1}{3}\right)(3x^2 + 8x + 4) = \frac{1}{4}\left(x + \frac{1}{3}\right)(3x + 2)(x + 2).$$

We see that $-\frac{2}{3}$ and -2 are also zeros of $f(x)$. Thus, the zeros of $f(x)$ are -2, $-\frac{2}{3}$, and $-\frac{1}{3}$.

21. We look first for any rational zeros by using direct substitution into $f(x)$ or synthetic division to test the possibilities

$$\frac{p}{s}: \quad \pm 1, \ \pm\frac{1}{2}, \ \pm\frac{1}{4}, \ \pm\frac{1}{8}, \ \pm 3, \ \pm\frac{3}{2}, \ \pm\frac{3}{4}, \ \pm\frac{3}{8}.$$

We find $f(-1) = 11$, $f(1) = 5$, $f(-\frac{1}{2}) = \frac{35}{4}$, $f(\frac{1}{2}) = -\frac{1}{4}$, $f(-\frac{1}{4}) = \frac{95}{16}$, $f(\frac{1}{4}) = \frac{11}{16}$, $f(-\frac{1}{8}) = \frac{71}{16}$, $f(\frac{1}{8}) = \frac{55}{32}$, $f(-3) = -135$, $f(3) = 231$, $f(-\frac{3}{2}) = \frac{15}{4}$, $f(\frac{3}{2}) = \frac{99}{4}$, $f(-\frac{3}{4}) = \frac{171}{16}$, $f(\frac{3}{4}) = \frac{15}{16}$, $f(-\frac{3}{8}) = \frac{237}{32}$, and (finally!) $f(\frac{3}{8}) = 0$. This last result follows from the synthetic division

$$
\begin{array}{r|rrrr}
\frac{3}{8} & 8 & 5 & -11 & 3 \\
  &   & 3 & 3 & -3 \\
\hline
  & 8 & 8 & -8 & \boxed{0} = r
\end{array}
$$

so

$$f(x) = \left(x - \frac{3}{8}\right)(8x^2 + 8x - 8) = (8x - 3)(x^2 + x - 1).$$

Using the quadratic formula to solve $x^2 + x - 1 = 0$, we find

$$x = \frac{-1 \pm \sqrt{1^2 - 4(1)(-1)}}{2(1)} = -\frac{1}{2} \pm \frac{1}{2}\sqrt{5}.$$

The zeros of $f(x)$ are $\frac{3}{8}$, $-\frac{1}{2} - \frac{1}{2}\sqrt{5}$, and $-\frac{1}{2} + \frac{1}{2}\sqrt{5}$, and a factorization is

$$f(x) = (8x - 3)\left[x - \left(-\frac{1}{2} - \frac{1}{2}\sqrt{5}\right)\right]\left[x - \left(-\frac{1}{2} + \frac{1}{2}\sqrt{5}\right)\right]$$

$$= (8x - 3)\left(x + \frac{1}{2} + \frac{1}{2}\sqrt{5}\right)\left(x + \frac{1}{2} - \frac{1}{2}\sqrt{5}\right).$$

24. We look first for any rational zeros by using direct substitution in $f(x)$ or synthetic division to test the possibilities

$$\frac{p}{s} : \ \pm 1, \ \pm 2, \ \pm 3, \ \pm 4, \ \text{etc.}$$

We find $f(-1) = 100$, $f(1) = 144$, $f(-2) = 36$, $f(2) = 100$, and $f(-3) = 0$. This result follows from the synthetic division

$$
\begin{array}{r|rrrrr}
-3 & 1 & -2 & -23 & 24 & 144 \\
   &   & -3 & 15 & 24 & -144 \\
\hline
   & 1 & -5 & -8 & 48 & \boxed{0 = r}
\end{array}
$$

so

$$f(x) = (x+3)(x^3 - 5x^2 - 8x + 48).$$

We continue by testing $q(x) = x^3 - 5x^2 - 8x + 48$ for rational zeros. We already know that ± 1 and ± 2 cannot be zeros, so we start with $x = -3$ (in case it is a multiple zero of $f(x)$). From

$$
\begin{array}{r|rrrr}
-3 & 1 & -5 & -8 & 48 \\
   &   & -3 & 24 & -48 \\
\hline
   & 1 & -8 & 16 & \boxed{0 = r}
\end{array}
$$

we see that -3 is a zero of $q(x)$, and hence at least a double root of $f(x)$. We now have, by simple factoring, that

$$f(x) = (x+3)^2(x - 8x + 16) = (x+3)^2(x-4)^2.$$

The zeros of $f(x)$ are -3 (multiplicity 2) and 4 (multiplicity 2).

27. We begin by noting that $x = 0$ is a zero of $f(x)$ and

$$f(x) = 4x(x^4 - 2x^3 - 6x^2 + 10x - 3).$$

We now check

$$p/s : \ \pm 1, \ \pm 3$$

to see if there are any rational zeros of $q(x) = x^4 - 2x^3 - 6x^2 + 10x - 3$. Using synthetic division, we find $q(-1) = -16$, $q(1) = 0$ and $q(x) = (x-1)(x^3 - x^2 - 7x + 3)$. We continue by testing $q_1(x) = x^3 - x^2 - 7x + 3$ for rational zeros. We already know that -1 cannot be a zero, so we start with $x = 1$ (in case it is a multiple zero of $q(x)$). We find $q_1(1) = 4$, $q_1(-3) = -12$, $q_1(3) = 0$ and $q_1(x) = (x-3)(x^2 + 2x - 1)$. Since $x^2 + 2x - 1$ does not factor, we use the quadratic formula:

$$x = \frac{-2 \pm \sqrt{2^2 - 4(1)(-1)}}{2(1)} = -1 \pm \frac{1}{2}\sqrt{4+4} = -1 \pm \sqrt{2}.$$

The zeros of $f(x)$ are 0, 1, 3, $-1 - \sqrt{2}$ and $-1 + \sqrt{2}$, and

$$f(x) = 4x(x-1)(x-3)\left[x - (-1-\sqrt{2})\right]\left[x - (-1+\sqrt{2})\right]$$

$$= 4x(x-1)(x-3)(x+1+\sqrt{2})(x+1-\sqrt{2}).$$

30. We first look for rational zeros by using synthetic division to test the possibilities

$$\frac{p}{s} : \ \pm 1, \ \pm 2, \ \pm 8, \ \pm 16, \ \pm 32, \ \pm 64.$$

We find $f(-1) = -27$, $f(1) = -27$, $f(-2) = 0$ and

$$f(x) = (x+2)(x^5 - 2x^4 - 8x^3 + 16x^2 + 16x - 32).$$

We now test $q(x) = x^5 - 2x^4 - 8x^3 + 16x^2 + 16x - 32$ for rational zeros starting with -2 (in case it is a multiple zero of $f(x)$). In this case, -2 is a zero of $q(x)$ so it is a multiple zero of $f(x)$ and

$$f(x) = (x+2)^2(x^4 - 4x^3 + 16x - 16).$$

Next, we test $q_1(x) = x^4 - 4x^3 + 16x - 16$ for rational zeros starting again with $x = -2$. We find $q_1(-2) = 0$ and

$$f(x) = (x+2)^3(x^3 - 6x^2 + 12x - 8).$$

Testing $q_2(x) = x^3 - 6x^2 + 12x - 8$ for rational zeros starting again with $x = -2$ we find $q_2(-2) = -64$, $q_2(2) = 0$ and

$$f(x) = (x+2)^3(x-2)(x^2 - 4x + 4) = (x+2)^3(x-2)(x-2)^2$$

$$= (x+2)^3(x-2)^3.$$

The zeros of $f(x)$ are -2 and 2, each with multiplicity 3.

33. To find the real zeros of

$$f(x) = 2x^4 + 7x^3 - 8x^2 - 25x - 6,$$

(which are the same as the real solutions of $2x^4 + 7x^3 - 8x^2 - 25x - 6 = 0$), we first find any rational zeros. Using synthetic division to test the possibilities

$$\frac{p}{s} : \ \pm 1, \ \pm \frac{1}{2}, \ \pm 2, \ \pm 3, \ \pm \frac{3}{2}, \ \pm 6$$

we find $f(-1) = 6$, $f(1) = -30$, $f(-\frac{1}{2}) = \frac{15}{4}$, $f(\frac{1}{2}) = -\frac{39}{2}$, $f(-2) = -12$, $f(2) = 0$, and

$$f(x) = (x-2)(2x^3 + 11x^2 + 14x + 3).$$

Next, we look for rational zeros of $q(x) = 2x^3 + 11x^2 + 14x + 3$ using synthetic division. We start with $x = 2$ (in case it is a multiple zero of $f(x)$) and find $q(2) = 91$, $q(-3) = 6$, $q(3) = 198$, $q(-\frac{3}{2}) = 0$ and

$$q(x) = \left(x + \frac{3}{2} \right)(2x^2 + 8x + 2) = 2\left(x + \frac{3}{2} \right)(x^2 + 4x + 1).$$

We use the quadratic formula to solve $x^2 + 4x + 1 = 0$:

$$x = \frac{-4 \pm \sqrt{4^2 - 4(1)(1)}}{2(1)} = -2 \pm \frac{1}{2}\sqrt{16 - 4} = -2 \pm \sqrt{3}.$$

The real solutions of the original equation are 2, $-\frac{3}{2}$, $-2 - \sqrt{3}$, and $-2 + \sqrt{3}$.

36. To find the real zeros of

$$f(x) = 8x^4 - 6x^3 - 7x^2 + 6x - 1,$$

(which are the same as the real solutions of $8x^4 - 6x^3 - 7x^2 + 6x - 1 = 0$), we first find any rational zeros. Using synthetic division to test the possibilities

$$\frac{p}{s} : \pm 1, \ \pm\frac{1}{2}, \ \pm\frac{1}{4}, \ \pm\frac{1}{8}$$

we find $f(-1) = 0$ and

$$f(x) = (x + 1)(8x^3 - 14x^2 + 7x - 1).$$

Next, we look for rational zeros of $q(x) = 8x^3 - 14x^2 + 7x - 1$ using synthetic division. We start with $x = -1$ (in case it is a multiple zero of $f(x)$) and find $q(-1) = -30$, $q(1) = 0$ and

$$q(x) = (x - 1)(8x^2 - 6x + 1) = (x - 1)(2x - 1)(4x - 1).$$

The real solutions of $8x^4 - 6x^3 - 7x^2 + 6x - 1 = 0$ are -1, 1, $\frac{1}{2}$, and $\frac{1}{4}$.

39. Since $f(0) = -6 < 0$ and $f(1) = 1 > 0$, and f is continuous, the graph of f must cross the x-axis somewhere within the interval $[0, 1]$. The zero is $\frac{3}{4}$.

42. A cubic polynomial with the given zeros is

$$g(x) = \left(x - \frac{1}{2} \right)\left[x - (1 + \sqrt{3}) \right]\left[x - (1 - \sqrt{3}) \right]$$

$$= \left(x - \frac{1}{2} \right)\left[(x - 1) - \sqrt{3} \right]\left[(x - 1) + \sqrt{3} \right]$$

$$= \left(x - \frac{1}{2} \right)[(x - 1)^2 + 3] = \left(x - \frac{1}{2} \right)(x^2 - 2x + 1 + 3)$$

$$= \left(x - \frac{1}{2} \right)(x^2 - 2x + 4) = x^3 - \frac{5}{2}x^2 + 5x - 2.$$

Any nonzero constant multiple of $g(x)$ has the same zeros, so let

$$f(x) = ag(x) = ax^3 - \frac{5}{2}ax^2 + 5ax - 2a.$$

We want $5a = 2$, so $a = \frac{2}{5}$ and

$$f(x) = \frac{2}{5}x^3 - x^2 + 2x - \frac{4}{5}.$$

3.5   Approximating Real Zeros

3. From the graph in the text, there is one zero between -2 and -1. We must have an error less than 0.0005. The first approximation is

$$m_1 = \frac{-2 - 1}{2} = -1.5 \quad \text{with} \quad \text{error} < \frac{1}{2}(-1 + 2) = 0.5.$$

The second approximation is

$$m_2 = \frac{-1.5 - 1}{2} = -1.25 \quad \text{with} \quad \text{error} < \frac{1}{2}(-1 + 1.5) = 0.25.$$

After 11 iterations we find

$$m_{11} = -1.315 \quad \text{with} \quad \text{error} < 0.0005.$$

6. From the graph in the text, there are three zeros, the first between -2 and -1, the second between 0 and 1, and the third between 1 and 2. We must have an error less than 0.0005. The first approximation is

$$m_1 = \frac{-2 - 1}{2} = -1.5 \quad \text{with} \quad \text{error} < \frac{1}{2}(-1 + 2) = 0.5.$$

The second approximation is

$$m_2 = \frac{-1.5 - 1}{2} = -1.25 \quad \text{with} \quad \text{error} < \frac{1}{2}(-1 + 1.5) = 0.25.$$

After 11 iterations we find

$$m_{11} = -1.343 \quad \text{with} \quad \text{error} < 0.0005.$$

The first approximation to the second zero is

$$m_1 = \frac{0 + 1}{2} = 0.5 \quad \text{with} \quad \text{error} < \frac{1}{2}(1 - 0) = 0.5.$$

The second approximation is

$$m_2 = \frac{0.5 + 1}{2} = 0.75 \quad \text{with} \quad \text{error} < \frac{1}{2}(1 - 0.5) = 0.25.$$

After 11 iterations we find

$$m_{11} = 0.643 \quad \text{with} \quad \text{error} < 0.0005.$$

The first approximation to the third zero is

$$m_1 = \frac{1+2}{2} = 1.5 \quad \text{with} \quad \text{error} < \frac{1}{2}(2-1) = 0.5.$$

The second approximation is

$$m_2 = \frac{1+1.5}{2} = 1.25 \quad \text{with} \quad \text{error} < \frac{1}{2}(1.5-1) = 0.25.$$

After 11 iterations we find

$$m_{11} = 1.217 \quad \text{with} \quad \text{error} < 0.0005.$$

Miscellaneous Applications

9. Using $\rho_b = 0.4\rho_w$ and $r = 2$ we have

$$\frac{\pi}{3}\rho_w h^2(6-h) = \frac{4\pi}{3}(0.4\rho_w)(8)$$

or

$$h^3 - 6h^2 + 12.8 = 0.$$

To find a zero of $f(h) = h^3 - 6h^2 + 12.8$ we note that $f(1) = 7.8 > 0$ and $f(2) = -3.2 < 0$. We need an error less then 0.005. The first approximation is

$$m_1 = \frac{1+2}{2} = 1.5 \quad \text{with} \quad \text{error} < \frac{1}{2}(2-1) = 0.5.$$

The second approximation is

$$m_2 = \frac{1.5+2}{2} = 1.75 \quad \text{with} \quad \text{error} < \frac{1}{2}(2-1.5) = 0.25.$$

After 8 iterations we find

$$m_8 = 1.73 \text{ in.} \quad \text{with} \quad \text{error} < 0.005.$$

3.6 | Rational Functions |

3. *Vertical Asymptotes:* Setting $x - 2 = 0$ we see that $x = 2$ is a vertical asymptote.

Horizontal Asymptote: The degree of the numerator is less than the degree of the denominator, so $y = 0$ is a horizontal asymptote.

Intercepts: Since $f(0) = -\frac{1}{2}$, the y-intercept is $(0, -\frac{1}{2})$. The numerator is never 0, so there are no x-intercepts.

Graph: The graph is the graph of $y = 1/x$ shifted 2 units to the right.

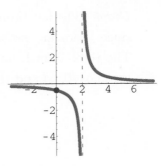

6. *Vertical Asymptotes:* Setting $2x - 5 = 0$ we see that $x = \frac{5}{2}$ is a vertical asymptote.

Horizontal Asymptote: The degree of the numerator equals the degree of the denominator, so $y = \frac{1}{2}$ is the horizontal asymptote.

Intercepts: Since $f(0) = 0$, the y-intercept is $(0, 0)$. Since the numerator is simply x, $(0, 0)$ is also the only x-intercept.

Graph: The asymptotes are shown as dashed lines. The left branch has to be below the horizontal asymptote in order to pass through the intercept. The right branch has to be above the horizontal asymptote since there is no x-intercept to the right of $x = \frac{5}{2}$ and the graph does not cross $y = \frac{1}{2}$ since $x/(2x - 5) = 1/2$ has no solution.

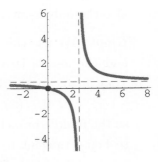

9. *Vertical Asymptotes:* Setting $x + 1 = 0$ we see that $x = -1$ is a vertical asymptote.

Horizontal Asymptote: The degree of the numerator equals the degree of the denominator, so $y = -1/1 = -1$ is the horizontal asymptote.

Intercepts: Since $f(0) = 1$, the y-intercept is $(0, 1)$. Setting $1 - x = 0$ we see that $x = 1$, or $(1, 0)$, is the x-intercept.

Graph: Using long division, we find that

$$\frac{1 - x}{x + 1} = -1 + \frac{2}{x + 1}.$$

Thus, the graph of $y = (1 - x)/(x + 1)$ is the graph of $y = 1/x$ stretched vertically by a factor of 2, shifted left 1 unit, and finally shifted 1 unit downward.

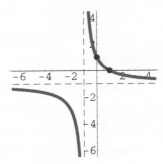

12. *Vertical Asymptotes:* Setting $(x + 2)^3 = 0$ we see that $x = -2$ is a vertical asymptote.

Horizontal Asymptote: The degree of the numerator is less than the degree of the denominator, so $y = 0$ is the horizontal asymptote.

Intercepts: Since $f(0) = 4/2^3 = \frac{1}{2}$, the y-intercept is $(0, \frac{1}{2})$. The numerator is never 0, so there are no x-intercepts.

Graph: The left branch must lie entirely in the third quadrant since there are no x-intercepts, and $f(x) < 0$ for $x < -2$. The right branch must lie above the x-axis because it passes through $(0, \frac{1}{2})$ and there are no x-intercepts.

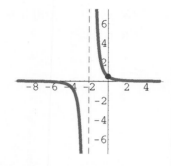

15. *Vertical Asymptotes:* Setting $x^2 - 1 = (x + 1)(x - 1) = 0$ we see that $x = -1$ and $x = 1$ are vertical asymptotes.

Horizontal Asymptote: The degree of the numerator is less than the degree of the denominator, so $y = 0$ is the horizontal asymptote.

Intercepts: Since $f(0) = 0$, the y-intercept is $(0, 0)$. Since the numerator is simply x, $(0, 0)$ is also the only x-intercept.

Graph: We use the facts that the only x-intercept is $(0, 0)$ and the x-axis is a horizontal asymptote. For $x < -1$, $f(x) < 0$, so the left branch is below the x-axis. For $-1 < x < 0$, $f(x) > 0$, and for $0 < x < 1$, $f(x) < 0$, so the middle branch passes through the origin (as opposed to being tangent to the origin and lying strictly above or below the x-axis). For $x > 1$, $f(x) > 0$, so the right branch is above the x-axis.

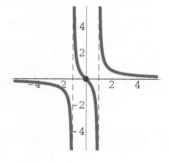

18. *Vertical Asymptotes:* Setting $x^2 - 2x - 8 = (x + 2)(x - 4) = 0$ we see that $x = -2$ and $x = 4$ are vertical asymptotes.

Horizontal Asymptote: The degree of the numerator is less than the degree of the denominator, so $y = 0$ is the horizontal asymptote.

Intercepts: Since $f(0) = -\frac{1}{8}$, the y-intercept is $(0, -\frac{1}{8})$. The numerator is never 0, so there are no x-intercepts.

Graph: We use the facts that there are no x-intercepts and the x-axis is a horizontal asymptote. For $x < -2$, $f(x) > 0$, so the left branch is above the x-axis. For $-2 < x < 4$ the graph must lie entirely below the x-axis because it passes through $(0, -\frac{1}{8})$ and there are no x-intercepts. For $x > 4$, $f(x) > 0$, so the right branch is above the x-axis.

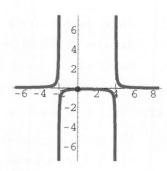

21. *Vertical Asymptotes:* Setting $(x - 1)^2 = 0$ we see that $x = 1$ is a vertical asymptote.

Horizontal Asymptote: The degree of the numerator equals the degree of the denominator, so $y = -2/1 = -2$ is the horizontal asymptote.

Intercepts: Since $f(0) = 8$, the y-intercept is $(0, 8)$. Setting $x^2 - 4 = 0$ we see that $x = -2$ and $x = 2$ or $(-2, 0)$ and $(2, 0)$ are the x-intercepts.

Graph: Solving

$$\frac{-2x^2 + 8}{x^2 - 2x + 1} = 2$$

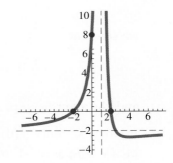

we obtain $x = \frac{5}{2}$. Since the graph passes through $(-2, 0)$, but not through $y = -2$ for $x < 1$, we see that the left branch must lie above $y = -2$. On the other hand, for $x > 1$, the graph does intersect $y = -2$ at $x = \frac{5}{2}$. Thus, the right branch drops down from $x = 1$, through $(2, 0)$ and $(\frac{5}{2}, -2)$, and then rises back toward $y = -2$.

24. *Vertical Asymptotes:* Setting the denominator equal to 0, we see that $x = 0$ or the y-axis is a vertical asymptote.

Slant Asymptote: Since the degree of the numerator is one greater than the degree of the denominator, the graph of $f(x)$ possesses a slant asymptote. From

$$f(x) = \frac{x^2 - 3x + 10}{x} = x - 3 + \frac{10}{x}$$

we see that $y = x - 3$ is a slant asymptote.

Intercepts: Since the y-axis is a vertical asymptote, the graph has no y-intercept. Setting $(x + 2)(x - 5) = 0$ we see that $x = -2$ and $x = 5$ or $(-2, 0)$ and $(5, 0)$ are the x-intercepts.

Graph: We can just about find the graph from the asymptotes and intercepts, but we need to determine if the graph crosses the slant asymptote. To do this we solve

$$\frac{x^2 - 3x - 10}{x} = x - 3$$
$$x^2 - 3x - 10 = x^2 - 3x$$
$$-10 = 0.$$

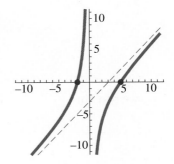

Since there is no solution, the graph does not cross its slant asymptote.

27. *Vertical Asymptotes:* Setting $x - 1 = 0$ we see that $x = 1$ is a vertical asymptote.

Slant Asymptote: Since the degree of the numerator is one greater than the degree of the denominator, the graph of $f(x)$ possesses a slant asymptote. Using synthetic division,

$$
\begin{array}{r|rrr}
1 & 1 & -2 & -3 \\
  &   & 1 & -1 \\
\hline
  & 1 & -1 & \underline{-4} = r
\end{array}
$$

we see that

$$f(x) = x - 1 - \frac{4}{x - 1}$$

and the slant asymptote is $y = x - 1$.

Intercepts: Since $f(0) = -3/(-1) = 3$, the y-intercept is $(0, 3)$. Setting $(x + 1)(x - 3) = 0$, we see that $x = -1$ and $x = 3$ or $(-1, 0)$ and $(3, 0)$ are the x-intercepts.

Graph: We can just about find the graph from the asymptotes and intercepts, but we need to determine if the graph crosses the slant asymptote. To do this we solve

$$\frac{x^2 - 2x - 3}{x - 1} = x - 1$$
$$x^2 - 2x - 3 = x^2 - 2x + 1$$
$$-3 = 1.$$

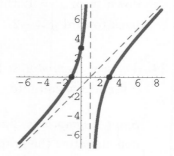

Since there is no solution, the graph does not cross its slant asymptote.

30. *Vertical Asymptotes:* The denominator is never 0, so there are no vertical asymptotes.

Slant Asymptote: Since the degree of the numerator is one greater than the degree of the denominator, the graph of $f(x)$ possesses a slant asymptote. Using polynomial long division

$$
\begin{array}{r}
5x - 15 \\
x^2 + 1 \overline{\smash{)}\ 5x^3 - 5x^2 - 20x + 0} \\
\underline{5x^3 + 0x^2 + 5x} \\
-15x^2 - 25x + 0 \\
\underline{-15x^2 + 0x - 15} \\
-25x + 15
\end{array}
$$

we see that

$$f(x) = 5x - 15 + \frac{-25x + 15}{x^2 + 1}$$

and the slant asymptote is $y = 5x - 15$.

Intercepts: Since $f(0) = 0$, the y-intercept is $(0, 0)$. Setting $5x(x + 1)(x - 4) = 0$ we see that, in addition to $(0, 0)$, the other x-intercepts are $(-1, 0)$ and $(4, 0)$.

Graph: We first see if the graph of $f(x)$ intersects the slant asymptote:

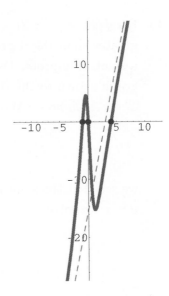

$$\frac{5x^3 - 15x^2 - 20x}{x^2 + 1} = 5x - 15$$

$$5x^3 - 15x^2 - 20x = 5x^3 - 15x^2 + 5x - 15$$

$$-20x = 5x - 15$$

$$15 = 25x$$

$$x = \frac{15}{25} = \frac{3}{5}.$$

Thus, the graph of $f(x)$ intersects the slant asymptote at $x = \frac{3}{5}$. Using the four points we have found and the slant asymptote, we obtain the graph of $f(x)$.

33. *Vertical Asymptotes:* Setting $(x - 3)(x + 4) = 0$ we see that $x = 3$ and $x = -4$ are vertical asymptotes.

Horizontal Asymptote: The degree of the numerator equals the degree of the denominator, so $y = 4/1 = 4$ is the horizontal asymptote.

Intercepts: Since $f(0) = 0$, the y-intercept is $(0, 0)$. Setting $4x(x - 2) = 0$, we see that in addition to $(0, 0)$, the other x-intercept is $(0, 2)$.

Graph: To determine if the graph of $f(x)$ crosses the horizontal asymptote $y = 4$ we solve

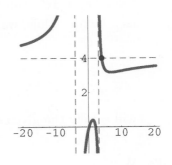

$$\frac{4x(x - 2)}{(x - 3)(x + 4)} = 4$$

$$4x^2 - 8x = 4(x^2 + x - 12) = 4x^2 + 4x - 48$$

$$-8x = 4x - 48$$

$$48 = 12x$$

$$x = 4.$$

Since the graph of $f(x)$ does not intersect the horizontal asymptote $y = 4$ for $x < -4$ and there are no x-intercepts for $x < -4$, the left branch of the graph lies entirely above $y = 4$ and to the left of $x = -4$. The graph intersects the x-axis at $x = 0$ and $x = 2$ and does not intersect the horizontal asymptote $y = 4$ for $-4 < x < 3$, so it must lie mostly in quadrants III and IV, with a small portion in quadrant I. Finally, there are no x-intersects for $x > 3$ and the graph intersects the horizontal asymptote $y = 4$ at $x = 4$, so the right branch must fall from just right of $x = 3$, through $(4, 4)$ and then rise up again toward the horizontal asymptote $y = 4$.

36. For $f(x) = (x^3 + 2x - 4)/x^2$ the degree of the numerator is one greater than the degree of the denominator, so the graph has a slant asymptote. From a calculator or computer we find the graph shown to the right along with the slant asymptote. *Slant Asymptote:* Writing

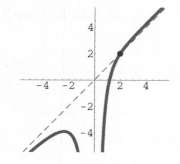

$$f(x) = \frac{x^3 + 2x - 4}{x^2} = x + \frac{2x - 4}{x^2}$$

we see that the slant asymptote is $y = x$. To find where the graph of $f(x)$ intersects the line $y = x$ we solve

$$\frac{x^3 + 2x - 4}{x^2} = x$$
$$x^3 + 2x - 4 = x^3$$
$$2x - 4 = 0$$
$$x = 2.$$

Since the point is on the line $y = x$, we have $y = 2$, and the point of intersection is $(2, 2)$.

39. If the function is $f(x) = P(x)/Q(x)$ and the vertical asymptotes are $x = -1$ and $x = 2$, then we can take $Q(x) = (x + 1)(x - 2)$. For the horizontal asymptote to be $y = 3$ we want the numerator to be a second degree polynomial with leading coefficient 3. For example, we can use $P(x) = 3x^2$. But we also need an x-intercept to be $(3, 0)$, so we need $P(x)$ to have $x - 3$ as a factor. Thus, we use $P(x) = 3x(x - 3)$. The function

$$f(x) = \frac{P(x)}{Q(x)} = \frac{3x(x - 3)}{(x + 1)(x - 2)}$$

will satisfy the given conditions.

42. Writing

$$f(x) = \frac{x - 1}{(x + 1)(x - 1)}$$

we see that both the numerator and denominator have a factor of $x - 1$. Thus, there is a hole in the graph of $f(x)$ at $x = 1$ and we can write

$$f(x) = \frac{x - 1}{(x + 1)(x - 1)} = \frac{1}{x + 1}, \quad x \neq 1.$$

We see that the y-coordinate of the hole is $1/(1+1) = \frac{1}{2}$.

Intercepts: Since $f(0) = 1$, the y-intercept is $(0,1)$. There are no x-intercepts because the numerator of $f(x)$ is never 0.

Graph: The graph of $f(x)$ is the graph of $y = 1/x$ shifted one unit left, with a hole at $(1, \frac{1}{2})$.

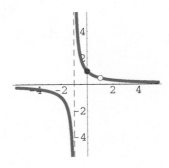

Miscellaneous Applications

45. The vertical asymptote is $r = -5$ and the horizontal asymptote is $R = 5$. Since $r > 0$, there are no intercepts. From the graph we see that the resistance R, as r becomes very large, approaches 5 from below.

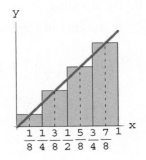

3.7 $\int$ **Calculus PREVIEW** **The Area Problem**

3. The width of each subinterval is $\Delta x = \frac{1-0}{4} = \frac{1}{4}$. The midpoints of the subintervals are $x_1^* = \frac{1}{8}$, $x_2^* = \frac{3}{8}$, $x_3^* = \frac{5}{8}$, and $x_4^* = \frac{7}{8}$. Then

$$A \approx f\left(\frac{1}{8}\right)\frac{1}{4} + f\left(\frac{3}{8}\right)\frac{1}{4} + f\left(\frac{5}{8}\right)\frac{1}{4} + f\left(\frac{7}{8}\right)\frac{1}{4}$$

$$= \frac{1}{8}\cdot\frac{1}{4} + \frac{3}{8}\cdot\frac{1}{4} + \frac{5}{8}\cdot\frac{1}{4} + \frac{7}{8}\cdot\frac{1}{4}$$

$$= \frac{1+3+5+7}{8\cdot 4} = \frac{16}{32} = \frac{1}{2}.$$

6. By dividing $[-1, 2]$ into twelve subintervals, the width of each subinterval is $\Delta x = \frac{2-(-1)}{12} = \frac{3}{12} = \frac{1}{4}$.

(a) The left-hand endpoints are $x_1^* = -1$, $x_2^* = -\frac{3}{4}$, $x_3^* = -\frac{1}{2}$, $x_4^* = -\frac{1}{4}$, $x_5^* = 0$, $x_6^* = \frac{1}{4}$, $x_7^* = \frac{1}{2}$, $x_8^* = \frac{3}{4}$, $x_9^* = 1$, $x_{10}^* = \frac{5}{4}$, $x_{11}^* = \frac{3}{2}$, and $x_{12}^* = \frac{7}{4}$. Then

$$A \approx f(-1)\frac{1}{4} + f\left(-\frac{3}{4}\right)\frac{1}{4} + f\left(-\frac{1}{2}\right)\frac{1}{4} + f\left(-\frac{1}{4}\right)\frac{1}{4} + f(0)\frac{1}{4} + f\left(\frac{1}{4}\right)\frac{1}{4}$$

$$+ f\left(\frac{1}{2}\right)\frac{1}{4} + f\left(\frac{3}{4}\right)\frac{1}{4} + f(1)\frac{1}{4} + f\left(\frac{5}{4}\right)\frac{1}{4} + f\left(\frac{3}{2}\right)\frac{1}{4} + f\left(\frac{7}{4}\right)\frac{1}{4}$$

$$= 1 \cdot \frac{1}{4} + \frac{5}{4} \cdot \frac{1}{4} + \frac{3}{2} \cdot \frac{1}{4} + \frac{7}{4} \cdot \frac{1}{4} + 2 \cdot \frac{1}{4} + \frac{9}{4} \cdot \frac{1}{4} + \frac{5}{2} \cdot \frac{1}{4} + \frac{11}{4} \cdot \frac{1}{4} + 3 \cdot \frac{1}{4}$$

$$+ \frac{13}{4} \cdot \frac{1}{4} + \frac{7}{2} \cdot \frac{1}{4} + \frac{15}{4} \cdot \frac{1}{4}$$

$$= \frac{1}{4} + \frac{5}{16} + \frac{3}{8} + \frac{7}{16} + \frac{1}{2} + \frac{9}{16} + \frac{5}{8} + \frac{11}{16} + \frac{3}{4} + \frac{13}{16} + \frac{7}{8} + \frac{15}{16} = \frac{57}{8}.t$$

(b) The right-hand endpoints are $x_1^* = -\frac{3}{4}$, $x_2^* = -\frac{1}{2}$, $x_3^* = -\frac{1}{4}$, $x_4^* = 0$, $x_5^* = \frac{1}{4}$, $x_6^* = \frac{1}{2}$, $x_7^* = \frac{3}{4}$, $x_8^* = 1$, $x_9^* = \frac{5}{4}$, $x_{10}^* = \frac{3}{2}$, $x_{11}^* = \frac{7}{4}$, and $x_{12}^* = 2$. Then

$$A \approx f\left(-\frac{3}{4}\right)\frac{1}{4} + f\left(-\frac{1}{2}\right)\frac{1}{4} + f\left(-\frac{1}{4}\right)\frac{1}{4} + f(0)\frac{1}{4} + f\left(\frac{1}{4}\right)\frac{1}{4} + f\left(\frac{1}{2}\right)\frac{1}{4}$$

$$+ f\left(\frac{3}{4}\right)\frac{1}{4} + f(1)\frac{1}{4} + f\left(\frac{5}{4}\right)\frac{1}{4} + f\left(\frac{3}{2}\right)\frac{1}{4} + f\left(\frac{7}{4}\right)\frac{1}{4} + f(2)\frac{1}{4}$$

$$= \frac{5}{4} \cdot \frac{1}{4} + \frac{3}{2} \cdot \frac{1}{4} + \frac{7}{4} \cdot \frac{1}{4} + 2 \cdot \frac{1}{4} + \frac{9}{4} \cdot \frac{1}{4} + \frac{5}{2} \cdot \frac{1}{4} + \frac{11}{4} \cdot \frac{1}{4} + 3 \cdot \frac{1}{4} + \frac{13}{4} \cdot \frac{1}{4}$$

$$+ \frac{7}{2} \cdot \frac{1}{4} + \frac{15}{4} \cdot \frac{1}{4} + 2 \cdot \frac{1}{4}$$

$$= \frac{5}{16} + \frac{3}{8} + \frac{7}{16} + \frac{1}{2} + \frac{9}{16} + \frac{5}{8} + \frac{11}{16} + \frac{3}{4} + \frac{13}{16} + \frac{7}{8} + \frac{15}{16} + \frac{1}{2} = \frac{63}{8}.$$

9. By dividing $[0, 5]$ into 5 subintervals, the width of each subinterval is $\Delta x = \frac{5-0}{5} = 1$. The midpoints of the subintervals are $x_1^* = \frac{1}{2}$, $x_2^* = \frac{3}{2}$, $x_3^* = \frac{5}{2}$, $x_4^* = \frac{7}{2}$, $x_5^* = \frac{9}{2}$. Then

$$A \approx f\left(\frac{1}{2}\right)1 + f\left(\frac{3}{2}\right)1 + f\left(\frac{5}{2}\right)1 + f\left(\frac{7}{8}\right)1 + f\left(\frac{9}{2}\right)1$$

$$= \left[-\left(\frac{1}{2}\right)^2 + 5\left(\frac{1}{2}\right)\right] + \left[-\left(\frac{3}{2}\right)^2 + 5\left(\frac{3}{2}\right)\right] + \left[-\left(\frac{5}{2}\right)^2 + 5\left(\frac{5}{2}\right)\right]$$

$$+ \left[-\left(\frac{7}{2}\right)^2 + 5\left(\frac{7}{2}\right)\right] + \left[-\left(\frac{9}{2}\right)^2 + 5\left(\frac{9}{2}\right)\right]$$

$$= \left(-\frac{1}{4} + \frac{5}{2}\right) + \left(-\frac{9}{4} + \frac{15}{2}\right) + \left(-\frac{25}{4} + \frac{25}{2}\right) + \left(-\frac{49}{4} + \frac{35}{2}\right) + \left(-\frac{81}{4} + \frac{45}{2}\right) = \frac{85}{4}.$$

12. First, we divide $[0, 9]$ into 9 subintervals, each of width 1. The left-hand endpoints are $x_1^* = 0$, $x_2^* = 1$, $x_3^* = 2$, $x_4^* = 3$, $x_5^* = 4$, $x_6^* = 5$, $x_7^* = 6$, $x_8^* = 7$, and $x_9^* = 8$. The approximate area in this case is

$$A_L = f(0)1 + f(1)1 + f(2)1 + f(3)1 + f(4)1 + f(5)1 + f(6)1 + f(7)1 + f(8)1$$
$$= 0 + 3 + 1 + 3 + 5 + 3 + 7 + 4 + 6 = 32.$$

Second, we again divide $[0, 9]$ into 9 subintervals, each of width 1. The right-hand endpoints are $x_1^* = 1$, $x_2^* = 2$, $x_3^* = 3$, $x_4^* = 4$, $x_5^* = 5$, $x_6^* = 6$, $x_7^* = 7$, $x_8^* = 8$, and $x_9^* = 9$. The approximate area in this case is

$$A_R = f(1)1 + f(2)1 + f(3)1 + f(4)1 + f(5)1 + f(6)1 + f(7)1 + f(8)1 + f(9)1$$
$$= 3 + 1 + 3 + 5 + 3 + 7 + 4 + 6 + 3 = 35.$$

Chapter 3 Review Exercises

A. Fill in the Blanks

3. Writing

$$f(x) = x^2(x + 3)(x - 5) = x^4 - 2x^3 - 15x^2,$$

we see that the end behavior of the graph of $f(x)$ resembles the graph of the power function $f(x) = x^4$.

6. The y-intercept of the graph of $f(x)$ is the point $(0, f(0))$, or, in this case, $(0, \frac{8}{4})$ or $(0, 2)$.

9. Since

$$f(x) = \frac{x^3 - x}{4 - 2x^3},$$

we see that $y = 1/(-2) = -\frac{1}{2}$ is a horizontal asymptote.

12. Since the graph of a polynomial function of degree n can have at most $n - 1$ turning points, the graph of $f(x) = 3x^5 - 4x^2 + 5x - 2$ can have at most 4 turning points.

15. When a polynomial function f of degree 5 is divided by $x - 3$ we get

$$\frac{f(x)}{x - 3} = q(x) + \frac{7}{x - 3}.$$

The degree of the quotient $q(x)$ must be 4.

18. If $f(x) = \dfrac{x + 10}{2 - x}$, then $f(x) \to -\infty$ as $x \to 2^+$.

B. True/False

3. This is true; *rational functions* can have graphs with holes, but polynomial functions cannot.

6. This is true, because if a, b, c, d, and x are all positive, then so is $f(x) = ax^3 + bx^2 + cx + d$ and $f(x)$ cannot be 0.

9. This is true because the powers of x with nonzero coefficients, namely 6 and 2, are both even.

12. Computing

$$f\left(\frac{1}{3}\right) = \frac{2(\frac{1}{3}) + 4}{3 - \frac{1}{3}} = \frac{2(\frac{1}{3}) + 4}{3 - \frac{1}{3}}\left(\frac{3}{3}\right) = \frac{2 + 12}{9 - 1} = \frac{14}{8} = \frac{7}{4}$$

we see that $(\frac{1}{3}, \frac{7}{4})$ is on the graph of $f(x)$. Thus, the statement is true.

15. This is false. The polynomial $f(x) = x^2 + 1$ has no rational zeros.

18. True. See Figure 3.1.5 in the text.

C. Review Exercises

3. Using synthetic division, we have

$$
\begin{array}{r|rrrrr}
2 & 7 & 0 & -6 & 9 & 3 \\
  &   & 14 & 28 & 44 & 106 \\
\hline
  & 7 & 14 & 22 & 53 & \boxed{109} = r
\end{array}
$$

Thus,

$$\frac{7x^4 - 6x^2 + 9x + 3}{x - 2} = 7x^3 + 14x^2 + 22x + 53 + \frac{109}{x - 2}.$$

6. By the Remainder Theorem, $f(2)$ is equal to the remainder r when $f(x)$ is divided by $x - 2$. Using synthetic division,

$$
\begin{array}{r|rrrrrrr}
2 & 1 & -3 & 2 & 3 & -1 & 5 & -1 \\
  &   & 2 & -2 & 0 & 6 & 10 & 30 \\
\hline
  & 1 & -1 & 0 & 3 & 5 & 15 & \boxed{29} = r
\end{array}
$$

Thus, $f(2) = 29$.

9. The possible rational zeros of $f(x)$ are

$$\pm 1,\ \pm 3,\ \pm 5,\ \pm 15,\ \pm\frac{1}{2},\ \pm\frac{3}{2},\ \pm\frac{5}{2},\ \pm\frac{15}{2},$$

$$\pm\frac{1}{4},\ \pm\frac{3}{4},\ \pm\frac{5}{4},\ \pm\frac{15}{4},\ \pm\frac{1}{8},\ \pm\frac{3}{8},\ \pm\frac{5}{8},\ \pm\frac{15}{8}.$$

12. Since

$$f(-1) = (-1+2)^4 - 1 = 1^4 - 1 = 1 - 1 = 0,$$

-1 is a zero of $f(x)$. Writing

$$f(x) = (x+2)^4 - 1 = x^4 + 8x^3 + 24x^2 + 32x + 16 - 1$$

$$= x^4 + 8x^3 + 24x^2 + 32x + 15,$$

and using synthetic division to divide $f(x)$ by $x + 1$,

$$
\begin{array}{r|rrrrr}
-1 & 1 & 8 & 24 & 32 & 15 \\
   &   & -1 & -7 & -17 & -15 \\
\hline
   & 1 & 7 & 17 & 15 & \boxed{0 = r}
\end{array}
$$

we see that $f(x) = (x+1)(x^3+7x^2+17x+15)$. We continue by testing $q(x) = x^3+7x^2+17x+15$ for rational zeros. Using synthetic division, we find $q(-3) = 0$ and

$$q(x) = (x+3)(x^2 + 4x + 5).$$

Since $x^2 + 4x + 5$ does not factor, we use the quadratic formula to solve $x^2 + 4x + 5 = 0$:

$$x = \frac{-4 \pm \sqrt{4^2 - 4(1)(5)}}{2(1)} = -2 \pm \frac{1}{2}\sqrt{-4} = -2 \pm i.$$

The complete factorization of $f(x)$ is

$$(x + 2)^4 - 1 = x^4 + 8x^3 + 24x^2 + 32x + 15$$

$$= (x + 1)(x + 3)(x + 2 + i)(x + 2 - i).$$

15. For $x - k$ to be a factor of $f(x) = 2x^3 + x^2 + 2x - 12$ we need

$$f(k) = 2k^3 + k^2 + 2k - 12 = 0.$$

Using synthetic division to test possible zeros, we find that $f(\frac{3}{2}) = 0$. Thus, $k = \frac{3}{2}$.

18. Since the graph is tangent to the x-axis at $x = -2$, $x = 1$, and $x = 3$, the function has zeros of even multiplicity at those points. The degree of $f(x)$ is 6, so

$$f(x) = a(x + 2)^2(x - 1)^2(x - 3)^2.$$

Since $(0, 9)$ lies on the graph, we see that $9 = 36a$, so $a = \frac{1}{4}$ and

$$f(x) = \frac{1}{4}(x + 2)^2(x - 1)^2(x - 3)^2.$$

21. Since

$$f(x) = \frac{2x}{x^2 + 1}$$

has no vertical asymptotes and its horizontal asymptote is $y = 0$, its graph must be either (a) or (f). Because $f(x)$ is negative for $x < 0$, it must be (f).

24. Since

$$f(x) = 2 - \frac{1}{x^2} = \frac{2x^2 - 1}{x^2}$$

has vertical asymptote $x = 0$ (the y-axis), its graph must be (g).

27. Since

$$f(x) = \frac{x^2 - 10}{2x - 4} = \frac{1}{2} \cdot \frac{x^2 - 10}{x - 2} = \frac{1}{2}\left(x + 2 - \frac{6}{x - 2}\right)$$

using synthetic division, we see that the graph of $f(x)$ has vertical asymptote $x = 2$ and slant asymptote $y = \frac{1}{2}x + 1$. Thus, the graph of f must be (c).

30. Since

$$f(x) = \frac{3}{x^2 + 1}$$

has no vertical asymptotes and its horizontal asymptote is $y = 0$, the graph must be either (a) or (f). Because $f(x)$ is always positive, it must be (a).

Chapter 4

Trigonometric Functions

4.1 | **Angles and Their Measurement**

3.
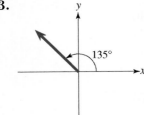

6.
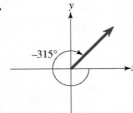

9.
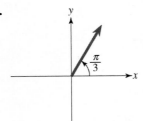

12.
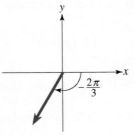

15.
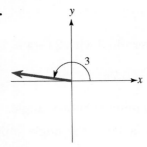

In Problem 18 *we use the fact that* $1' = \left(\dfrac{1}{60}\right)^{\circ}$ *and* $1'' = \dfrac{1}{60}\left(\dfrac{1}{60}\right)^{\circ} = \left(\dfrac{1}{3600}\right)^{\circ}$.

18. $143°7'2'' = 143° + 7(1') + 2(1'') = 143° + 7\left(\dfrac{1}{60}\right)^{\circ} + 2\left(\dfrac{1}{3600}\right)^{\circ} = \left(143 + \dfrac{7}{60} + \dfrac{2}{3600}\right)^{\circ}$

$$= 143.11722°$$

In Problems 21 *and* 24 *we use the fact that* $1° = 60'$ *and* $1' = 60''$.

21. $210.78° = 210° + 0.78(60') = 210° + 46.8' = 210°46' + 0.8(60'') = 210°46' + 48''$

$\quad\quad = 210°46'48''$

24. $110.5° = 110° + 0.5(60') = 110° + 30' = 110°30'$

27. To convert from degrees to radians we use

$$45° = 45(1°) = 45\left(\frac{\pi}{180}\text{ radian}\right) = \frac{\pi}{4}\text{ radian}.$$

30. To convert from degrees to radians we use

$$-120° = -120(1°) = -120\left(\frac{\pi}{180}\text{ radian}\right) = -\frac{2\pi}{3}\text{ radians}.$$

33. To convert from radians to degrees we use

$$\frac{2\pi}{9}\text{ radians} = \frac{2\pi}{9}(1\text{ radian}) = \frac{2\pi}{9}\left(\frac{180}{\pi}\right)^° = 40°.$$

36. To convert from radians to degrees we use

$$\frac{5\pi}{12}\text{ radians} = \frac{5\pi}{12}(1\text{ radian}) = \frac{5\pi}{12}\left(\frac{180}{\pi}\right)^° = 75°.$$

39. To convert from radians to degrees we use

$$3.1\text{ radians} = 3.1(1\text{ radian}) = 3.1\left(\frac{180}{\pi}\right)^° \approx 177.62°.$$

42. Since $400 = 1(360) + 40$, we see that $400°$ is coterminal with $40°$.

45. Since $155 - 360 = -205$, we see that $875°$ is also coterminal with $-205°$.

48. Since $17\pi/2 = 4(2\pi) + \pi/2$, we see that $17\pi/2$ is coterminal with $\pi/2$.

51. Since $-4 = -1(2\pi) + (2\pi - 4) \approx -1(2\pi) + 2.28$, we see that -4 is coterminal with $2\pi - 4 \approx 2.28$.

54. Since $1.3\pi - 2\pi = -0.7\pi$, we see that 5.3π is also coterminal with -0.7π.

57. (a) Since $98.4° > 90°$, it is not an acute angle and hence, has no complementary angle.

    (b) Since the two positive angles are supplementary if their sum is $180°$, the angle supplementary to $98.4°$ is $180° - 98.4° = 81.6°$.

60. (a) Since two acute angles are complementary if their sum is $\pi/2$ radians, the angle complementary to $\pi/6$ is

$$\frac{\pi}{2} - \frac{\pi}{6} = \frac{3\pi}{6} - \frac{\pi}{6} = \frac{2\pi}{6} = \frac{\pi}{3}\text{ radians}.$$

(b) Since two positive angles are supplementary if their sum is π radians, the angle supplementary to $\pi/6$ is

$$\pi - \frac{\pi}{6} = \frac{6\pi}{6} - \frac{\pi}{6} = \frac{5\pi}{6} \text{ radians.}$$

63. (a) A counter-clockwise rotation is $360°$, so three-fifths of a clockwise rotation is

$$\frac{3}{5}(360°) = 216° \quad \text{or} \quad 216(1°) = 216\left(\frac{\pi}{180} \text{ radian}\right) = \frac{6\pi}{5} \text{ radians.}$$

(b) Since $5\frac{1}{8} = \frac{41}{8}$, five and one-eighth clockwise rotations is

$$\frac{41}{8}(-360°) = -1845°$$

or

$$(-1845)(1°) = (-1845)\left(\frac{\pi}{180} \text{ radians}\right) = -\frac{41\pi}{4} \text{ radians.}$$

66. In 2 hours, the minute hand on a clock makes 2 complete rotations. This is

$$2(360°) = 720° \quad \text{or} \quad 2(2\pi) = 4\pi \text{ radians.}$$

If the clockwise nature of the rotation is taken into account, the angles are $-720°$ and -4π radians.

69. The arc length subtended by an angle of 3 radians is $s = r(3) = 3r$, where r is the radius of the circle.

(a) When $r = 3$, $s = 3(3) = 9$.

(b) When $r = 5$, $s = 3(5) = 15$.

72. Using $s = r\theta$ where θ is measured in radians, we have $\pi/6 = 1 \cdot \theta$ when the radius is 1 and the arc length is $\pi/6$. Thus,

(a) $\theta = \dfrac{\pi}{6}$ radian

(b) $\left(\dfrac{\pi}{6}\right)(1 \text{ radian}) = \left(\dfrac{\pi}{6}\right)\left(\dfrac{180}{\pi}\right)° = 30°.$

Miscellaneous Applications

75. The time 23 hours, 56 minutes, and 4 seconds converts to 86,164 seconds.

(a) The angular speed is

$$\omega = \frac{\theta}{t} = \frac{2\pi}{86,164} \approx 0.000072921 \text{ rad/s.}$$

(b) Since the value of r is the sum of the Earth's radius and the height of the satellite above the Earth, $r = 6{,}378 + 35{,}786 = 42{,}164$ km. Thus, the linear speed of the satellite is

$$v = r\omega = (42{,}164) \cdot (0.000072921) \approx 3.074641 \text{ km/s}.$$

78. (a) We first convert $7.2°$ to radians:

$$7.2° = 7.2(1°) = 7.2 \left(\frac{\pi}{180} \text{ radian} \right) = 0.125664 \text{ radian}.$$

Since arc length is given by $s = r\theta$, we have

$$5000 = r(0.125664) \quad \text{or} \quad r = 39788.7 \text{ stades}.$$

(b) Since 1 stade $= 559$ feet, we have that Eratosthenes' measure of the radius of the Earth is

$$39788.7(1 \text{ stade}) = 39788.7(559 \text{ feet}) = 22{,}241{,}900 \text{ feet}.$$

The circumference of the Earth would then be

$$2\pi(22{,}241{,}900 \text{ feet}) = 139{,}750{,}000 \text{ feet} = 139{,}750{,}000(1 \text{ foot})$$

$$= 139{,}750{,}000 \left(\frac{1}{5280} \text{ mi} \right) = 26{,}467.8 \text{ mi}.$$

If the actual diameter of the Earth is 7900 miles, then its circumference is $7900\pi = 24{,}818.6$ miles. Thus, Eratosthenes' value is in error by $26{,}467.8 - 24{,}818.6 = 1649.22$ miles. The error as a percentage of the actual circumference is

$$\frac{1649.22}{24{,}818.6} \times 100\% = 6.6\%.$$

81. We first convert 55 mph to $\frac{55}{60} = \frac{11}{12}$ miles per minute, which in turn is $\frac{11}{12}(63{,}360)$ inches per minute. The circumference of the tire is 26π inches.

(a) The number of revolutions per minute is $58{,}080/26\pi \approx 711.06$.

(b) The angular speed is $711.06(2\pi) \approx 4467.7$ radians per minute.

4.2 | **The Sine and Cosine Functions**

3. Using the Pythagorean identity gives $(-2/3)^2 + \cos^2 t = 1$ or $\cos^2 t = 5/9$. Since $P(t)$ is in the third quadrant, $\cos t < 0$, so $\cos t = -\sqrt{5}/3$.

6. Using the Pythagorean identity gives $\sin^2 t + (3/10)^2 = 1$ or $\sin^2 t = 91/100$. Thus, $\sin t = \pm\sqrt{91}/10$.

9. Since $2\sin t - \cos t = 0$, $\cos t = 2\sin t$, and from the Pythagorean identity, we have

$$\sin^2 t + \cos^2 t = 1$$
$$\sin^2 t + (2\sin t)^2 = 1$$
$$5\sin^2 t = 1$$
$$\sin^2 t = \frac{1}{5}.$$

Thus, $\sin t = \pm 1/\sqrt{5}$ and $\cos t = 2\sin t = \pm 2/\sqrt{5}$.

12. **(a)** Since $\sin 3\pi = \sin(\pi + 2\pi) = \sin \pi$, we have from Figure 4.2.4 in the text that $\sin 3\pi = 0$.

(b) Similarly, $\cos 3\pi = \cos \pi = -1$.

15. The angle $t = 2\pi/3$ is in the second quadrant and the reference angle is $t' = \pi - 2\pi/3 = \pi/3$, so

$$\sin \frac{2\pi}{3} = \sin \frac{\pi}{3} = \frac{\sqrt{3}}{2} \qquad \text{and} \qquad \cos \frac{2\pi}{3} = -\cos \frac{\pi}{3} = -\frac{1}{2}.$$

18. The angle $t = 3\pi/4$ is in the second quadrant and the reference angle is $t' = \pi - 3\pi/4 = \pi/4$, so

$$\sin \frac{3\pi}{4} = \sin \frac{\pi}{4} = \frac{\sqrt{2}}{2} \qquad \text{and} \qquad \cos \frac{3\pi}{4} = -\cos \frac{\pi}{4} = -\frac{\sqrt{2}}{2}.$$

21. The angle $t = -\pi/4$ is in the fourth quadrant and the reference angle is $t' = -(-\pi/4) = \pi/4$, so

$$\sin\left(-\frac{\pi}{4}\right) = -\sin \frac{\pi}{4} = -\frac{\sqrt{2}}{2} \qquad \text{and} \qquad \cos\left(-\frac{\pi}{4}\right) = \cos \frac{\pi}{4} = \frac{\sqrt{2}}{2}.$$

24. The angle $t = -11\pi/6$ is in the first quadrant and the reference angle is $t' = 2\pi - 11\pi/6 = \pi/6$, so

$$\sin\left(-\frac{11\pi}{6}\right) = \sin \frac{\pi}{6} = \frac{1}{2} \qquad \text{and} \qquad \cos\left(-\frac{11\pi}{6}\right) = \cos \frac{\pi}{6} = \frac{\sqrt{3}}{2}.$$

27. The angle $t = -11\pi/3$ is in the first quadrant and the reference angle is $t' = 4\pi - 11\pi/3 = \pi/3$, so

$$\sin\left(-\frac{11\pi}{3}\right) = \sin \frac{\pi}{3} = \frac{\sqrt{3}}{2}.$$

30. Since $-19\pi/2 = -20\pi/2 + \pi/2$,

$$\sin\left(-\frac{19\pi}{2}\right) = \sin\left(-\frac{20\pi}{2} + \frac{\pi}{2}\right) = \sin\left(-10\pi + \frac{\pi}{2}\right) = \sin \frac{\pi}{2} = 1.$$

33. Using periodicity,

$$\sin 3\pi = \sin(\pi + 2\pi) = \sin \pi.$$

36. Using periodicity,

$$\cos 16.8\pi = \cos(14.8\pi + 2\pi) = \cos 14.8\pi.$$

39. The angle $t = 135°$ is in the second quadrant and the reference angle is $t = 180° - 135° = 45°$, so

$$\sin 135° = \sin 45° = \frac{\sqrt{2}}{2}.$$

42. The terminal side of $t = 270°$ lies along the negative y-axis, so the point $P(t)$ is $(0, -1)$ and $\sin 270° = -1$.

45. The angle $t = -60°$ is in the fourth quadrant and the reference angle is $60°$, so

$$\sin(-60°) = -\sin 60° = -\frac{\sqrt{3}}{2}.$$

48. Since $\cos t$ is the x-coordinate of $P(t)$, $\cos t = -1$ only for $t = \pi$ in $[0, 2\pi)$.

51. Since $\cos 30° = \sqrt{3}/2$, the only other angle in $[0°, 360°)$ whose cosine is $\sqrt{3}/2$ lies in the fourth quadrant and has reference angle $30°$. This angle is $330°$.

54. Since $\cos \theta$ is the x-coordinate of $P(\theta)$, $\cos \theta = 1$ only for $\theta = 0°$ in $[0°, 360°)$.

Miscellaneous Applications

57. (a) When $\theta = 0$, $\sin \theta = 0$ and $\sin 2\theta = 0$, so

$$g_{\text{sat}} = 978.0309 \text{ cm}/s^2.$$

(b) At the north pole, $\theta = 90°$, and $\sin \theta = 1$ and $\sin 2\theta = \sin 180° = 0$, so

$$g_{\text{sat}} = 978.0309 + 5.18552 = 983.2164 \text{ cm}/s^2.$$

(c) When $\theta = 45°$, $\sin \theta = \sqrt{2}/2$ and $\sin 2\theta = \sin 90° = 1$, so

$$g_{\text{sat}} = 978.0309 + 5.18552(\sqrt{2}/2)^2 - 0.00570(1)^2$$

$$= 978.0309 + 2.59276 - 0.00570 = 980.618 \text{ cm}/s^2.$$

4.3 | Graphs of Sine and Cosine Functions

3.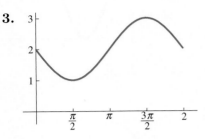

6.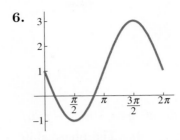

9. Since the graph flattens out at $x = 0$, we use $y = A\cos x + D$. The amplitude is $\frac{1}{2}[4 - (-2)] = 3$ and the graph has been reflected through the line $y = 1$, so $A = -3$ and $D = 1$. Thus, $y = -3\cos x + 1$.

12. Setting $-\cos 2x = 0$, we have by (2) in this section of the text that the x-intercepts are determined by $2x = (2n + 1)\pi/2$, n an integer. Thus, the x-intercepts are at $x = (2n + 1)\pi/4$, so the intercepts are $(\pi/4 + n\pi/2, 0)$, where n is an integer.

15. Setting $\sin(x - \pi/4) = 0$ we have by (1) in this section of the text that the x-intercepts are determined by $x - \pi/4 = n\pi$, n an integer. Thus, the x-intercepts are at $x = n\pi + \pi/4 = (n + \frac{1}{4})\pi$, so the intercepts are $(\pi/4 + n\pi, 0)$, where n is an integer.

18. Setting $1 - 2\cos x = 0$ we have $2\cos x = 1$ or $\cos x = \frac{1}{2}$. Since $\cos x$ is positive, x is in the first or fourth quadrant. Since $\cos x = \frac{1}{2}$ for $x = \pi/3$, we see that $1 - 2\cos x = 0$ for $x = \pi/3$ and $x = 5\pi/3$ in $[0, 2\pi]$. By periodicity, then, the x-intercepts of $1 - 2\cos 2x$ are at

$$\frac{\pi}{3} + 2n\pi = \left(\frac{1}{3} + 2n\right)\pi \qquad \text{and} \qquad \frac{5\pi}{3} + 2n\pi = \left(\frac{5}{3} + 2n\right)\pi,$$

so the intercepts are

$$\left(\frac{\pi}{3} + 2n\pi, 0\right) \qquad \text{and} \qquad \left(\frac{5\pi}{3} + 2n\pi, 0\right), \text{ where } n \text{ is an integer.}$$

21. Since the y-intercept is $\frac{1}{2}$ (not 0), the equation has the form $y = A\cos Bx$. The amplitude of the graph is $A = \frac{1}{2}$ and the period is $2 = 2\pi/B$, so $B = \pi$ and $y = \frac{1}{2}\cos \pi x$.

24. Since the y-intercept is 3 (not 0), the equation has the form $y = A\cos Bx$. The amplitude of the graph is $A = 3$ and the period is $8 = 2\pi/B$, so $B = \pi/4$ and $y = 3\cos \pi x/4$.

27. The amplitude of $y = -3\cos 2\pi x$ is $A = |-3| = 3$ and the period is $2\pi/2\pi = 1$.

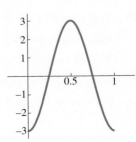

30. The amplitude of $y = 2 - 2\sin \pi x$ is $A = |-2| = 2$ and the period is $2\pi/\pi = 2$.

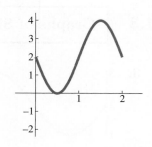

33. The amplitude of $y = \sin(x - \pi/6)$ is $A = 1$ and the period is $2\pi/1 = 2\pi$. The phase shift is $|-\pi/6|/1 = \pi/6$. Since $C = -\pi/6 < 0$, the shift is to the right.

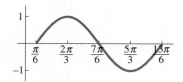

36. The amplitude of $y = -2\cos(2x - \pi/6)$ is $A = |-2| = 2$ and the period is $2\pi/2 = \pi$. The phase shift is $|-\pi/6|/2 = -\pi/12$. Since $C = -\pi/6 < 0$, the shift is to the right.

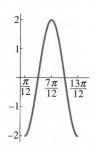

39. The amplitude of $y = 3\sin(x/2 - \pi/3)$ is $A = 3$ and the period is $2\pi/(1/2) = 4\pi$. The phase shift is $|-\pi/3|/(1/2) = 2\pi/3$. Since $C = -\pi/3 < 0$, the shift is to the right.

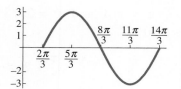

42. We use the fact that $\cos(-\theta) = \cos\theta$ to write

$$y = 2\cos(-2\pi x - 4\pi/3) = 2\cos(2\pi x + 4\pi/3).$$

Then the amplitude of the function is $A = 2$ and the period is $2\pi/2\pi = 1$. The phase shift is $|4\pi/3|/2\pi = \frac{2}{3}$. Since $C = 4\pi/3 > 0$, the shift is to the left.

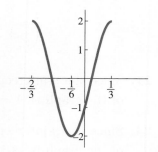

45. We identify $A = 3$ and $2\pi/B = 2\pi/3$, so $B = 3$. Since the graph is shifted to the right, $C < 0$. The phase shift is $|C|/B = |C|/3 = \pi/3$, so $C = -\pi$. The functions are $y = 3\sin(3x - \pi)$, shown as the blue curve, and $y = 3\cos(3x - \pi)$, shown as the red curve.

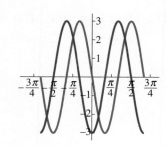

48. We identify $A = \frac{5}{4}$ and $2\pi/B = 4$, so $B = \pi/2$. Since the graph is shifted to the left, $C > 0$. The phase shift is $|C/B| = C/(\pi/2) = 1/2\pi$, so $C = \frac{1}{4}$. The functions are $y = \frac{5}{4}\sin(\pi x/2 + 1/2\pi)$, shown as the blue curve, and $y = \frac{5}{4}\cos(\pi x/2 + 1/2\pi)$, shown as the red curve.

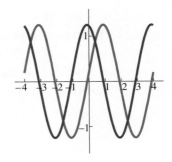

Miscellaneous Applications

51. For $\omega = 2$ and $\theta_0 = \pi/10$ the function is $\theta(t) = (\pi/10)\cos 2t$. The amplitude is $A = 2$ and the period is $2\pi/2 = \pi$. Thus, two cycles will extend from $t = 0$ to $t = 2\pi$.

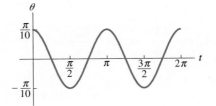

54. (a) Since 8 A.M. is 8 hours after midnight,

$$T(8) = 50 + 10\sin\frac{\pi}{12}(8-8) = 50$$

and the temperature at 8 A.M. is 50°F.

   (b) We solve

$$50 + 10\sin\frac{\pi}{12}(t-8) = 60$$

$$\sin\frac{\pi}{12}(t-8) = 1$$

$$\frac{\pi}{12}(t-8) = \frac{\pi}{2}$$

$$t - 8 = 6$$

$$t = 14.$$

Since 14 hours after midnight is noon plus 2 hours, the temperature will be 60°F at 2 P.M.

   (c) The amplitude of the graph is $A = 10$ and the graph is vertically centered at $T = 50$. The period is $2\pi/(\pi/12) = 24$ and the phase shift is $|C|/B = (8\pi/12)/(\pi/12) = 8$. Since $C = 8\pi/12 > 0$, the graph of $y = 50 + 10\sin \pi t/12$ is shifted 8 units to the left to obtain the graph of

$$T = 50 + 10\sin\left(\frac{\pi}{12}t - \frac{8\pi}{12}\right).$$

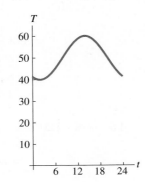

   (d) The minimum temperature will be 40°F and this will occur when $\pi(t-8)/12 = 3\pi/2$, since $x = 3\pi/2$ where $y = \sin x$ first reaches its minimum. Solving for t, we obtain

$t = 26$. Since the period is 24, the temperature is first minimum when $t = 2$ or at 2 A.M. The maximum temperature will be 60°F and this will occur when $\pi(t - 8)/12 = \pi/2$, since $x = \pi/2$ where $y = \sin x$ first reaches its maximum. Solving for t, we obtain $t = 14$. Thus, the temperature is first maximum when $t = 14$ or at 2 P.M. We could have reached the same conclusion about the first occurrence of the maximum temperature from the information in parts **(b)** and **(c)**.

4.4   Other Trigonometric Functions

3. $\cot \dfrac{13\pi}{6} = \cot\left(\dfrac{13\pi}{6} - 2\pi\right)$ (because the cotangent is π-periodic)

$\qquad = \cot \dfrac{\pi}{6} = \dfrac{\cos \pi/6}{\sin \pi/6}$ (by the definition of the cotangent)

$\qquad = \dfrac{\sqrt{3}/2}{1/2} = \sqrt{3}$

6. $\sec 7\pi = \sec(7\pi - 6\pi)$ (because the secant is 2π-periodic)

$\qquad = \sec \pi = \dfrac{1}{\cos \pi}$ (by the definition of the secant)

$\qquad = \dfrac{1}{-1} = -1$

9. $\tan \dfrac{23\pi}{4} = \tan\left(\dfrac{23\pi}{4} - 5\pi\right)$ (because the tangent is π-periodic)

$\qquad = \tan \dfrac{3\pi}{4} = \dfrac{\sin 3\pi/4}{\cos 3\pi/4}$ (by the definition of the tangent)

$\qquad = \dfrac{\sqrt{2}/2}{-\sqrt{2}/2} = -1$

12. $\cot \dfrac{17\pi}{6} = \cot\left(\dfrac{17\pi}{6} - 2\pi\right)$ (because the cotangent is π-periodic)

$\qquad = \cot \dfrac{5\pi}{6} = \dfrac{\cos 5\pi/6}{\sin 5\pi/6}$ (by the definition of the cotangent)

$\qquad = \dfrac{-\sqrt{3}/2}{1/2} = -\sqrt{3}$

15. $\sec(-120°) = \sec(-120° + 360°)$ (because the secant has period 360°)

$\qquad = \sec 240° = \dfrac{1}{\cos 240°}$ (by the definition of the secant)

$\qquad = \dfrac{1}{-1/2} = -2$

18. $\cot(-720°) = \cot(-720° + 2 \cdot 360°)$ (because the cotangent has period $360°$)

$$= \cot 0° = \frac{\cos 0°}{\sin 0°}$$ (by the definition of the cotangent)

Since $\sin 0° = 0$, the cotangent of $-720°$ is not defined.

21. We first compute $\cot x$ using the Pythagorean identity:

$$\cot^2 x = \csc^2 x - 1 = \left(\frac{4}{3}\right)^2 + 1 = \frac{16}{9} - 1 = \frac{7}{9}.$$

Since $0 < x < \pi/2$, x is in the first quadrant and $\cot x > 0$. Thus, $\cot x = \sqrt{7}/3$. Then

$$\tan x = \frac{1}{\cot x} = \frac{1}{\sqrt{7}/3} = \frac{3}{\sqrt{7}}.$$

Finally,

$$\sin x = \frac{1}{\csc x} = \frac{1}{4/3} = \frac{3}{4},$$

$$\cos x = \frac{\cos x \sin x}{\sin x} = \cot x \sin x = \frac{\sqrt{7}}{3}\left(\frac{3}{4}\right) = \frac{\sqrt{7}}{4},$$

and

$$\sec x = \frac{1}{\cos x} = \frac{1}{\sqrt{7}/4} = \frac{4}{\sqrt{7}}.$$

24. We first compute $\sin x$ using the Pythagorean identity:

$$\sin^2 x = 1 - \cos^2 x = 1 - \left(-\frac{1}{\sqrt{5}}\right)^2 = 1 - \frac{1}{5} = \frac{4}{5}.$$

Since $\pi < x < 3\pi/2$, x is in the quadrant and $\sin x < 0$. Thus, $\sin x = -2/\sqrt{5}$. Then

$$\tan x = \frac{\sin x}{\cos x} = \frac{-2/\sqrt{5}}{-1/\sqrt{5}} = 2.$$

Finally,

$$\cot x = \frac{1}{\tan x} = \frac{1}{2},$$

$$\sec x = \frac{1}{\cos x} = \frac{1}{-1/\sqrt{5}} = -\sqrt{5},$$

and

$$\csc x = \frac{1}{\sin x} = \frac{1}{-2/\sqrt{5}} = -\frac{\sqrt{5}}{2}.$$

27. From $3\cos x = \sin x$ we see that $\tan x = \sin x / \cos x = 3$. Since $\tan x = 3 > 0$, x is in either the first or third quadrant. (This is consistent with the fact that $\sin x$ and $\cos x$ have the same algebraic sign, as can be inferred from $3\cos x = \sin x$.) From $\tan x = 3$ we see that $\cot x = 1/\tan x = 1/3$. Using the Pythagorean identity, we have

$$\sec^2 x = 1 + \tan^2 x = 1 + 3^2 = 10,$$

so $\sec x = \pm\sqrt{10}$. From

$$\csc^2 x = 1 + \cot^2 = 1 + \left(\frac{1}{3}\right)^2 = \frac{10}{9}$$

we see that $\csc x = \pm\sqrt{10}/3$. Finally, $\sin x = 1/\csc x = 1/(\pm\sqrt{10}/3) = \pm 3/\sqrt{10}$, and $\cos x = 1/\sec x = \pm 1/\sqrt{10}$.

30. The period of $y = \tan(x/2)$ is $\pi/(1/2) = 2\pi$. Since

$$\tan\frac{x}{2} = \frac{\sin(x/2)}{\cos(x/2)},$$

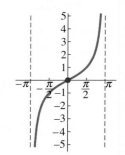

the x-intercepts of $\tan(x/2)$ occur at the zeros of $\sin(x/2)$; namely, at $x/2 = n\pi$ or $x = 2n\pi$ for n an integer. The vertical asymptotes occur at the zeros of $\cos(x/2)$; namely, at

$$\frac{x}{2} = \frac{(2n+1)\pi}{2} \qquad \text{or} \qquad x = (2n+1)\pi = \pi + 2n\pi$$

for n an integer. Since the graph has vertical asymptotes at $-\pi$ and π (using $n = -1$ and $n = 0$), we graph one cycle on the interval $(-\pi, \pi)$.

33. The period of $y = \tan(x/2 - \pi/4)$ is $\pi/(1/2) = 2\pi$. Since

$$\tan\left(\frac{x}{2} - \frac{\pi}{4}\right) = \frac{\sin(x/2 - \pi/4)}{\cos(x/2 - \pi/4)},$$

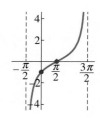

the x-intercepts of $\tan(x/2-\pi/4)$ occur at the zeros of $\sin(x/2-\pi/4)$; namely, at

$$\frac{x}{2} - \frac{\pi}{4} = n\pi \qquad \text{or} \qquad x = 2n\pi + \frac{\pi}{2},$$

for n an integer. The vertical asymptotes occur at the zeros of $\cos(x/2 - \pi/4)$; namely, at

$$\frac{x}{2} - \frac{\pi}{4} = \frac{(2n+1)\pi}{2} \qquad \text{or} \qquad x = (2n+1)\pi + \frac{\pi}{2} = \frac{3\pi}{2} + 2n\pi,$$

for n an integer. Since the graph has vertical asymptotes at $-\pi/2$ and $3\pi/2$ (using $n = -1$ and $n = 0$), we graph one cycle on the interval $(-\pi/2, 3\pi/2)$.

36. The period of $y = \tan(x + 5\pi/6)$ is π. Since

$$\tan(x + 5\pi/6) = \frac{\sin(x + 5\pi/6)}{\cos(x + 5\pi/6)},$$

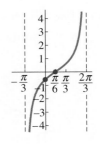

the x-intercepts of $\tan(x + 5\pi/6)$ occur at the zeros of $\sin(x + 5\pi/6)$; namely, at

$$x + 5\pi/6 = n\pi$$

or

$$x = n\pi - 5\pi/6,$$

for n an integer. The vertical asymptotes occur at the zeros of $\cos(x + 5\pi/6)$; namely, at

$$x + 5\pi/6 = (2n + 1)\pi/2 \qquad \text{or} \qquad x = (2n + 1)\pi/2 - 5\pi/6 = -\pi/3 + n\pi,$$

for n an integer. Since the graph has vertical asymptotes at $-\pi/3$ and $2\pi/3$ (using $n = 0$ and $n = 1$), we graph one cycle on the interval $(-\pi/3, 2\pi/3)$.

39. The period of $y = 3\csc \pi x$ is $2\pi/\pi = 2$. Since $3\csc \pi x = 3/\sin \pi x$, the vertical asymptotes occur at the zeros of $\sin \pi x$; namely, at $\pi x = n\pi$ or $x = n$, for n an integer. We plot one cycle of the graph on $(-1, 1)$, since the period of the function is $2 = 1 - (-1)$ and vertical asymptotes occur at $x = -1$ and $x = 1$ (taking $n = -1$ and 1).

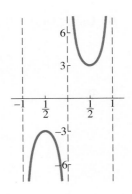

42. The period of $y = \csc(4x + \pi)$ is $2\pi/4 = \pi/2$. Since

$$\csc(4x + \pi) = \frac{1}{\sin(4x + \pi)},$$

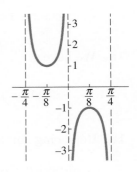

the vertical asymptotes occur at the zeros of $\sin(4x + \pi)$, namely, at

$$4x + \pi = n\pi \qquad \text{or} \qquad x = (n - 1)\frac{\pi}{4} = -\frac{\pi}{4} + \frac{n\pi}{4},$$

for n an integer. We plot one cycle of the graph on $(-\pi/4, \pi/4)$ since the period of the function is $\pi/2 = \pi/4 - (-\pi/4)$ and vertical asymptotes occur at $x = -\pi/4$ and $x = \pi/4$ (taking $n = 0$ and 2).

45. The graph of $y = \cot x$ is shown in the figures below as a red curve, while the graphs of $y = A\tan(x + C)$, for various choices of A and C, are shown as blue curves.

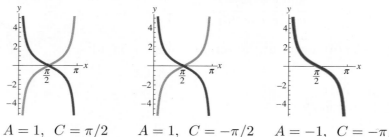

$A = 1, \ C = \pi/2$ $A = 1, \ C = -\pi/2$ $A = -1, \ C = -\pi/2$

We see from the third graph that $\cot x = -\tan(x - \pi/2)$.

4.5 | Verifying Trigonometric Identities

3. We have

$$\frac{\sin\theta}{\csc\theta} + \frac{\cos\theta}{\sec\theta} = \frac{\sin\theta}{1/\sin\theta} + \frac{\cos\theta}{1/\cos\theta} = \sin^2\theta + \cos^2\theta = 1.$$

6. Using $\tan(-\theta) = -\tan\theta$ and $1 + \tan^2\theta = \sec^2\theta$ we have

$$1 + \tan^2(-\theta) = 1 + \big[\tan(-\theta)\big]^2 = 1 + \big[-\tan\theta\big]^2 = 1 + \tan^2\theta = \sec^2\theta.$$

9. We have

$$\sec(-x)\cos x = \sec x\cos x = \frac{1}{\cos x}\cos x = 1.$$

12. Since $1 + \tan^2 x = \sec^2 x$ and $1/\cos x = \sec x$ we have

$$\cos x + \cos x\tan^2 x = \cos x(1 + \tan^2 x) = \cos x(\sec^2 x) = (\cos x\sec x)\sec x = \sec x.$$

15. We have

$$\sin x + \cos x\cot x = \sin x + \cos x\left(\frac{\cos x}{\sin x}\right) = \frac{\sin^2 x + \cos^2 x}{\sin x} = \frac{1}{\sin x} = \csc x.$$

18. Replacing $\sin^2\theta$ with $1 - \cos^2\theta$ we have

$$\frac{\sin^2\theta\cos\theta + \cos^3\theta - \cos\theta + \sin\theta}{\cos\theta} = \frac{(1 - \cos^2\theta)\cos\theta + \cos^3\theta - \cos\theta + \sin\theta}{\cos\theta}$$

$$= \frac{\cos\theta - \cos^3\theta + \cos^3\theta - \cos\theta + \sin\theta}{\cos\theta} = \frac{\sin\theta}{\cos\theta} = \tan\theta.$$

21. $\dfrac{1}{1+\sin t} + \dfrac{1}{1-\sin t} = \dfrac{1-\sin t + 1 + \sin t}{(1+\sin t)(1-\sin t)} = \dfrac{2}{1-\sin^2 t} = \dfrac{2}{\cos^2 t} = 2\sec^2 t$

24. We show that the left-hand side is equivalent to the right-hand side of the equation.

$$\frac{1+\sin x}{\cos x} = \frac{1}{\cos x} + \frac{\sin x}{\cos x} = \sec x + \tan x$$

27. We show that the left-hand side is equivalent to the right-hand side of the equation.

$$1 - 2\sin^2 t = 1 - \sin^2 t - \sin^2 t$$
$$= \cos^2 t - \sin^2 t$$
$$= \cos^2 t - (1 - \cos^2 t)$$
$$= 2\cos^2 t - 1$$

30. Converting to sines and cosines we show that the left-hand side is equivalent to the right-hand side of the equation.

$$\frac{\sin t + \tan t}{1 + \cos t} = \frac{\sin t + \dfrac{\sin t}{\cos t}}{1 + \cos t} \frac{\cos t}{\cos t} = \frac{\sin t \cos t + \sin t}{(1 + \cos t)\cos t}$$

$$= \frac{\sin t\,(\cos t + 1)}{(1 + \cos t)\cos t} = \frac{\sin t}{\cos t} = \tan t$$

33. We show that the left-hand side is equivalent to the right-hand side of the equation.

$$\sin^2 x \cot^2 x + \cos^2 x \tan^2 x = \sin^2 x \frac{\cos^2 x}{\sin^2 x} + \cos^2 x \frac{\sin^2 x}{\cos^2 x}$$

$$= \cos^2 x + \sin^2 x = 1$$

36. Using $1 + \tan^2 t = \sec^2 t$ we show that the left-hand side is equal to the right-hand side of the equation.

$$\frac{1}{\sec t - \tan t}\frac{\sec t + \tan t}{\sec t + \tan t} = \frac{\sec t + \tan t}{\sec^2 t - \tan^2 t} = \frac{\sec t + \tan t}{1} = \sec t + \tan t$$

39. We show that the left-hand side is equivalent to the right-hand side of the equation.

$$(\csc t - \cot t)^2 = \left(\frac{1}{\sin t} - \frac{\cos t}{\sin t}\right)^2 = \left(\frac{1 - \cos t}{\sin t}\right)^2$$

$$= \frac{(1 - \cos t)^2}{\sin^2 t} = \frac{(1 - \cos t)^2}{1 - \cos^2 t}$$

$$= \frac{(1 - \cos t)(1 - \cos t)}{(1 - \cos t)(1 + \cos t)} = \frac{1 - \cos t}{1 + \cos t}$$

42. Converting to sines and cosines, we show that the left-hand side of the equation is equal to the right-hand side.

$$\frac{\tan t + \cot t}{\cos^2 t} - \sin t \sec^3 t = \frac{\dfrac{\sin t}{\cos t} + \dfrac{\cos t}{\sin t}}{\cos^2 t} - \sin t \frac{1}{\cos^3 t}$$

$$= \frac{\sin t}{\cos^3 t} + \frac{1}{\sin t \cos t} - \frac{\sin t}{\cos^3 t} = \csc t \sec t$$

45. We show that the left-hand side is equivalent to the right-hand side of the equation.

$$\cos(-t)\csc(-t) = \cos t(-\csc t) = -\cos t \left(\frac{1}{\sin t}\right) = -\cot t$$

48. In this problem we first use the fact that $\sin^2 \alpha + \cos^2 \alpha = 1$ and then the fact that $\cos \alpha \leq 1$ so that $1 - \cos \alpha \geq 0$ and $|1 - \cos \alpha| = 1 - \cos \alpha$, to show that the left-hand side of the equation is equal to the right-hand side.

$$\sqrt{\frac{1 + \cos \alpha}{1 - \cos \alpha}} \sqrt{\frac{1 - \cos \alpha}{1 - \cos \alpha}} = \sqrt{\frac{(1 + \cos \alpha)(1 - \cos \alpha)}{(1 - \cos \alpha)^2}} = \sqrt{\frac{1 - \cos^2 \alpha}{(1 - \cos \alpha)^2}}$$

$$= \sqrt{\frac{\sin^2 \alpha}{(1 - \cos \alpha)^2}} = \left|\frac{\sin \alpha}{1 - \cos \alpha}\right| = \frac{|\sin \alpha|}{1 - \cos \alpha}$$

51. We show that the left-hand side is equivalent to the right-hand side of the equation.

$$(\tan^2 t + 1)(\cos^2 t - 1) = (\sec^2 t)(-\sin^2 t)$$

$$= \frac{\sin^2 t}{\cos^2 t} = -\tan^2 t = 1 - \sec^2 t$$

54. $\dfrac{1 - \cos \alpha}{1 + \cos \alpha} = \dfrac{\sec \alpha - 1}{\sec \alpha + 1} \Rightarrow$ LHS is

$$\frac{1 - \cos \alpha}{1 + \cos \alpha} = \frac{1 - \cos \alpha}{1 + \cos \alpha} \frac{\dfrac{1}{\cos \alpha}}{\dfrac{1}{\cos \alpha}} = \frac{\dfrac{1}{\cos \alpha} - 1}{\dfrac{1}{\cos \alpha} + 1} = \frac{\sec \alpha - 1}{\sec \alpha + 1}$$

57. We show that the left-hand side is equivalent to the right-hand side of the equation.

$$\frac{\sin\theta}{1-\cot\theta}+\frac{\cos\theta}{1-\tan\theta}=\frac{\sin\theta}{1-\dfrac{\cos\theta}{\sin\theta}}+\frac{\cos\theta}{1-\dfrac{\sin\theta}{\cos\theta}}$$

$$=\frac{\sin\theta}{\dfrac{\sin\theta-\cos\theta}{\sin\theta}}+\frac{\cos\theta}{\dfrac{\cos\theta-\sin\theta}{\cos\theta}}$$

$$=\frac{\sin^2\theta}{\sin\theta-\cos\theta}+\frac{\cos^2\theta}{\cos\theta-\sin\theta}$$

$$=\frac{\sin^2\theta}{\sin\theta-\cos\theta}-\frac{\cos^2\theta}{\sin\theta-\cos\theta}$$

$$=\frac{\sin^2\theta-\cos^2\theta}{\sin\theta-\cos\theta}$$

$$=\frac{(\sin\theta-\cos\theta)(\sin\theta+\cos\theta)}{\sin\theta-\cos\theta}$$

$$=\cos\theta+\sin\theta$$

60. Recalling that $\dfrac{1/a}{b}=\dfrac{1}{ab}$ and converting to sines and cosines, we show that the left-hand side of the equation is equal to the right-hand side.

$$\frac{\tan x-\cot x}{\sin x\cos x}=\frac{\dfrac{\sin x}{\cos x}-\dfrac{\cos x}{\sin x}}{\sin x\cos x}=\frac{\sin x}{\sin x\cos^2 x}-\frac{\cos x}{\sin^2 x\cos x}=\frac{1}{\cos^2 x}-\frac{1}{\sin^2 x}=\sec^2 x-\csc^2 x$$

63. $\dfrac{\tan\alpha-\tan\beta}{1+\tan\alpha\tan\beta}=\dfrac{\cot\beta-\cot\alpha}{\cot\alpha\cot\beta+1}\Rightarrow$ LHS is

$$\frac{\tan\alpha-\tan\beta}{1+\tan\alpha\tan\beta}\frac{\dfrac{1}{\tan\alpha\tan\beta}}{\dfrac{1}{\tan\alpha\tan\beta}}=\frac{\dfrac{1}{\tan\beta}-\dfrac{1}{\tan\alpha}}{\dfrac{1}{\tan\alpha\tan\beta}+1}=\frac{\cot\alpha-\cot\beta}{\cot\alpha\cot\beta+1}$$

66. If we let $x=a\tan\theta$, then

$$\sqrt{a^2+x^2}=\sqrt{a^2+a^2\tan^2\theta}=\sqrt{a^2(1+\tan^2\theta)}=\sqrt{a^2\sec^2\theta}=a\sec\theta$$

for $a>0$ and $-\pi/2<\theta<\pi/2$.

69. If we let $x=3\sin\theta$, then

$$\frac{x}{\sqrt{9-x^2}}=\frac{3\sin\theta}{\sqrt{9-9\sin^2\theta}}=\frac{3\sin\theta}{\sqrt{9(1-\sin^2\theta)}}=\frac{3\sin\theta}{\sqrt{9\cos^2\theta}}=\frac{3\sin\theta}{3\cos\theta}=\tan\theta$$

for $-\pi/2<\theta<\pi/2$.

72. If we let $x = 6\tan\theta$, then

$$\left(36 + x^2\right)^{3/2} = \left(36 + 36\tan^2\theta\right)^{3/2} = 36^{3/2}\left(1 + \tan^2\theta\right)^{3/2} = 216\left(\sec^2\theta\right)^{3/2} = 216\sec^3\theta$$

for $-\pi/2 < \theta < \pi/2$.

4.6 | Sum and Difference Formulas

3. $\sin 75° = \sin(30° + 45°) = \sin 30° \sin 45° + \cos 30° \sin 45°$

$$= \left(\frac{1}{2}\right)\left(\frac{\sqrt{2}}{2}\right) + \left(\frac{\sqrt{3}}{2}\right)\left(\frac{\sqrt{2}}{2}\right) = \frac{\sqrt{2}}{4}(1 + \sqrt{3})$$

6. $\cos\dfrac{11\pi}{12} = \cos\left(\dfrac{3\pi}{4} + \dfrac{\pi}{6}\right) = \cos\dfrac{3\pi}{4}\cos\dfrac{\pi}{6} - \sin\dfrac{3\pi}{4}\sin\dfrac{\pi}{6}$

$$= \left(-\frac{\sqrt{2}}{2}\right)\left(\frac{\sqrt{3}}{2}\right) - \left(\frac{\sqrt{2}}{2}\right)\left(\frac{1}{2}\right) = -\frac{\sqrt{2}}{4}(\sqrt{3} + 1)$$

9. $\sin\left(-\dfrac{\pi}{12}\right) = \sin\left(-\dfrac{\pi}{4} - \dfrac{\pi}{3}\right) = \sin\dfrac{\pi}{4}\cos\dfrac{\pi}{3} - \cos\dfrac{\pi}{4}\sin\dfrac{\pi}{3}$

$$= \left(\frac{\sqrt{2}}{2}\right)\left(\frac{1}{2}\right) - \left(\frac{\sqrt{2}}{2}\right)\left(\frac{\sqrt{3}}{2}\right) = \frac{\sqrt{2}}{4}(1 - \sqrt{3})$$

12. $\tan\dfrac{7\pi}{12} = \tan\left(\dfrac{\pi}{3} + \dfrac{\pi}{4}\right) = \dfrac{\tan\dfrac{\pi}{3} + \tan\dfrac{\pi}{4}}{1 - \tan\dfrac{\pi}{3}\tan\dfrac{\pi}{4}}$

$$= \frac{\sqrt{3} + 1}{1 - (\sqrt{3})(1)} = \frac{\sqrt{3} + 1}{1 - \sqrt{3}}$$

15. $\tan 165° = \tan(135° + 30°) = \dfrac{\tan 135° + \tan 30°}{1 - \tan 135° \tan 30°} = \dfrac{-1 + 1/\sqrt{3}}{1 - (-1)(1/\sqrt{3})}$

$$= \frac{-1 + 1/\sqrt{3}}{1 + 1/\sqrt{3}} \cdot \frac{\sqrt{3}}{\sqrt{3}} = \frac{1 - \sqrt{3}}{1 + \sqrt{3}} = \frac{1 - \sqrt{3}}{1 + \sqrt{3}} \cdot \frac{1 - \sqrt{3}}{1 - \sqrt{3}} = \frac{1 - 2\sqrt{3} + 3}{1 - 3} = -2 + \sqrt{3}$$

18. $\tan 195° = \tan(135° + 60°) = \dfrac{\tan 135° + \tan 60°}{1 - \tan 135° \tan 60°}$

$$= \frac{-1 + \sqrt{3}}{1 - (-1)\sqrt{3}} = \frac{\sqrt{3} - 1}{\sqrt{3} + 1}$$

21. $\cos\dfrac{13\pi}{12} = \cos\left(\dfrac{3\pi}{4} + \dfrac{\pi}{3}\right) = \cos\dfrac{3\pi}{4}\cos\dfrac{\pi}{3} - \sin\dfrac{3\pi}{4}\sin\dfrac{\pi}{3}$

$$= \left(-\frac{\sqrt{2}}{2}\right)\left(\frac{1}{2}\right) - \left(\frac{\sqrt{2}}{2}\right)\left(\frac{\sqrt{3}}{2}\right) = -\frac{\sqrt{2}}{4}(1 + \sqrt{3})$$

24. $\cos^2 2t - \sin^2 2t = \cos 2(2t) = \cos 4t$

27. From (13) of Theorem 4.6.4 with x replaced by $3t$,

$$\frac{\tan 3t}{1 - \tan^2 3t} = \frac{1}{2}\frac{2\tan 3t}{1 - \tan^2 3t} = \tfrac{1}{2}\tan 2(3t) = \tfrac{1}{2}\tan 6t$$

30. Since $3\pi/2 < x < 2\pi$, x is in the fourth quadrant, so $\sin x$ is negative. Using $\sin^2 x + \cos^2 x = 1$ we have

$$\sin x = -\sqrt{1 - \cos^2 x} = -\sqrt{1 - \frac{3}{25}} = -\frac{\sqrt{22}}{5}.$$

(a)   From the double-angle formula (11) in the text,

$$\cos 2x = \cos^2 x - \sin^2 x = \left(\frac{\sqrt{3}}{5}\right)^2 - \left(-\frac{\sqrt{22}}{5}\right)^2 = \frac{3}{25} - \frac{22}{25} = -\frac{19}{25}.$$

(b)   From the double-angle formula (12) in the text,

$$\sin 2x = 2\sin x\cos x = 2\left(-\frac{\sqrt{22}}{5}\right)\left(\frac{\sqrt{3}}{5}\right) = -\frac{2\sqrt{66}}{25}.$$

(c)   Since $\tan 2x = \sin 2x / \cos 2x$, we have

$$\tan 2x = \frac{-2\sqrt{66}/25}{-19/25} = \frac{2\sqrt{66}}{19}.$$

33. Since $\pi/2 < x < \pi$, x is in the second quadrant and $\sin x$ is positive. Using $\cos x = 1/\sec x$ we have

$$\cos x = \frac{1}{-13/5} = -\frac{5}{13},$$

and, using $\sin^2 x + \cos^2 x = 1$, we have

$$\sin x = \sqrt{1 - \cos^2 x} = \sqrt{1 - \left(-\frac{5}{13}\right)^2} = \frac{1}{13}\sqrt{169 - 25} = \frac{12}{13}.$$

(a)   From the double-angle formula (11) in the text,

$$\cos 2x = \cos^2 x - \sin^2 x = \left(-\frac{5}{13}\right)^2 - \left(\frac{12}{13}\right)^2 = -\frac{119}{169}.$$

(b)   From the double-angle formula (12) in the text,

$$\sin 2x = 2\sin x\cos x = 2\left(\frac{12}{13}\right)\left(-\frac{5}{13}\right) = -\frac{120}{169}.$$

(c)   Since $\tan 2x = \sin 2x / \cos 2x$, we have

$$\tan 2x = \frac{-120/169}{-119/169} = \frac{120}{119}.$$

36. If we let $x = \pi/4$, then $x/2 = \pi/8$ and formula (18) in the text yields

$$\sin^2 \frac{\pi}{8} = \frac{1}{2}\left(1 - \cos\frac{\pi}{4}\right) = \frac{1}{2}\left(1 - \frac{\sqrt{2}}{2}\right) = \frac{2 - \sqrt{2}}{4}.$$

Since $\pi/8$ is in the first quadrant, $\sin(\pi/8) > 0$ and $\sin(\pi/8) = \frac{1}{2}\sqrt{2 - \sqrt{2}}$.

39. If we let $\theta = 135°$, then $\theta/2 = 67.5°$ and formula (17) in the text yields

$$\cos^2 67.5° = \frac{1}{2}(1 + \cos 135°) = \frac{1}{2}\left[1 + \left(-\frac{\sqrt{2}}{2}\right)\right] = \frac{2 - \sqrt{2}}{4}.$$

Since $67.5°$ is in the first quadrant, $\cos 67.5° > 0$ and $\cos 67.5° = \frac{1}{2}\sqrt{2 - \sqrt{2}}$.

42. To find $\cot 157.5$ we use (19) of Theorem 4.6.5 with x replaced by $315°$,

$$\tan^2 157.5° = \tan^2\left(\frac{315°}{2}\right) = \frac{1 - \cos 315°}{1 + \cos 315°} = \frac{1 - (\frac{\sqrt{2}}{2})}{1 + (\frac{\sqrt{2}}{2})} = \frac{2 - \sqrt{2}}{2 + \sqrt{2}} \Rightarrow$$

$$\tan 157.5° = -\sqrt{\frac{2 - \sqrt{2}}{2 + \sqrt{2}}} \Rightarrow \cot 157.5° = -\sqrt{\frac{2 + \sqrt{2}}{2 - \sqrt{2}}} = -\sqrt{3 + 2\sqrt{2}}$$

after rationalizing the denominator. Alternatively, if we rationalize before taking the square root:

$$\tan^2 157.5° = \frac{2 - \sqrt{2}}{2 + \sqrt{2}} = \tfrac{1}{2}(2 - \sqrt{2})^2 \Rightarrow \tan 157.5° = -(\sqrt{2} - 1) \Rightarrow$$

$$\cot 157.5° = -\sqrt{2} - 1.$$

45. Since $\cos^2 x = 1 - \sin^2 x$ and $\sin x = \frac{12}{13}$, we have $\cos^2 x = 1 - \frac{144}{169} = \frac{25}{169}$. Because $\pi/2 < x < \pi$, x is in the second quadrant and $\cos x < 0$, so $\cos x = -\frac{5}{13}$. Using the half-angle formulas, we have

$$\cos^2 \frac{x}{2} = \frac{1}{2}\left(1 + \left(-\frac{5}{13}\right)\right) = \frac{1}{2}\left(\frac{8}{13}\right) = \frac{4}{13}$$

and

$$\sin^2 \frac{x}{2} = \frac{1}{2}\left(1 - \left(-\frac{5}{13}\right)\right) = \frac{1}{2}\left(\frac{18}{13}\right) = \frac{9}{13}.$$

Then, since $\pi/2 < x < \pi$, we have $\pi/4 < x/2 < \pi/2$, so $x/2$ is in the first quadrant where $\cos(x/2) > 0$ and $\sin(x/2) > 0$. Thus,

(a) $\cos\dfrac{x}{2} = \sqrt{\dfrac{4}{13}} = \dfrac{2\sqrt{13}}{13}$

(b) $\sin\dfrac{x}{2} = \sqrt{\dfrac{9}{13}} = \dfrac{3\sqrt{13}}{13}$

(c) $\tan \dfrac{x}{2} = \dfrac{\sin(x/2)}{\cos(x/2)} = \dfrac{3\sqrt{13}/13}{2\sqrt{13}/13} = \dfrac{3}{2}.$

48. From $\csc x = 9$ we infer that $\sin x = 1/\csc x = \frac{1}{9}$. Since $\cos^2 x = 1 - \sin^2 x$, we have $\cos^2 x = 1 - \frac{1}{81} = \frac{80}{81}$. Because $0 < x < \pi/2$, x is in the first quadrant and $\cos x > 0$, so $\cos x = 4\sqrt{5}/9$. Using the half-angle formulas, we have

$$\cos^2 \frac{x}{2} = \frac{1}{2}\left(1 + \frac{4\sqrt{5}}{9}\right) = \frac{9 + 4\sqrt{5}}{18}$$

and

$$\sin^2 \frac{x}{2} = \frac{1}{2}\left(1 - \frac{4\sqrt{5}}{9}\right) = \frac{9 - 4\sqrt{5}}{18}.$$

Then, since $0 < x < \pi/2$, we have $0 < x/2 < \pi/4$, so $x/2$ is in the first quadrant where $\cos(x/2) > 0$ and $\sin(x/2) > 0$. Thus,

(a) $\cos \dfrac{x}{2} = \sqrt{\dfrac{9 + 4\sqrt{5}}{18}}$

(b) $\sin \dfrac{x}{2} = \sqrt{\dfrac{9 - 4\sqrt{5}}{18}}$

(c) $\tan \dfrac{x}{2} = \dfrac{\sin(x/2)}{\cos(x/2)} = \dfrac{\sqrt{\dfrac{9 - 4\sqrt{5}}{18}}}{\sqrt{\dfrac{9 + 4\sqrt{5}}{18}}} = \dfrac{\sqrt{9 - 4\sqrt{5}}}{\sqrt{9 + 4\sqrt{5}}} = \sqrt{161 - 72\sqrt{5}}.$

51. Using the formula for the sine of a double angle we have

$$\sin 4x = 2\sin 2x \cos 2x = 2(2\sin x \cos x)(1 - 2\sin^2 x) = 4\sin x \cos x - 8\cos x \sin^3 x$$

$$= 4\cos x(\sin x - 2\sin^3 x)$$

54. Using the formula for the cosine of a double angle we have

$$\cos 2x = (\cos^2 x - \sin^2 x)1 = (\cos^2 x - \sin^2 x)(\cos^2 x + \sin^2 x) = \cos^4 x - \sin^4 x.$$

57. Using the double-angle formula for the sine we have

$$\frac{2\tan x}{1 + \tan^2 x} = \frac{2\tan x}{\sec^2 x} = 2\tan x \cos^2 x = 2\left(\frac{\sin x}{\cos x}\right)(\cos^2 x) = 2\sin x \cos x = \sin 2x$$

60. To obtain $\tan \dfrac{x}{2} = \dfrac{\sin x}{1 + \cos x}$ we first write

$$\tan \frac{x}{2} = \frac{\sin \dfrac{x}{2}}{\cos \dfrac{x}{2}} = \frac{2\cos \dfrac{x}{2} \sin \dfrac{x}{2}}{2\cos \dfrac{x}{2} \cos \dfrac{x}{2}} = \frac{\sin x}{2\cos^2 \dfrac{x}{2}} = \frac{\sin x}{2\left[\frac{1}{2}(1 + \cos x)\right]} = \frac{\sin x}{1 + \cos x}$$

63. $y = 2 \sin 2x \cos 2x = \sin 4x \Rightarrow$ amplitude is 1 and period is $\pi/2$

66. Since x_1 is in the second quadrant, $\cos x_1 < 0$. Using $\sin^2 x + \cos x^2 = 1$, we have

$$\cos x_1 = -\sqrt{1 - \frac{64}{289}} = -\frac{15}{17}.$$

Since x_2 is in the third quadrant, $\sec x_2 = 1/\cos x_2 < 0$. Using $1 + \tan^2 x = \sec^2 x$, we have

$$\sec x_2 = -\sqrt{1 + \frac{9}{16}} = -\frac{5}{4},$$

so $\cos x_2 = 1/\sec x_2 = -4/5$. Using $\tan x_2 = \sin x_2/\cos x_2$, we find that $\sin x_2 = -3/5$.

(a) $\sin(x_1 + x_2) = \sin x_1 \cos x_2 + \cos x_1 \sin x_2 = \left(\dfrac{8}{17}\right)\left(-\dfrac{4}{5}\right) + \left(-\dfrac{15}{17}\right)\left(-\dfrac{3}{5}\right) = \dfrac{13}{85}$

(b) $\sin(x_1 - x_2) = \sin x_1 \cos x_2 - \cos x_1 \sin x_2 = \left(\dfrac{8}{17}\right)\left(-\dfrac{4}{5}\right) - \left(-\dfrac{15}{17}\right)\left(-\dfrac{3}{5}\right) = -\dfrac{77}{85}$

(c) $\cos(x_1 + x_2) = \cos x_1 \cos x_2 - \sin x_1 \sin x_2 = \left(-\dfrac{15}{17}\right)\left(-\dfrac{4}{5}\right) - \left(\dfrac{8}{17}\right)\left(-\dfrac{3}{5}\right) = \dfrac{84}{85}$

(d) $\cos(x_1 - x_2) = \cos x_1 \cos x_2 + \sin x_1 \sin x_2 = \left(-\dfrac{15}{17}\right)\left(-\dfrac{4}{5}\right) + \left(\dfrac{8}{17}\right)\left(-\dfrac{3}{5}\right) = \dfrac{36}{85}$

Miscellaneous Applications

69. Referring to Figure 4.6.4 in the text, let h be the height of the triangle and let b represent the length of the base. Then, by the Pythagorean Theorem,

$$h = \sqrt{x^2 - \frac{b^2}{4}} = \sqrt{\frac{4x^2 - b^2}{4}} = \frac{1}{2}\sqrt{4x^2 - b^2}.$$

Then, letting $b = 2x \cos(\theta/2)$, we have

$$h = \frac{1}{2}\sqrt{4x^2 - \left(2x \cos \frac{\theta}{2}\right)^2} = \frac{1}{2}\sqrt{4x^2 \left(1 - \cos^2 \frac{\theta}{2}\right)}$$

$$= x\sqrt{\sin^2 \frac{\theta}{2}} = x \sin \frac{\theta}{2}.$$

Thus, the area of the triangle is

$$A = \frac{1}{2}bh = \frac{1}{2}\left(2x \cos \frac{\theta}{2}\right)\left(x \sin \frac{\theta}{2}\right) = \frac{1}{2}x^2 \left(2 \cos \frac{\theta}{2} \sin \frac{\theta}{2}\right)$$

$$= \frac{1}{2}x^2 \sin 2\left(\frac{\theta}{2}\right) = \frac{1}{2}x^2 \sin \theta.$$

4.7 | Product-to-Sum and Sum-to-Product Formulas

3. $\sin 2x \sin 5x = \frac{1}{2} \left[\cos(2x - 5x) - \cos(2x + 5x) \right] = \frac{1}{2} \left[\cos 3x - \cos 7x \right]$

6. $-\sin t \sin 2t = -\frac{1}{2} \left[\cos(t - 2t) - \cos(t + 2t) \right] = \frac{1}{2} \left[\cos 3t - \cos t \right]$

9. $2 \cos 3\beta \sin \beta = 2 \sin \beta \cos 3\beta = 2 \cdot \frac{1}{2} \left[\sin(\beta + 3\beta) + \sin(\beta - 3\beta) \right] = \sin 4\beta - \sin 2\beta$

12. $2 \sin \left(t + \dfrac{\pi}{2} \right) \cos \left(t - \dfrac{\pi}{2} \right) = 2 \cdot \dfrac{1}{2} \left[\sin \left(t + \dfrac{\pi}{2} + t - \dfrac{\pi}{2} \right) + \sin \left(t + \dfrac{\pi}{2} - t + \dfrac{\pi}{2} \right) \right]$

$\qquad = \sin 2t - \sin \pi = \sin 2t$

15. $\sin 75° \sin 15° = \frac{1}{2} \left[\cos 60° - \cos 90° \right] = \frac{1}{4}$

18. $\sin 105° \cos 195° = \frac{1}{2} \left[\sin(105° + 195°) + \sin(105° - 195°) \right]$

$\qquad = \frac{1}{2} \left[\sin 300° + \sin(-90°) \right] = -\frac{\sqrt{3}}{4} - \frac{1}{2}$

21. $\cos \dfrac{9x}{2} - \cos \dfrac{x}{2} = -2 \sin \dfrac{\dfrac{9x}{2} + \dfrac{x}{2}}{2} \sin \dfrac{\dfrac{9x}{2} - \dfrac{x}{2}}{2} = -2 \sin \dfrac{5x}{2} \sin 2x$

24. $\sin 5t + \sin 3t = 2 \sin \dfrac{5t + 3t}{2} \cos \dfrac{5t - 3t}{2} = 2 \sin 4t \sin t$

27. $-\frac{1}{2} \cos t + \frac{1}{2} \cos 5t = \frac{1}{2} \cos 5t - \frac{1}{2} \cos t = -\sin \dfrac{5t + t}{2} \sin \dfrac{5t - t}{2} = -\sin 3t \sin 2t$

30. $\cos \left(t + \dfrac{\pi}{2} \right) - \cos \left(t - \dfrac{\pi}{2} \right) = -2 \sin \dfrac{t + \dfrac{\pi}{2} + t - \dfrac{\pi}{2}}{2} \sin \dfrac{t + \dfrac{\pi}{2} - t + \dfrac{\pi}{2}}{2}$

$\qquad = -2 \sin t \sin \dfrac{\pi}{2} = -2 \sin t$

33. $\cos 105° - \cos 15° = -2 \sin \dfrac{105° + 15°}{2} \sin \dfrac{105° - 15°}{2} = -2 \cdot \dfrac{\sqrt{3}}{2} \cdot \dfrac{\sqrt{2}}{2} = -\dfrac{\sqrt{6}}{2}$

36. $2 \cos 195° - 2 \cos 105° = -4 \sin \dfrac{195° + 105°}{2} \sin \dfrac{195° - 105°}{2}$

$\qquad = -4 \sin 150° \sin 45° = -4 \cdot \dfrac{1}{2} \cdot \dfrac{\sqrt{2}}{2} = -\sqrt{2}$

Miscellaneous Applications

39. $\sin \omega t \sin(\omega t + \phi) = \frac{1}{2} \left[\cos(\omega t - \omega t - \phi) - \cos(\omega t + \omega t + \phi) \right] = \frac{1}{2} \left[\cos \phi - \cos(2\omega t + \phi) \right]$

4.8 | Inverse Trigonometric Functions

3. Letting $y = \arccos(-1)$, we have that $\cos y = -1$, and we must find y such that $0 \le y \le \pi$. Since $\cos \pi = -1$,

$$y = \arccos(-1) = \pi.$$

6. Letting $y = \arctan(-\sqrt{3})$, we have that $\tan y = -\sqrt{3}$, and we must find y such that $-\pi/2 < y < \pi/2$. Since $\tan(-\pi/3) = -\sqrt{3}$,

$$y = \arctan(-\sqrt{3}) = -\frac{\pi}{3}.$$

9. Letting $y = \tan^{-1} 1$, we have that $\tan y = 1$, and we must find y such that $-\pi/2 < y < \pi/2$. Since $\tan(\pi/4) = 1$,

$$y = \tan^{-1} 1 = \frac{\pi}{4}.$$

12. Letting $y = \arccos(-\frac{1}{2})$, we have that $\cos y = -\frac{1}{2}$, and we must find y such that $0 \le y \le \pi$. Since $\cos(2\pi/3) = -\frac{1}{2}$,

$$y = \arccos\left(-\frac{1}{2}\right) = \frac{2\pi}{3}.$$

15. Letting $t = \cos^{-1} \frac{3}{5}$, we want to find $\sin t$. Using $\cos t = \frac{3}{5}$ and $\sin^2 t + \cos^2 t = 1$ we see that

$$\sin^2 t + \left(\frac{3}{5}\right)^2 = 1 \quad \text{or} \quad \sin t = \sqrt{1 - \frac{9}{25}} = \sqrt{\frac{16}{25}} = \frac{4}{5}.$$

We used the positive square root because $\cos^{-1} \frac{3}{5}$ is in the first quadrant, so $\sin(\cos^{-1} \frac{3}{5})$ is positive. Thus, $\sin(\cos^{-1} \frac{3}{5}) = \frac{4}{5}$.

18. Letting $t = \arctan \frac{1}{4}$, we want to find $\sin t$. Using $\tan t = \frac{1}{4}$ and $1 + \tan^2 t = \sec^2 t$, we see that

$$1 + \left(\frac{1}{4}\right)^2 = \sec^2 t \quad \text{or} \quad \sec^2 t = \frac{17}{16},$$

so $\cos^2 t = 1/(17/16) = \frac{16}{17}$. Using $\sin^2 t + \cos^2 t = 1$, we have

$$\sin^2 t + \frac{16}{17} = 1 \quad \text{or} \quad \sin t = \sqrt{1 - \frac{16}{17}} = \frac{1}{\sqrt{17}}.$$

We used the positive square root because $\arctan \frac{1}{4}$ is in the first quadrant, so $\sin(\arctan \frac{1}{4})$ is positive. Thus, $\sin(\arctan \frac{1}{4}) = 1/\sqrt{17}$.

21. We use

$$\csc\left(\sin^{-1} \frac{3}{5}\right) = \frac{1}{\sin(\sin^{-1}(3/5))}.$$

Since $-1 \le \frac{3}{5} \le 1$, $\sin(\sin^{-1} \frac{3}{5}) = \frac{3}{5}$ by *(ii)* of the **Properties of Inverse Trigonometric Functions** in this section of the text and $\csc(\sin^{-1} \frac{3}{5}) = \frac{1}{3/5} = \frac{5}{3}$.

24. Since $-1 \le -\frac{4}{5} \le 1$, $\cos(\cos^{-1} -\frac{4}{5}) = -\frac{4}{5}$ by *(iv)* of the **Properties of Inverse Trigonometric Functions** in this section of the text.

27. Since $-\pi/2 \le \pi/16 \le \pi/2$, $\arcsin(\sin(\pi/16)) = \pi/16$ by (i) of the Properties of Inverse Trigono-metric Functions in this section of the text.

30. Since $-\pi/2 \le 5\pi/6 \le \pi/2$, $\sin^{-1}(\sin(5\pi/6)) = 5\pi/6$ by (i) of the Properties of Inverse Trigono-metric Functions in this section of the text.

33. Letting $t = \tan^{-1} x$, we want to find $\sin t$. Using $\tan t = x$ and $1 + \tan^2 t = \sec^2 t$, we see that $\sec^2 t = 1 + x^2$, so $\cos^2 t = 1/(1 + x^2)$. Using $\sin^2 t + \cos^2 t = 1$, we have

$$\sin^2 t + \frac{1}{1+x^2} = 1 \qquad \text{or} \qquad \sin t = \sqrt{1 - \frac{1}{1+x^2}} = \sqrt{\frac{x^2}{1+x^2}} = \frac{x}{\sqrt{1+x^2}}.$$

We used x in the numerator rather than $|x|$ because the range of $\tan^{-1} t$ is $(-\pi/2, \pi/2)$, and the sine function is positive for arguments in the first quadrant but negative for arguments in the fourth quadrant.

36. Using $\sec t = 1/\cos t$, we have

$$\sec(\arccos x) = \frac{1}{\cos(\arccos x)} = \frac{1}{x}$$

by (iv) of the Properties of Inverse Trigonometric Functions in this section of the text.

39. Letting $t = \arctan x$, we want to find $\csc t$. Since $\tan t = x$, $\cot t = 1/x$. Using $1 + \cot^2 t = \csc^2 t$, we have

$$1 + \frac{1}{x^2} = \csc^2 t \qquad \text{or} \qquad \csc t = \frac{\sqrt{1+x^2}}{x}.$$

We used x in the denominator rather than $|x|$ because the range of $\arctan t$ is $(-\pi/2, \pi/2)$, and the cosecant function is positive for arguments in the first quadrant but negative for arguments in the fourth quadrant.

42. This is the graph of $y = \arctan x$, first reflected through the x-axis, and then shifted upward by $\pi/2$ units.

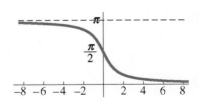

45. This is the graph of $y = \cos^{-1} x$, stretched vertically by a factor of two units.

48. Letting $t = \arcsin x$, we want to find $\cos t$. Using $\sin t = x$ and $\sin^2 t + \cos^2 t = 1$, we have

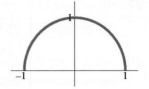

$$x^2 + \cos^2 t = 1 \qquad \text{or} \qquad \cos t = \sqrt{1 - x^2}.$$

We used the positive square root because $-\pi/2 \le t \le \pi/2$ and the cosine of an angle in the first or fourth quadrant is positive. Thus, $\cos(\arcsin x) = \sqrt{1 - x^2}$, which is the upper half of the unit circle centered at the origin.

51. The domain of $\operatorname{arcsec} x$ is $(-\infty, -1] \cup [1, \infty)$ and the range is $[0, \pi/2) \cup (\pi/2, \pi]$.

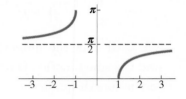

54. (a) $\csc(\operatorname{arccsc} x) = x$ for $x \le -1$ or $x \ge 1$
(b) $\operatorname{arccsc}(\csc x) = x$ for $-\pi/2 \le x < 0$ or $0 < x \le \pi/2$

These are obtained by simply using the domain of $\operatorname{arccsc} x$ in part **(a)** and the domain of $\csc x$ in part **(b)**.

57. Let $y = \arcsin(1/x)$, for $|x| \ge 1$. Then, $\sin y = 1/x$, where $-\pi/2 \le y \le \pi/2$, and $\csc y = x$. It follows that

$$\operatorname{arccsc} x = y = \arcsin\left(\frac{1}{x}\right).$$

60. $\csc^{-1}(-1.3) = \csc^{-1}\left(-\frac{13}{10}\right) = \sin^{-1}\left(-\frac{1}{13/10}\right) = \sin^{-1}\left(-\frac{10}{13}\right) \approx -0.8776$

63. $\operatorname{arcsec}(-1.2) = \operatorname{arcsec}\left(-\frac{6}{5}\right) = \arccos\left(-\frac{1}{6/5}\right) = \arccos\left(-\frac{5}{6}\right) \approx 2.5559$

Miscellaneous Applications

66. We have

$$\cos 2\theta = \frac{(9.81)(2.25)}{(13.7)^2 + (9.81)(2.25)} = \frac{22.0725}{209.7625} \approx 0.10523.$$

Then, $2\theta = \cos^{-1} 0.10523 \approx 1.46$, and

$$\theta \approx 0.73 \text{ radian} = 42°.$$

69. Referring to Figure 4.8.10 in the text, the triangle with height $H_{\text{CO}} = 0.18 W_{\text{CO}} = 0.18$ and angle ϕ has base $(W_{\text{CO}} - W_{\text{CR}})/2 = (1 - 0.4)/2 = 0.3$. Thus, $\tan \phi = 0.18/0.3$ and

$$\phi = \tan^{-1} 0.6 \approx 0.5404 \text{ radian} = 31°.$$

4.9 | Trigonometric Equations

3. If $\sec x = \sqrt{2}$, the reference angle for x is $\pi/4$. Since the value of $\sec x$ is positive, the terminal side of the angle lies in either the first or fourth quadrant. Thus, as shown in the figure, the only solutions between 0 and 2π are $x = \pi/4$ or $x = 7\pi/4$. Since the secant function is 2π-periodic, all of the remaining solutions can be obtained by adding integer multiples of 2π to these solutions:

$$x = \frac{\pi}{4} + 2n\pi \qquad \text{or} \qquad x = \frac{7\pi}{4} + 2n\pi,$$

where n is an integer.

Illustrations, like the one shown in the first solution of this section, will not be drawn for the remaining problems in this section.

6. If $\csc x = 2$, the reference angle for x is $\pi/6$. Since the value of $\csc x$ is positive, the terminal side of the angle lies in either the first or second quadrant. Thus, the only solutions between 0 and 2π are $x = \pi/6$ or $x = 5\pi/6$. Since the cosecant function is 2π-periodic, all of the remaining solutions can be obtained by adding integer multiples of 2π to these solutions:

$$x = \frac{\pi}{6} + 2n\pi \qquad \text{or} \qquad x = \frac{5\pi}{6} + 2n\pi,$$

where n is an integer.

9. Since $\tan x = 0$ for $x = 0$ and $x = \pi$, these are the only solutions of $\tan x = 0$ in $[0, 2\pi)$. The tangent function is π-periodic, so all of the remaining solutions can be obtained by adding integer multiples of π to $x = 0$. Thus, all solutions are $x = n\pi$, where n is an integer.

12. If $\sqrt{3} \cot x = 1$, then $\cot x = 1/\sqrt{3}$ and the reference angle for x is $\pi/3$. Since the value of $\cot x$ is positive, the terminal side of the angle lies in either the first or third quadrant. Thus, the only solutions between 0 and 2π are $x = \pi/3$ or $x = 4\pi/3$. Since the cotangent function is π-periodic, all of the remaining solutions can be obtained by adding integer multiples of π to these solutions:

$$x = \frac{\pi}{3} + n\pi,$$

where n is an integer.

15. If $1 + \cot \theta = 0$, then $\cot \theta = -1$ and the reference angle for θ is $135°$. Since the value of $\cot \theta$ is negative, the terminal side of the angle lies in either the second or fourth quadrant. Thus, the only solutions between $0°$ and $360°$ are $\theta = 135°$ or $\theta = 315°$. Since the cotangent function has period $180°$, all of the remaining solutions can be obtained by adding integer multiples of $180°$ to these solutions:

$$\theta = 135° + 180°n,$$

where n is an integer.

18. If $2\cos\theta + \sqrt{2} = 0$, then $\cos\theta = -\sqrt{2}/2$ and the reference angle for θ is 135°. Since the value of $\cos\theta$ is negative, the terminal side of the angle lies in either the second or third quadrant. Thus, the only solutions between 0° and 360° are $\theta = 135°$ or $\theta = 225°$. Since the cosine function has period 360°, all of the remaining solutions can be obtained by adding integer multiples of 360° to these solutions:

$$x = 135° + 360°n \qquad \text{or} \qquad x = 225° + 360°n,$$

where n is an integer.

21. Since $\sec x \geq 1$ for all x, we can divide the equation by $\sec x$, obtaining $3\sec x = 1$ or $\sec x = \frac{1}{3}$. But, again, since $\sec x \geq 1$ for all x, this equation has no solutions. Thus, the original equation, $3\sec^2 x = \sec x$, has no solutions.

24. Factoring the equation, we have $(2\sin\theta + 1)(\sin\theta - 1) = 0$, from which we conclude that $\sin\theta = -\frac{1}{2}$ and $\sin\theta = 1$. We first consider $\sin\theta = -\frac{1}{2}$. The reference angle for θ is 30°. Since the value of $\sin\theta$ is negative, the terminal side of the angle lies in either the third or fourth quadrants. Thus, the only solutions between 0° and 360° are $\theta = 210°$ or $\theta = 330°$. Since the sine function has period 360°, all of the remaining solutions can be obtained by adding integer multiples of 360° to these solutions:

$$\theta = 210° + 360°n \qquad \text{or} \qquad \theta = 330° + 360°n,$$

where n is an integer. Since $\sin\theta = 1$ for $0° \leq \theta \leq 360°$ only when $\theta = 90°$, the solutions of $\sin\theta = 1$ are $90° + 360°n$, where n is an integer. Therefore, all solutions of $2\sin^2\theta - \sin\theta - 1 = 0$ are

$$\theta = 210° + 360°n, \quad 330° + 360°n, \qquad \text{or} \qquad \theta = 90° + 360°n,$$

where n is an integer.

27. Since $\cos 2x = -1$ on $[0, 2\pi)$ only when $2x = \pi$, the solutions of $\cos 2x = -1$ are

$$2x = \pi + 2n\pi \qquad \text{or} \qquad x = \pi/2 + n\pi,$$

where n is an integer.

30. If $\tan 4\theta = -1$, the reference angle for 4θ is 135°. Since the value of $\tan 4\theta$ is negative, the terminal side of the angle lies in either the second or fourth quadrant. Thus, the only solutions between 0° and 360° are $4\theta = 135°$ or $4\theta = 315°$. Since the tangent function has period 180°, all of the remaining solutions can be obtained by adding integer multiples of 180° to these solutions:

$$4\theta = 135° + 180°n \qquad \text{or} \qquad \theta = 33.75° + 45°n,$$

where n is an integer.

33. Using the double-angle formula and then factoring, we have

$$\sin 2x + \sin x = 0$$
$$2 \sin x \cos x + \sin x = 0$$
$$\sin x (2 \cos x + 1) = 0.$$

Thus, $\sin x = 0$, so $x = nx$, where n is an integer, and $\cos x = -\frac{1}{2}$. In this case, the reference angle for x is $\pi/3$. Since the value of $\cos x$ is negative, the terminal side of the angle lies in either the second or third quadrant. Therefore, the only solutions between 0 and 2π are $x = 2\pi/3$ or $x = 4\pi/3$. Since the cosine function is 2π-periodic, all of the remaining solutions can be obtained by adding integer multiples of 2π to these solutions:

$$x = \frac{2\pi}{3} + 2n\pi \qquad \text{or} \qquad x = \frac{4\pi}{3} + 2n\pi,$$

where n is an integer. Hence, all solutions of $\sin 2x + \sin x = 0$ are

$$x = n\pi, \quad x = \frac{2\pi}{3} + 2n\pi, \quad \text{or} \quad x = \frac{4\pi}{3} + 2n\pi,$$

where n is an integer.

36. Using the double-angle formula and then factoring, we have

$$\sin 2\theta + 2 \sin \theta - 2 \cos \theta = 2$$
$$2 \sin \theta \cos \theta + 2 \sin \theta - 2 \cos \theta - 2 = 0$$
$$2 \sin \theta (\cos \theta + 1) - 2 (\cos \theta + 1) = 0$$
$$2 (\sin \theta - 1)(\cos \theta + 1) = 0.$$

Thus, $\sin \theta = 1$ and $\theta = 90° + 360°n$ or $\cos \theta = -1$ and $\theta = 180° + 360°n$, where n is an integer.

39. Using $\sec x = 1/\cos x$ and $\tan x = \sin x/\cos x$, we have

$$\sec x \sin^2 x = \tan x$$
$$\frac{\sin^2 x}{\cos x} = \frac{\sin x}{\cos x}$$
$$\sin^2 x = \sin x$$
$$\sin^2 x - \sin x = 0$$
$$\sin x (\sin x - 1) = 0.$$

We note that $\cos x \neq 0$, and hence $x \neq \pi/2 + n\pi$ where n is an integer, because the secant and tangent functions are not defined when $\cos x = 0$. Then $\sin x = 0$, so $x = n\pi$, and $\sin x = 1$, so $x = \pi/2 + 2n\pi$, where n is an integer. We have already seen that $x = \pi/2 + 2n\pi$ cannot be a solution, so the only solutions are $x = n\pi$, where n is an integer.

42. We rewrite $\sin x + \cos x = 0$ as $\sin x = -\cos x$. Since $\cos x$ and $\sin x$ are never 0 for the same value of x, we conclude that $\cos x \neq 0$ for any values of x that are solutions of the equation. Then $\sin x + \cos x = 0$ is equivalent to $\sin x / \cos x = \tan x = -1$. Thus, $x = 3\pi/4$ and, since the tangent function is π-periodic, the solutions are

$$x = \frac{3\pi}{4} + n\pi,$$

where n is an integer.

45. We begin by writing the equation in the form $\cos \theta = \sqrt{\cos \theta}$ and squaring both sides:

$$\cos \theta = \sqrt{\cos \theta}$$
$$\cos^2 \theta = \cos \theta$$
$$\cos^2 \theta - \cos \theta = 0$$
$$\cos \theta (\cos \theta - 1) = 0.$$

This implies $\cos \theta = 0$, so $\theta = 90° + 180°n$, and $\cos \theta = 1$, so $\theta = 360°n$, where n is an integer. Since we squared both sides of the equation, it is imperative that we check the answers. When $\theta = 90° + 180°n$,

$$\cos(90° + 180°n) = 0 = \sqrt{\cos(90° + 180°n)};$$

and when $\theta = 360°n$,

$$\cos(360°n) = 1 = \sqrt{\cos(360°n)}.$$

The solutions of $\cos \theta - \sqrt{\cos \theta} = 0$ are

$$\theta = 90° + 180°n \qquad \text{or} \qquad \theta = 360°n,$$

where n is an integer.

48. $\cos 2t + \cos 3t = 0, \quad [0, \ 2\pi)$

$\cos 2t + \cos 3t = 2 \cos \dfrac{5}{2} t \cos \dfrac{1}{2} t = 0 \Rightarrow \cos \dfrac{5}{2} t = 0 \Rightarrow$

$\dfrac{5}{2} t = \pi/2, \ 3\pi/2, \ 5\pi/2, \ 7\pi/2, \ 9\pi/2 \Rightarrow$

$t = \pi/5, \ 3\pi/5, \ \pi, \ 7\pi/5, \ 9\pi/5$

$\cos \dfrac{1}{2} t = 0 \Rightarrow \dfrac{1}{2} t = \pi/2 \Rightarrow t = \pi \Rightarrow$ solution set is

$\left\{ \pi/5, \ 3\pi/5, \ \pi, \ 7\pi/5, \ 9\pi/5 \right\}$

51. $\sin 7x - \sin x - 2 \sin 3x = 0, \quad (-\pi, \ \pi)$

$\sin 7x - \sin x - 2 \sin 3x = 2 \cos 4x \sin 3x - 2 \sin 3x = 2 \sin 3x (\cos 4x - 1) = 0 \Rightarrow$

$\sin 3x = 0 \Rightarrow 3x = 0, \ \pm\pi, \ \pm 2\pi \Rightarrow x = 0, \ \pm\pi/3, \ \pm 2\pi/3$ or $\cos 4x = 1 \Rightarrow$

$4x = 0, \ \pm 2\pi \Rightarrow x = 0, \ \pm\pi/2 \Rightarrow$ solution set is

$\{0, \ \pm\pi/3, \ \pm 2\pi/3, \ \pm\pi/2 \}$

54. Solving $f(x) = 2\cos(x + \pi/4) = 0$ we obtain $x + \pi/4 = \pi/2 + n\pi$, or $x = \pi/4 + n\pi$, where n is an integer. The first three positive x-intercepts are at $\pi/4$, $\pi/4 + \pi$, and $\pi/4 + 2\pi$, so the intercepts are $(\pi/4, 0)$, $(5\pi/4, 0)$, and $(9\pi/4, 0)$.

57. We need to solve $f(x) = \sin x + \tan x = 0$. Writing $\tan x = \sin x/\cos x$ and factoring, we have

$$\sin x + \frac{\sin x}{\cos x} = 0$$

$$\sin x \left(1 + \frac{1}{\cos x}\right) = 0$$

so $\sin x = 0$ and $\cos x = -1$. Thus, solutions are $x = n\pi$ or $x = (2n + 1)\pi$, where n is an integer. That is, the set of solutions is $x = n\pi$, where n is an integer. The first three positive x-intercepts are $(\pi, 0)$, $(2\pi, 0)$, and $(3\pi, 0)$.

60. Using $\cos 3x = \cos(x + 2x) = \cos x \cos 2x - \sin x \sin 2x$, we have

$$f(x) = \cos x + \cos 3x = \cos x + \cos x \cos 2x - \sin x \sin 2x$$
$$= \cos x(1 + \cos 2x) - \sin x(2 \sin x \cos x)$$
$$= \cos x(1 + \cos 2x - 2\sin^2 x) = \cos[1 + \cos 2x - 2(1 - \cos^3 x)]$$
$$= \cos x(\cos 2x + 2\cos^2 x - 1) = \cos x(2\cos 2x) = 2\cos x \cos 2x.$$

Thus, $\cos x = 0$, so $x = \pi/2 + n\pi$, or $\cos 2x = 0$, so $2x = \pi/2 + m\pi$ and $x = \pi/4 + m\pi/2$, where n and m are integers. This means that the first three positive zeros of $f(x)$ are $\pi/4$, $\pi/2$, and $3\pi/4$. Here we use $m = 0$, $n = 0$, and $m = 1$, respectively.

63. Since $\cot x - x = 0$ is equivalent to $\cot x = x$, we see from the graphs of $y = \cot x$ and $y = x$ that there are infinitely many solutions of $\cot x - x = 0$.

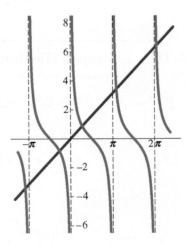

66. The equation $3\sin^2 x - 8\sin x + 4 = 0$ is quadratic in $\sin x$. Factoring, we obtain

$$(3\sin x - 2)(\sin x - 2) = 0,$$

so $\sin x = \frac{2}{3}$ or $\sin x = 2$, which we disregard since $\sin x$ cannot be greater than 1. Thus, $x = \sin^{-1}(\frac{2}{3}) \approx 0.73$.

69. Factoring, we obtain

$$\cos x(5\cos^2 x - 3\cos x - 1) = 0.$$

The equation $5\cos^2 x - 3\cos x - 1 = 0$ is quadratic in $\cos x$. Since it does not factor with integer coefficients, we apply the quadratic formula to obtain

$$\cos x = \frac{3 \pm \sqrt{(-3)^2 - 4(5)(-1)}}{2(5)} = \frac{3}{10} \pm \frac{\sqrt{29}}{10}.$$

The solutions of the original equation are

$$\cos^{-1}(0) = \frac{\pi}{2} \approx 1.57, \quad \cos^{-1}\left(\frac{3 + \sqrt{29}}{10}\right) \approx 0.58, \quad \text{and} \quad \cos^{-1}\left(\frac{3 - \sqrt{29}}{10}\right) \approx 1.81.$$

Miscellaneous Applications

72. We need to determine when $y = 8\cos(\pi t - \pi/12) = 0$. This will occur when

$$\pi t - \pi/12 = \pi t - \frac{\pi}{12} = \frac{\pi}{2} + n\pi \qquad \text{or} \qquad t = \frac{7}{12} + n \text{ seconds,}$$

where n is a nonnegative integer (assuming t must be positive).

75. We need to solve $V = 110\sin(120\pi t - \pi/6) = 0$. This will occur when

$$120\pi t - \frac{\pi}{6} = n\pi \qquad \text{or} \qquad t = \frac{1}{720} + \frac{n}{120} \text{ seconds,}$$

where n is a nonnegative integer (assuming t must be positive).

4.10   Simple Harmonic Motion

3. For $y = \sqrt{3}\sin 2x + \cos 2x$ we identify in (2) in the text, $c_1 = 1$, $c_2 = \sqrt{3}$, and $B = 2$. Then, we have

$$A = \sqrt{c_1^2 + c_2^2} = \sqrt{1^2 + (\sqrt{3})^2} = 2$$

and

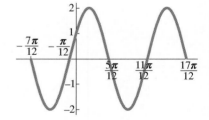

$$\sin \phi = \frac{1}{2}, \qquad \cos \phi = \frac{\sqrt{3}}{2}.$$

From the last two equations, we conclude that ϕ can be taken to be the first quadrant angle $\pi/6$. Therefore,

$$y = \sqrt{3}\sin 2x + \cos 2x = 2\sin\left(2x + \frac{\pi}{6}\right).$$

The amplitude is 2, the period is $2\pi/2 = \pi$, and the phase shift is $\pi/12$ units to the left.

6. For $y = \sin x + \cos x$ we identify in (2) in the text, $c_1 = 1$, $c_2 = 1$, and $B = 1$. Then, we have

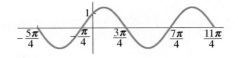

$$A = \sqrt{1^2 + 1^2} = \sqrt{2}$$

and

$$\sin\phi = \frac{1}{\sqrt{2}} = \frac{\sqrt{2}}{2}, \qquad \cos\phi = \frac{1}{\sqrt{2}} = \frac{\sqrt{2}}{2}.$$

From the last two equations, we conclude that ϕ can be taken to be the first quadrant angle $\pi/4$. Therefore,

$$y = \sqrt{2}\sin\left(x + \frac{\pi}{4}\right).$$

The amplitude is $\sqrt{2}$, the period is 2π, and the phase shift is $\pi/4$ units to the left.

9. We begin with the equation for simple harmonic motion (5) in the text. Since $y_0 = -1$, $\omega = \sqrt{k/m} = \sqrt{16/1} = 4$, and $v_0/\omega = -2/4 = -\frac{1}{2}$, equation (5) becomes

$$y(t) = -\cos 4t - \frac{1}{2}\sin 4t.$$

The period of motion is $2\pi/4 = \pi/2$ seconds and the frequency is $4/2\pi = 2/\pi$ cycles per second. With $c_1 = -1$ and $c_2 = -\frac{1}{2}$, we find the amplitude of motion is

$$A = \sqrt{(-1)^2 + \left(-\frac{1}{2}\right)^2} = \sqrt{\frac{5}{4}} = \frac{\sqrt{5}}{2} \text{ ft.}$$

Then $\tan\phi = -1/(-\frac{1}{2}) = 2$. Because $c_1 < 0$ and $c_2 < 0$, $\sin\phi < 0$ and $\cos\phi < 0$, so we see that ϕ lies in the third quadrant. However, because the range of the inverse tangent function is $(-\pi/2, \pi/2)$, $\tan^{-1} 2 = 1.10715$ is a first quadrant angle. The correct angle is found by using the reference angle 1.10715 for $\tan^{-1} 2$ to find the third quadrant angle

$$\phi = \pi + 1.10715 = 4.24874 \text{ radians.}$$

Using (7) in the text, the equation of motion is then

$$y(t) = \frac{\sqrt{5}}{2}\sin(4t + 4.24874).$$

12. Using Problem 11, the mass will pass through the equilibrium position when

$$y(t) = \frac{5}{2}\sin\left(2t - \frac{\pi}{3}\right) = 0 \qquad \text{or} \qquad 2t - \frac{\pi}{3} = n\pi, \; n \text{ an integer.}$$

Solving for t we have $t = \pi/6 + n\pi/2$, where n is an integer.

4.11 $\displaystyle\int$ **Calculus** **The Limit Concept Revisited**
PREVIEW

3. Since the sine is an odd function, $\sin(-\theta) = -\sin\theta$ and

$$\lim_{\theta\to 0}\frac{\sin(-\theta)}{\theta} = \lim_{\theta\to 0}\left(-\frac{\sin\theta}{\theta}\right) = -\lim_{\theta\to 0}\frac{\sin\theta}{\theta} = -1.$$

6. Since the sine function is continuous at $x = \pi/4$,

$$\lim_{x\to\pi/4}\sin x = \sin\frac{\pi}{4} = \frac{1}{\sqrt{2}}.$$

9. By (9) in the text,

$$\lim_{x\to 0}\frac{\cos x - 1}{10x} = \frac{1}{10}\lim_{x\to 0}\frac{\cos x - 1}{x} = \frac{1}{10}(0) = 0.$$

12. By (7) and (9) in the text,

$$\lim_{x\to 0}\frac{2\sin 4x + 1 - \cos x}{x} = 2\lim_{x\to 0}\frac{\sin 4x}{x} + \lim_{x\to 0}\frac{1 - \cos x}{x} = 2(4) + 0 = 8.$$

15. We use a trigonometric identity and (2) in the text:

$$\lim_{x\to\pi/2}\frac{\cos x}{\cot x} = \lim_{x\to\pi/2}\frac{\cos x}{\cos x/\sin x} = \lim_{x\to\pi/2}\sin x = \sin\frac{\pi}{2} = 1.$$

18. We use the double-angle formula for the cosine followed by (2) and (3) in the text:

$$\lim_{x\to\pi/4}\frac{\cos 2x}{\cos x - \sin x} = \lim_{x\to\pi/4}\frac{\cos^2 x - \sin^2 x}{\cos x - \sin x} = \lim_{x\to\pi/4}\frac{(\cos x + \sin x)(\cos x - \sin x)}{\cos x - \sin x}$$

$$= \lim_{x\to\pi/4}(\cos x + \sin x) = \cos\frac{\pi}{4} + \sin\frac{\pi}{4} = \frac{1}{\sqrt{2}} + \frac{1}{\sqrt{2}} = \frac{2}{\sqrt{2}}.$$

21. Since $f(\pi/6) = \sin(\pi/6) = \frac{1}{2}$, the point of tangency is $(\pi/6, 1/2)$. By (12) in the text, the derivative of $f(x) = \sin x$ is $f'(x) = \cos x$, so the slope of the tangent line at the point of tangency is

$$f'\left(\frac{\pi}{6}\right) = \cos\frac{\pi}{6} = \frac{\sqrt{3}}{2}.$$

Using the point-slope form of a line, an equation of the tangent line is

$$y - \frac{1}{2} = \frac{\sqrt{3}}{2}\left(x - \frac{\pi}{6}\right) \qquad\text{or}\qquad y = \frac{\sqrt{3}}{2}x - \frac{\sqrt{3}\pi}{12} + \frac{1}{2}.$$

24. Since $f(\pi/3) = \cos(\pi/3) = \frac{1}{2}$, the point of tangency is $(\pi/3, 1/2)$. By the result of Problem 23, the derivative of $f(x) = \cos x$ is $f'(x) = -\sin x$, so the slope of the tangent line at the point of tangency is

$$f'\left(\frac{\pi}{3}\right) = -\sin\left(\frac{\pi}{3}\right) = -\frac{\sqrt{3}}{2}.$$

Using the point-slope form of a line, an equation of the tangent line is

$$y - \frac{1}{2} = -\frac{\sqrt{3}}{2}\left(x - \frac{\pi}{3}\right) \qquad\text{or}\qquad y = -\frac{\sqrt{3}}{2}x + \frac{\sqrt{3}\pi}{6} + \frac{1}{2}.$$

Chapter 4 Review Exercises

A. Fill in the Blanks

3. The coordinates are $x = \cos t$ and $y = \sin t$. For $t = 5\pi/6$, $\cos(5\pi/6) = -\sqrt{3}/2$ and $\sin(5\pi/6) = \frac{1}{2}$. The point is $(-\sqrt{3}/2, 1/2)$.

6. Clearly $\pi < 8\pi/5 < 2\pi$, so the angle is in the third or fourth quadrant and we want to compare it to $3\pi/2$. Putting both numbers over a common denominator,

$$\frac{8\pi}{5} = \frac{16\pi}{10} \quad \text{and} \quad \frac{3\pi}{2} = \frac{15\pi}{10},$$

we see that $8\pi/5 > 3\pi/2$, so $8\pi/5$ is in the fourth quadrant.

9. The y-intercept occurs where $x = 0$, and

$$2\sec(0 + \pi) = 2\sec\pi = 2(-1) = -2,$$

so the y-intercept is $(0, -2)$.

12. Since $\cos^2(t/2) = \frac{1}{2}(1 + \cos t)$, for $\cos t = -\frac{2}{3}$ we have

$$\cos^2\frac{t}{2} = \frac{1}{2}\left(1 - \frac{2}{3}\right) = \frac{1}{6},$$

so $\cos t/2 = 1/\sqrt{6}$ or $\cos t/2 = -1/\sqrt{6}$. Since t is in the third quadrant where $\cos t$ is negative, $\cos t/2 = -1/\sqrt{6}$.

15. Using the formula for the sum of the sine of two angles, we have

$$\sin\left(t + \frac{\pi}{4}\right) = \sin t \cos\frac{\pi}{4} + \cos t \sin\frac{\pi}{4} = (\sin t)\frac{1}{\sqrt{2}} + (\cos t)\frac{1}{\sqrt{2}}$$

$$= \frac{1}{\sqrt{2}}(\sin t + \cos t).$$

Thus,

$$\sin t + \cos t = \sqrt{2}\sin\left(t + \frac{\pi}{4}\right).$$

18. Using the formula for the difference of the cosine of two angles, we have

$$\cos\left(\frac{\pi}{6} - \frac{5\pi}{4}\right) = \left(\cos\frac{\pi}{6}\right)\left(\cos\frac{5\pi}{4}\right) + \left(\sin\frac{\pi}{6}\right)\left(\sin\frac{5\pi}{4}\right)$$

$$= \frac{\sqrt{3}}{2}\left(-\frac{1}{\sqrt{2}}\right) + \frac{1}{2}\left(-\frac{1}{\sqrt{2}}\right) = -\frac{\sqrt{3}+1}{2\sqrt{2}}.$$

21. The amplitude of the function $y = \sin x + 2\sqrt{2}\cos x$ is 3.

$$A = \sqrt{1^2 + (2\sqrt{2})^2} = 3$$

24. The exact value of $\cos 70° \cos 40° + \sin 70° \sin 40°$ is $\frac{\sqrt{3}}{2}$.

$$\cos(70° - 40°) = \cos 70° \cos 40° + \sin 70° \sin 40° = \cos 30° = \frac{\sqrt{3}}{2}$$

B. True/False

3. True, because

$$\sec(-\pi) = \sec(-\pi + 2\pi) \qquad \text{(the secant function is } 2\pi\text{-periodic)}$$
$$= \sec \pi = \frac{1}{\cos \pi} = -1,$$

and

$$\csc\left(\frac{3\pi}{2}\right) = \frac{1}{\sin(3\pi/2)} = \frac{1}{-1} = -1.$$

6. False; $\tan^2 \theta = \sec^2 \theta - 1$

9. True, because $\csc x = 1/\sin x$, and the range of $\sin x$ is $[1, 1]$.

12. True; since the tangent function is π-periodic.

15. False. The graph of $y = \sin(2x - \pi/3)$ is the graph of $y = \sin x$ shifted $(\pi/3)/2 = \pi/6$ units to the right.

18. True; it's domain is $[-1, 1]$, and a periodic function must have a domain of infinite length.

21. This is true because both are 0.

24. False because $\cos \frac{x}{2} \neq \frac{1}{2} \cos x$.

C. Review Exercises

3. An example is $3 \sin(Bx + C)$. Since the period of this function is $2\pi/B$, we want $2\pi/B = \pi/2$ or $B = 4$. The phase shift is $|C|/B = |C|/4 = \pi/4$ or $|C| = \pi$. Thus, two examples are $y = \pm 3 \sin(4x + \pi)$.

6. Writing the equation as $\cos t = \sin t$ we see that $1 = \sin t / \cos t = \tan t$. Thus, $t = \pi/4$ or $t = 5\pi/4$ in $[0, 2\pi]$ since $\tan t$ is positive in the first and third quadrants.

9. From $\cos 4t = -1$ we have $4t = \pi$ or $t = \pi/4$. Since $\cos 4t$ has period $\pi/2$, it will also be -1 when t is $\pi/4 + \pi/2 = 3\pi/4$.

12. $\cos 9t - \cos 3t = 0 \Rightarrow$
$\cos 9t - \cos 3t = -2 \sin 6t \sin 3t = 0 \Rightarrow \sin 6t = 0 \Rightarrow$
$6t = 0,\ \pi,\ 2\pi,\ 3\pi,\ 4\pi,\ 5\pi,\ 6\pi, 7\pi,\ 8\pi,\ 9\pi,\ 10\pi,\ 11\pi,\ 12\pi \Rightarrow$
$t = 0,\ \pi/6,\ \pi/3,\ \pi/2,\ 2\pi/3,\ 5\pi/6,\ \pi, 7\pi/6,\ 4\pi/3,\ 3\pi/2,\ 5\pi/3,\ 11\pi/6,\ 2\pi$
or $\sin 3t = 0 \Rightarrow 3t = 0,\ \pi,\ 2\pi,\ 3\pi,\ 4\pi,\ 5\pi,\ 6\pi \Rightarrow t = 0, \pi/3, \pi, 4\pi/3, 5\pi/3,\ 2\pi$
so the solution set is
$\{0,\ \pi/6,\ \pi/3,\ \pi/2,\ 2\pi/3,\ 5\pi/6,\ \pi, 7\pi/6,\ 4\pi/3,\ 3\pi/2,\ 5\pi/3,\ 11\pi/6,\ 2\pi\}$

15. The range of the arccos function is $[0, \pi]$ and $\cos t$ is negative in the second quadrant, so

$$\cos^{-1}\left(-\frac{1}{2}\right) = \pi - \frac{\pi}{3} = \frac{2\pi}{3}.$$

18. Letting $t = \arcsin\frac{2}{5}$, we want to find $\cos t$. Using $\sin t = \frac{2}{5}$ and $\sin^2 t + \cos^2 t = 1$ we see that

$$\left(\frac{2}{5}\right)^2 + \cos^2 t = 1 \qquad \text{or} \qquad \cos t = \sqrt{1 - \frac{4}{25}} = \sqrt{\frac{21}{25}} = \frac{\sqrt{21}}{5}.$$

We used the positive square root because $\arcsin\frac{2}{5}$ is in the first quadrant, so $\cos(\arcsin\frac{2}{5})$ is positive. Thus, $\cos(\arcsin\frac{2}{5}) = \sqrt{21}/5$.

21. Letting $t = \arccos\frac{5}{13}$, we want to find $\sin t$. Using $\cos t = \frac{5}{13}$ and $\sin^2 t + \cos^2 t = 1$, we see that

$$\sin^2 t + \left(\frac{5}{13}\right)^2 = 1 \qquad \text{or} \qquad \sin t = \sqrt{1 - \frac{25}{169}} = \sqrt{\frac{144}{169}} = \frac{12}{13}.$$

We used the positive square root because $\arccos\frac{5}{13}$ is in the first quadrant, so $\sin(\arccos\frac{5}{13})$ is positive. Thus, $\sin(\arccos\frac{5}{13}) = \frac{12}{13}$.

24. Letting $t = \tan^{-1} x$, we want to find $\sec t$. Using $\tan t = x$ and $1 + \tan^2 t = \sec^2 t$, we see that

$$1 + x^2 = \sec^2 t \qquad \text{or} \qquad \sec t = \sqrt{1 + x^2}.$$

Thus, $\sec(\tan^{-1} x) = \sqrt{1 + x^2}$. (We were able to use the positive square root because for $x \geq 0$, $0 \leq \tan^{-1} x < \pi/2$, and $t = \tan^{-1} x$ is in the first quadrant where $\sec t > 0$. On the other hand, for $x < 0$, $-\pi/2 < \tan^{-1} x < 0$, and $t = \tan^{-1} x$ is in the fourth quadrant where $\sec t > 0$.)

27. We note that the graph has amplitude $\frac{1}{2}$ and period 2π. We also note that it looks like the graph of $y = \cos x$. Since the graph is shifted up by 1 unit, it is $y = 1 + \frac{1}{2}\cos x$. Since $\cos x = \sin(\pi/2 - x)$, [see (16) in Section 4.2 of the text] we have

$$y = 1 + \frac{1}{2}\sin\left(x + \frac{\pi}{2}\right).$$

30. $\cos^4 t + 1 - \sin^4 t = 2\cos^2 t \Rightarrow$ LHS is

$$(1 - \sin^2 t)^2 + 1 - \sin^4 t = 1 - 2\sin^2 t + \sin^4 t + 1 - \sin^4 t = 2(1 - \sin^2 t) = 2\cos^2 t$$

Chapter 5

Triangle Trigonometry

5.1 | **Right Triangle Trigonometry**

3. The hypotenuse c is given by

$$c^2 = 2^2 + 6^2 = 40 \qquad \text{or} \qquad c = \sqrt{40} = 2\sqrt{10}.$$

Then,

$$\sin\theta = \frac{\text{opp}}{\text{hyp}} = \frac{6}{2\sqrt{10}} = \frac{3}{\sqrt{10}} \qquad\qquad \cos\theta = \frac{\text{adj}}{\text{hyp}} = \frac{2}{2\sqrt{10}} = \frac{1}{\sqrt{10}}$$

$$\tan\theta = \frac{\text{opp}}{\text{adj}} = \frac{6}{2} = 3 \qquad\qquad \cot\theta = \frac{\text{adj}}{\text{opp}} = \frac{2}{6} = \frac{1}{3}$$

$$\sec\theta = \frac{\text{hyp}}{\text{adj}} = \frac{2\sqrt{10}}{2} = \sqrt{10} \qquad\qquad \csc\theta = \frac{\text{hyp}}{\text{opp}} = \frac{2\sqrt{10}}{6} = \frac{\sqrt{10}}{3} \;.$$

6. The side a opposite the angle θ is given by

$$a^2 + 1^2 = (\sqrt{5})^2 \qquad \text{or} \qquad a = \sqrt{5-1} = 2.$$

Then

$$\sin\theta = \frac{\text{opp}}{\text{hyp}} = \frac{2}{\sqrt{5}} \qquad\qquad \cos\theta = \frac{\text{adj}}{\text{hyp}} = \frac{1}{\sqrt{5}}$$

$$\tan\theta = \frac{\text{opp}}{\text{adj}} = \frac{2}{1} = 2 \qquad\qquad \cot\theta = \frac{\text{adj}}{\text{opp}} = \frac{1}{2}$$

$$\sec\theta = \frac{\text{hyp}}{\text{adj}} = \frac{\sqrt{5}}{1} = \sqrt{5} \qquad\qquad \csc\theta = \frac{\text{hyp}}{\text{opp}} = \frac{\sqrt{5}}{2} \;.$$

9. The hypotenuse c is given by

$$c^2 = x^2 + y^2 \qquad \text{or} \qquad c = \sqrt{x^2 + y^2}.$$

194

Then

$$\sin\theta = \frac{\text{opp}}{\text{hyp}} = \frac{y}{\sqrt{x^2 + y^2}} \qquad\qquad \cos\theta = \frac{\text{adj}}{\text{hyp}} = \frac{x}{\sqrt{x^2 + y^2}}$$

$$\tan\theta = \frac{\text{opp}}{\text{adj}} = \frac{y}{x} \qquad\qquad \cot\theta = \frac{\text{adj}}{\text{opp}} = \frac{x}{y}$$

$$\sec\theta = \frac{\text{hyp}}{\text{adj}} = \frac{\sqrt{x^2 + y^2}}{x} \qquad\qquad \csc\theta = \frac{\text{hyp}}{\text{opp}} = \frac{\sqrt{x^2 + y^2}}{y}\,.$$

12. Since $c = 10$ and $\beta = 49°$, we have

$$\sin\beta = \frac{b}{c} \qquad \text{so} \qquad b = c\sin\beta = 10\sin 49° \approx 7.55$$

and

$$\cos\beta = \frac{a}{c} \qquad \text{so} \qquad a = c\cos\beta = 10\cos 49° \approx 6.56.$$

15. Since $b = 1.5$ and $c = 3$, we have

$$\cos\alpha = \frac{b}{c} = \frac{1.5}{3} = \frac{1}{2} \qquad \text{so} \qquad \alpha = \cos^{-1}\frac{1}{2} = 60°.$$

Then $\beta = 90° - \alpha = 90° - 60° = 30°$, and

$$a^2 + b^2 = c^2 \qquad \text{so} \qquad a = \sqrt{c^2 - b^2} = \sqrt{3^2 - (1.5)^2} \approx 2.60.$$

18. Since $b = 4$ and $\alpha = 58°$, we have

$$\tan\alpha = \frac{a}{b} \qquad \text{so} \qquad a = b\tan\alpha = 4\tan 58° \approx 6.40$$

and

$$\cos\alpha = \frac{b}{c} \qquad \text{so} \qquad c = \frac{b}{\cos\alpha} = \frac{4}{\cos 58°} \approx 7.55.$$

21. Since $b = 20$ and $\alpha = 23°$, we have

$$\tan\alpha = \frac{a}{b} \qquad \text{so} \qquad a = b\tan\alpha = 20\tan 23° \approx 8.49$$

and

$$\cos\alpha = \frac{b}{c} \qquad \text{so} \qquad c = \frac{b}{\cos\alpha} = \frac{20}{\cos 23°} \approx 21.73.$$

24. From the right triangle containing the $46°$ angle,

$$\tan 46° = \frac{x}{y} \Rightarrow \frac{y}{x} = \cot 46° \Rightarrow y = x\cot 46°$$

Also, from the right triangle containing the $34°$ angle,

$$\tan 34° = \frac{x}{y + 10} \Rightarrow x = y\tan 34° + 10\tan 34° = x\cot 46°\tan 34° + 10\tan 34° \Rightarrow$$

$$x(1 - \cot 46°\tan 34°) = 10\tan 34° \Rightarrow$$

$$x = \frac{10\tan 34°}{1 - \cot 46°\tan 34°} \approx 19.35$$

5.2 | Applications of Right Triangles

3. Let b denote the width of the river. Then

$$\tan 37° = \frac{50}{b} \quad \text{so} \quad b = \frac{50}{\tan 37°} \approx 66.4 \text{ ft.}$$

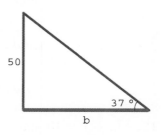

6. Let h denote the height of the mountain and let x denote the base of the $45°$ triangle in Figure 5.2.9 in the text. From this triangle we see that

$$\tan 45° = \frac{h}{x} \quad \text{so} \quad h = x \tan 45° = x(1) = x.$$

From the larger triangle, we have

$$\tan 28° = \frac{h}{x+1} = \frac{h}{h+1} \quad \text{so} \quad h = \frac{\tan 28°}{1 - \tan 28°} \approx 1.1 \text{ km.}$$

9. Using the fact that 1 mile is 5280 feet, we have that the height of the hill is $\frac{4000}{5280} = \frac{25}{33}$ mile. From the figure, we see that

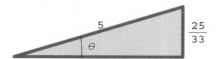

$$\sin \theta = \frac{25/33}{5} = \frac{5}{33} \approx 0.1515,$$

so $\theta = 8.7°$.

12. Let x denote the distance between the shore and the highest point of the bridge when it is completely open. (See Figure 5.2.12(a) in the text.) Then

$$\sin 43° = \frac{x}{7.5} \quad \text{so} \quad x = 7.5 \sin 43° \approx 5.11.$$

Then the distance from the shore to a point on the surface below is given by $d - 5.11$. From Figure 5.2.12(b) in the text, we see that

$$\tan 27° = \frac{d - 5.11}{7.5} \quad \text{or} \quad d = 7.5 \tan 27° + 5.11 \approx 8.93 \text{ m.}$$

15. Let h denote the altitude of the airplane in Figure 5.2.14 in the text. Then

$$\tan 1° = \frac{231/2}{h} \quad \text{so} \quad h = \frac{115.5}{\tan 1°} \approx 6617 \text{ ft.}$$

18. Let h denote the cloud ceiling. Then from Figure 5.2.15 in the text, we see that

$$\tan 8° = \frac{h}{1} \quad \text{so} \quad h = \tan 8° \approx 0.14 \text{ km.}$$

21. From Figure 5.2.17 in the text we see that

$$\tan 1° = \frac{3963}{d} \qquad \text{so} \qquad d = \frac{3963}{\tan 1°} \approx 227,000 \text{ miles.}$$

24. From Figure 5.2.20 in the text, we see that

$$\sin 28° = \frac{200}{d_1} \qquad \text{so} \qquad d_1 = \frac{200}{\sin 28°} \approx 426 \text{ ft.}$$

Let x denote the base of the triangle with hypotenuse d_1, as seen in the figure to the right. Then

$$\tan 28° = \frac{200}{x} \qquad \text{so} \qquad x = \frac{200}{\tan 28°} \approx 376 \text{ ft.}$$

Then, we see that

$$\tan 15° = \frac{200}{d_2 + x} = \frac{200}{d_2 + 376} \qquad \text{so} \qquad d_2 = \frac{200}{\tan 15°} - 376 \approx 370 \text{ ft.}$$

Thus, the total distance that the soapbox travels is

$$d_1 + d_2 = 426 + 370 = 796 \text{ ft.}$$

27. The vertical height of the helicopter is 12,442 m, and the distance from a point P on the ground to a point directly beneath the helicopter is 2000 m. Thus

$$\tan \theta = \frac{12,442}{2000} \qquad \text{and} \qquad \theta = \tan^{-1} \frac{12,442}{2000} \approx 80.87°.$$

30. Referring to Figure 5.2.24 in the text, we see that

$$\cos \theta = \frac{1/2}{d} \qquad \text{so} \qquad d = \frac{1/2}{\cos \theta} = \frac{1}{2} \sec \theta.$$

5.3 $\boxed{\textbf{Law of Sines}}$

3. With $\beta = 37°$ and $\gamma = 51°$ it follows that $\alpha = 180° - 37° - 51° = 92°$. Using the Law of Sines, we have

$$\frac{\sin \alpha}{a} = \frac{\sin \beta}{b} \qquad \text{or} \qquad \frac{\sin 92°}{5} = \frac{\sin 37°}{b}$$

so

$$b = \frac{5 \sin 37°}{\sin 92°} \approx 3.01.$$

Also,

$$\frac{\sin\alpha}{a} = \frac{\sin\gamma}{c} \qquad \text{or} \qquad \frac{\sin 92°}{5} = \frac{\sin 51°}{c}$$

so

$$c = \frac{5\sin 51°}{\sin 92°} \approx 3.89.$$

6. With $\alpha = 120°$, $a = 9$, and $c = 4$, and using the Law of Sines, we have

$$\frac{\sin\alpha}{a} = \frac{\sin\gamma}{c} \qquad \text{or} \qquad \frac{\sin 120°}{9} = \frac{\sin\gamma}{4},$$

so

$$\sin\gamma = \frac{4\sin 120°}{9} = \frac{4(\sqrt{3}/2)}{9} = \frac{2\sqrt{3}}{9} \approx 0.38.$$

Then $\gamma = \sin^{-1}(0.38) \approx 22.64°$. Thus, $\beta = 180° - 120° - 22.64° = 37.36°$. It remains to find b. Again using the Law of Sines, we have

$$\frac{\sin\alpha}{a} = \frac{\sin\beta}{b} \qquad \text{or} \qquad \frac{\sin 120°}{9} = \frac{\sin 37.36°}{b},$$

so

$$b = \frac{9\sin 37.36°}{\sin 120°} \approx 6.31.$$

9. With $\gamma = 15°$, $a = 8$, and $c = 5$, and using the Law of Sines, we have

$$\frac{\sin\alpha}{a} = \frac{\sin\gamma}{c} \qquad \text{or} \qquad \frac{\sin\alpha}{8} = \frac{\sin 15°}{5}$$

so

$$\sin\alpha = \frac{8\sin 15°}{5} \approx 0.41.$$

In this problem, we are given two sides and an angle opposite one of the sides, so there are potentially two solutions. From $\sin\alpha = 0.41$ we have $\alpha = 24.46°$ in the first quadrant and $\alpha = 180° - 24.46° = 155.54°$ in the second quadrant. For $\alpha = 24.46$ we have $\beta = 180° - 24.46° - 15° = 140.54°$. To find b in this case, we again use the Law of Sines:

$$\frac{\sin\beta}{b} = \frac{\sin\gamma}{c} \qquad \text{or} \qquad \frac{\sin 140.54°}{b} = \frac{\sin 15°}{c},$$

so

$$b = \frac{5\sin 140.54°}{\sin 15°} \approx 12.28.$$

Next, for $\alpha = 155.54°$ we have $\beta = 180° - 155.54° - 15° = 9.46°$. To find b in this case, we use the Law of Sines:

$$\frac{\sin\beta}{b} = \frac{\sin\gamma}{c} \qquad \text{or} \qquad \frac{\sin 9.46°}{b} = \frac{\sin 15°}{c},$$

so

$$b = \frac{5\sin 9.46°}{\sin 15°} \approx 3.18.$$

12. With $\alpha = 35°$, $a = 9$, and $b = 12$, and using the Law of Sines, we have

$$\frac{\sin \alpha}{a} = \frac{\sin \beta}{b} \qquad \text{or} \qquad \frac{\sin 35°}{9} = \frac{\sin \beta}{12},$$

so

$$\sin \beta = \frac{12 \sin 35°}{9} \approx 0.76.$$

In this problem we are given two sides and an angle opposite one of the sides, so there are potentially two solutions. From $\sin \beta = 0.76$ we have $\beta = 49.89°$ in the first quadrant and $\beta = 180° - 49.89° = 130.11°$ in the second quadrant. For $\beta = 49.89°$ we have $\gamma = 180° - 35° - 49.89° = 95.11°$. To find c in this case we again use the Law of Sines:

$$\frac{\sin \alpha}{a} = \frac{\sin \gamma}{c} \qquad \text{or} \qquad \frac{\sin 35°}{9} = \frac{\sin 95.11°}{c},$$

so

$$c = \frac{9 \sin 95.11°}{\sin 35°} \approx 15.63.$$

Next, for $\beta = 130.11°$ we have $\gamma = 180° - 35° - 130.11° = 14.89°$. To find c in this case we use the Law of Sines:

$$\frac{\sin \alpha}{a} = \frac{\sin \gamma}{c} \qquad \text{or} \qquad \frac{\sin 35°}{9} = \frac{\sin 14.89°}{c},$$

so

$$c = \frac{9 \sin 14.89°}{\sin 35°} \approx 4.03.$$

15. With $\alpha = 20°$, $a = 8$, and $c = 27$, and using the Law of Sines, we have

$$\frac{\sin \alpha}{a} = \frac{\sin \gamma}{c} \qquad \text{or} \qquad \frac{\sin 20°}{8} = \frac{\sin \gamma}{27},$$

so

$$\sin \gamma = \frac{27 \sin 20°}{8} \approx 1.15.$$

This is not possible, so the triangle has no solution.

Miscellaneous Applications

18. Using the figure to the right, we identify $\beta = 105°$, $\gamma = 20°$, and $a = 230$. We want to find the distance from A to B, which is the length of the side of the triangle labeled c and opposite $\alpha = 20°$. We note that $\alpha = 180° - 105° - 20° = 55°$, and then using the Law of Sines, we have

$$\frac{\sin \alpha}{a} = \frac{\sin \gamma}{c} \qquad \text{so} \qquad c = \frac{a \sin \gamma}{\sin \alpha} = \frac{230 \sin 20°}{\sin 55°} \approx 96.03 \text{ ft.}$$

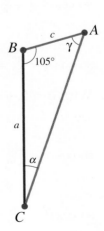

21. We let h denote the height of the street light, as shown in the figure. Then, using the results of Problem 20 and the Law of Sines, we have

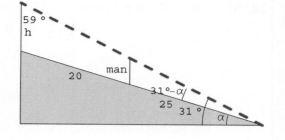

$$\frac{\sin(31° - \alpha)}{h} = \frac{\sin 59°}{20 + 25}$$

or

$$h = \frac{45(0.20)}{\sin 59°} \approx 10.35 \text{ ft.}$$

24. Using the solution and figure in Problem 23 above, we see from the Law of Sines that

$$\frac{\sin \alpha°}{a} = \frac{\sin 8.81°}{a} = \frac{\sin 160°}{370}.$$

Thus,

$$a = 370 \frac{\sin 8.81°}{\sin 160°} \approx 370 \frac{0.1531}{0.3420} \approx 156.6 \text{ yd,}$$

and the distance from the ball to the green is about 156.6 yards.

5.4   Law of Cosines

3. With $a = 8$, $b = 10$, and $c = 7$, and using the Law of Cosines, we have

$$b^2 = a^2 + c^2 - 2ac \cos \beta \qquad \text{or} \qquad 10^2 = 8^2 + 7^2 - 112 \cos \beta,$$

so $\cos \beta = \frac{13}{112}$ and $\beta = \cos^{-1} \frac{13}{112} \approx 83.33°$. To find α we consider

$$a^2 = b^2 + c^2 - 2bc \cos \alpha \qquad \text{or} \qquad 8^2 = 10^2 + 7^2 - 140 \cos \alpha,$$

so $\cos \alpha = \frac{85}{140} = \frac{17}{28}$ and $\alpha = \cos^{-1} \frac{17}{28} \approx 52.62°$, and $\gamma = 180° - 83.33° - 52.62° = 44.05°$.

6. With $a = 7$, $b = 9$, and $c = 4$, and using the Law of Cosines, we have

$$b^2 = a^2 + c^2 - 2ac \cos \beta \qquad \text{or} \qquad 9^2 = 7^2 + 4^2 - 56 \cos \beta,$$

so $\cos \beta = -\frac{16}{56} = -\frac{2}{7}$ and $\beta = \cos^{-1} \left(-\frac{2}{7}\right) \approx 106.60°$. To find α we consider

$$a^2 = b^2 + c^2 - 2bc \cos \alpha \qquad \text{or} \qquad 7^2 = 9^2 + 4^2 - 72 \cos \alpha,$$

so $\cos \alpha = \frac{48}{72} = \frac{2}{3}$ and $\alpha = \cos^{-1} \frac{2}{3} \approx 48.19°$, and $\gamma = 180° - 106.60° - 48.19° = 25.21°$.

9. With $a = 5$, $b = 7$, and $c = 10$, and using the Law of Cosines, we have

$$a^2 = b^2 + c^2 - 2bc \cos \alpha \qquad \text{or} \qquad 5^2 = 7^2 + 10^2 - 140 \cos \alpha,$$

so $\cos \alpha = \frac{124}{140} = \frac{31}{35}$ and $\alpha = \cos^{-1} \frac{31}{35} \approx 27.66°$. To find β we consider

$$b^2 = a^2 + c^2 - 2ac \cos \beta \qquad \text{or} \qquad 7^2 = 5^2 + 10^2 - 100 \cos \beta,$$

so $\cos \beta = \frac{76}{100} = \frac{19}{25}$ and $\beta = \cos^{-1} \frac{19}{25} \approx 40.54°$, and $\gamma = 180° - 27.66° - 40.54° = 111.80°$.

12. With $a = 5$, $b = 12$, and $c = 13$, and using the Law of Cosines, we have

$$a^2 = b^2 + c^2 - 2bc \cos \alpha \qquad \text{or} \qquad 5^2 = 12^2 + 13^2 - 312 \cos \alpha,$$

so $\cos \alpha = \frac{288}{312} = \frac{12}{13}$ and $\alpha = \cos^{-1} \frac{12}{13} \approx 22.62°$. To find β we consider

$$b^2 = a^2 + c^2 - 2ac \cos \beta \qquad \text{or} \qquad 12^2 = 5^2 + 13^2 - 130 \cos \beta,$$

so $\cos \beta = \frac{50}{130} = \frac{5}{13}$ and $\beta = \cos^{-1} \frac{5}{13} \approx 67.38°$, and $\gamma = 180° - 22.62° - 67.38° = 90°$.

15. With $\alpha = 22°$, $b = 3$, and $c = 9$, and using the Law of Cosines, we have

$$a^2 = b^2 + c^2 - 2bc \cos \alpha = 3^2 + 9^2 - 54 \cos 22° \approx 39.93,$$

so $a \approx 6.32$. We use the Law of Sines to find β:

$$\frac{\sin \alpha}{a} = \frac{\sin \beta}{b} \qquad \text{or} \qquad \sin \beta = \frac{3 \sin 22°}{6.32} \approx 0.18,$$

so $\beta = \sin^{-1}(0.18) \approx 10.24°$. Finally, $\gamma = 180° - 22° - 10.24° = 147.76°$.

Miscellaneous Applications

18. After one hour, each hiker will travel $5(1) = 5$ km. The angle α between the hikers' paths is

$$\alpha = 180° - 42° + 20° = 158°.$$

From the Law of Cosines,

$$a^2 = 5^2 + 5^2 - 2(5)(5) \cos 158° \approx 96.36$$

and $a \approx 9.82$ km.

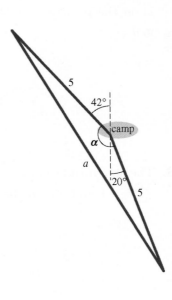

21. The distance from the origin to the point $(1, 2)$ is $\sqrt{1^2 + 2^2} = \sqrt{5}$. We see from the figure that $\alpha = \gamma + \delta$. Since $\tan \delta = 2/1 = 2$, we have $\delta = \tan^{-1} 2 \approx 63.43°$. Then, using the Law of Cosines on triangle ABC,

$$2^2 = 2^2 + (\sqrt{5})^2 - 2(2)\sqrt{5} \cos \gamma$$

or $\cos \gamma = 5/4\sqrt{5} \approx 0.56$, and $\gamma \approx 56.01°$. Thus, $\alpha = \delta + \gamma = 63.43° + 56.01° = 119.44°$. Again using the Law of Cosines on triangle ABC,

$$(\sqrt{5})^2 = 2^2 + 2^2 - 2(2)(2) \cos \beta$$

and $\cos \beta = 3/8 \approx 0.375$, so $\beta \approx 67.98°$.

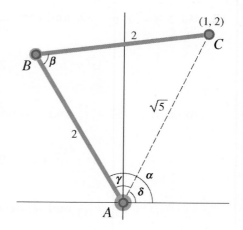

24. Using the values in Figure 5.4.11 in the text we have by the Law of Cosines

$$d^2 = 62^2 + 86^2 - 2(62)(86) \cos 123° \approx 17,048.$$

Thus, the distance from the top of one canyon wall to the other is $d = \sqrt{17,048} \approx 130.57$ ft.

5.5 $\displaystyle\int$ **Calculus PREVIEW** **Vectors and Dot Product**

3. The magnitude is $|\mathbf{v}| = \sqrt{5^2 + 0^2} = 5$. Since $\tan \theta = 0$, and θ is on the positive x-axis, the smallest *positive* direction angle is $\theta = 2\pi$.

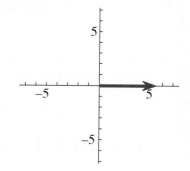

6. The magnitude is $|\mathbf{v}| = \sqrt{1^2 + (-1)^2} = \sqrt{2}$. Since $\tan \theta = \frac{-1}{1} = -1$ and θ is in the fourth quadrant, $\theta = 7\pi/4$.

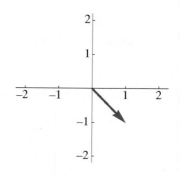

9. $\mathbf{u} + \mathbf{v} = \langle 2,\ 3 \rangle + \langle 1,\ -1 \rangle = \langle 2+1,\ 3-1 \rangle = \langle 3,\ 2 \rangle$

$\mathbf{u} - \mathbf{v} = \langle 2,\ 3 \rangle - \langle 1,\ -1 \rangle = \langle 2-1,\ 3-(-1) \rangle = \langle 1,\ 4 \rangle$

$-3\,\mathbf{u} = -3\langle 2,\ 3 \rangle = \langle -6,\ -9 \rangle$

$3\,\mathbf{u} - 4\,\mathbf{v} = 3\langle 2,\ 3 \rangle - 4\langle 1,\ -1 \rangle = \langle 6-4,\ 9+4 \rangle = \langle 2,\ 13 \rangle$

12. $\mathbf{u} + \mathbf{v} = \langle -1,\ -5 \rangle + \langle 8,\ 7 \rangle = \langle -1+8,\ -5+7 \rangle = \langle 7,\ 2 \rangle$

$\mathbf{u} - \mathbf{v} = \langle -1,\ -5 \rangle - \langle 8,\ 7 \rangle = \langle -1-8,\ -5-7 \rangle = \langle -9,\ -12 \rangle$

$-3\,\mathbf{u} = -3\langle -1,\ -5 \rangle = \langle 3,\ 15 \rangle$

$3\,\mathbf{u} - 4\,\mathbf{v} = 3\langle -1,\ -5 \rangle - 4\langle 8,\ 7 \rangle = \langle -3-32,\ -15-28 \rangle = \langle -35,\ -43 \rangle$

15. $\mathbf{u} - 4\,\mathbf{v} = (\mathbf{i} - 2\,\mathbf{j}) - 4(8\,\mathbf{i} + 3\,\mathbf{j}) = (1-32)\,\mathbf{i} + (-2-12)\,\mathbf{j} = -31\,\mathbf{i} - 14\,\mathbf{j}$

$2\,\mathbf{u} + 5\,\mathbf{v} = 2(\mathbf{i} - 2\,\mathbf{j}) + 5(8\,\mathbf{i} + 3\,\mathbf{j}) = (2+40)\,\mathbf{i} + (-4+15)\,\mathbf{j} = 42\,\mathbf{i} + 11\,\mathbf{j}$

18. $\mathbf{u} - 4\,\mathbf{v} = (2\,\mathbf{i} - 3\,\mathbf{j}) - 4(3\,\mathbf{i} - 2\,\mathbf{j}) = (2-12)\,\mathbf{i} + (-2+8)\,\mathbf{j} = -10\,\mathbf{i} + 5\,\mathbf{j}$

$2\,\mathbf{u} + 5\,\mathbf{v} = 2(2\,\mathbf{i} - 3\,\mathbf{j}) + 5(3\,\mathbf{i} - 2\,\mathbf{j}) = (4+15)\,\mathbf{i} + (-6-10)\,\mathbf{j} = 19\,\mathbf{i} - 16\,\mathbf{j}$

21.

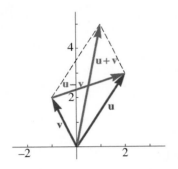

24.

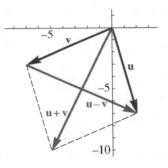

27. $2\,\mathbf{v} = 2(3\,\mathbf{i} - 5\,\mathbf{j}) = 6\,\mathbf{i} - 10\,\mathbf{j}$

$-2\,\mathbf{v} = -2(6\,\mathbf{i} - 10\,\mathbf{j}) = -6\,\mathbf{i} + 10\,\mathbf{j}$

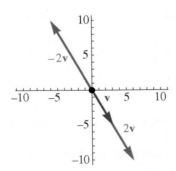

30. Since $3(\mathbf{u} + \mathbf{v}) = 3[(3\,\mathbf{i} - \mathbf{j}) + (2\,\mathbf{i} + 4\,\mathbf{j})] = 3(5\,\mathbf{i} + 3\,\mathbf{j}) = 15\,\mathbf{i} + 9\,\mathbf{j}$,   the horizontal component of $3(\mathbf{u} + \mathbf{v})$ is 15, and the vertical component is 9.

33. The magnitude of $\langle -\sqrt{2},\ \sqrt{2} \rangle$ is $\sqrt{\left(-\sqrt{2}\right)^2 + \left(\sqrt{2}\right)^2} = \sqrt{2+2} = \sqrt{4} = 2$.

Since $\tan\theta = \sqrt{2} / \left(-\sqrt{2}\right) = -1$ and θ is a second quadrant angle, we have $\theta = 3\pi/4$.

(a) In trigonometric form, $\langle -\sqrt{2},\ \sqrt{2} \rangle = 2\left(\cos\dfrac{3\pi}{4}\,\mathbf{i} + \sin\dfrac{3\pi}{4}\,\mathbf{j} \right).$

(b) As a linear combination of $\mathbf{i}$ and $\mathbf{j}$, $\langle -\sqrt{2},\ \sqrt{2} \rangle = -\sqrt{2}\,\mathbf{i} + \sqrt{2}\,\mathbf{j}.$

36. The magnitude of $\langle -4,\ -4 \rangle$ is $\sqrt{(-4)^2 + (-4)^2} = \sqrt{16 + 16} = \sqrt{32} = 4\sqrt{2}.$

Since $\tan\theta = -4/(-4) = 1$ and θ is a third quadrant angle, we have $\theta = 5\pi/4.$

(a) In trigonometric form, $\langle -4,\ -4 \rangle = 4\sqrt{2}\left(\cos\dfrac{5\pi}{4}\,\mathbf{i} + \sin\dfrac{5\pi}{4}\,\mathbf{j} \right).$

(b) As a linear combination of $\mathbf{i}$ and $\mathbf{j}$, $\langle -4,\ -4 \rangle = -4\,\mathbf{i} - 4\,\mathbf{j}.$

39. The magnitude of $\mathbf{v}$ is $|\langle 0,\ -5 \rangle| = \sqrt{0^2 + (-5)^2} = 5.$

(a) A vector in the same direction as $\mathbf{v}$ is $\dfrac{\mathbf{v}}{|\mathbf{v}|} = \dfrac{1}{5}\langle 0,\ -5 \rangle = \langle 0,\ -1 \rangle.$

(b) A vector in the opposite direction of $\mathbf{v}$ is $-\dfrac{\mathbf{v}}{|\mathbf{v}|} = -\dfrac{1}{5}\langle 0,\ -5 \rangle = \langle 0,\ 1 \rangle.$

42. $2\,\mathbf{v} - 3\,\mathbf{w} = 2\langle 2,\ 8 \rangle - 3\langle 3,\ 4 \rangle = \langle 4,\ 16 \rangle - \langle 9,\ 12 \rangle = \langle -5,\ 4 \rangle$

$|2\,\mathbf{v} - 3\,\mathbf{w}| = |\langle -5,\ 4 \rangle| = \sqrt{(-5)^2 + 4^2} = \sqrt{25 + 16} = \sqrt{41}$

$\dfrac{2\,\mathbf{v} - 3\,\mathbf{w}}{|2\,\mathbf{v} - 3\,\mathbf{w}|} = \dfrac{1}{41}\langle -5,\ 4 \rangle = \left\langle -\dfrac{5}{41},\ \dfrac{4}{41} \right\rangle$

45. $\mathbf{u} \cdot \mathbf{v} = \langle 4,\ 2 \rangle \cdot \langle 3,\ -1 \rangle = 4(3) + 2(-1) = 12 - 2 = 10$

48. $\mathbf{u} \cdot \mathbf{v} = 4\,\mathbf{i} \cdot (-3\,\mathbf{j}) = (4\,\mathbf{i} + 0\,\mathbf{j}) \cdot (0\,\mathbf{i} - 3\,\mathbf{j}) = 4(0) + 0(-3) = 0$

51. $\mathbf{u} \cdot \mathbf{w} = \langle 2,\ -3 \rangle \cdot \langle 3,\ -2 \rangle = 6 + 6 = 12$

54. $\mathbf{u} \cdot (\mathbf{v} + \mathbf{w}) = \langle 2,\ -3 \rangle \cdot \langle 2,\ 3 \rangle = 4 - 9 = -5$

57. $(-\mathbf{v}) \cdot \left(\dfrac{1}{2}\mathbf{w} \right) = \langle 1,\ -5 \rangle \cdot \left\langle \dfrac{3}{2},\ -1 \right\rangle = \dfrac{3}{2} + 5 = \dfrac{13}{2}$

60. $(2\,\mathbf{u}) \cdot (\mathbf{u} - 2\,\mathbf{v}) = \langle 4,\ -6 \rangle \cdot \langle 4,\ -13 \rangle = 16 + 78 = 94$

63. From (15) in the text, $\mathbf{u} \cdot \mathbf{v} = 10(5)\cos(\pi/4) = 25\sqrt{2}.$

66. From (17) in the text, $\cos\theta = (\mathbf{u} \cdot \mathbf{v})/|\mathbf{u}||\mathbf{v}|.$ Since $\langle 3,\ 5 \rangle \cdot \langle -4,\ -2 \rangle = 3(-4) + 5(-2) = -22,$

$$|\mathbf{u}| = \sqrt{3^2 + 5^2} = \sqrt{34}, \quad \text{and} \quad |\mathbf{v}| = \sqrt{(-4)^2 + (-2)^2} = 2\sqrt{5},$$

we have

$$\cos\theta = \frac{-22}{\sqrt{34}\,(2\sqrt{5})} = -\frac{11}{\sqrt{170}} \quad \text{and} \quad \theta = \cos^{-1}\left(-\frac{11}{\sqrt{170}} \right) \approx 147.53°.$$

69. Since $\mathbf{u} \cdot \mathbf{v} = \langle -5, 4 \rangle \cdot \langle -6, 8 \rangle = 30 - 32 = -2 \neq 0$, the vectors are not orthogonal.

72. Since $\mathbf{u} \cdot \mathbf{v} = \left\langle \dfrac{1}{2}, \dfrac{3}{4} \right\rangle \cdot \left\langle -\dfrac{2}{5}, \dfrac{4}{5} \right\rangle = -\dfrac{1}{5} + \dfrac{3}{4} = \dfrac{2}{5} \neq 0$, the vectors are not orthogonal.

75. We compute

$$\mathbf{u} \cdot \mathbf{w} = \mathbf{u} \cdot \left(\mathbf{v} - \dfrac{\mathbf{v} \cdot \mathbf{u}}{|\mathbf{u}|^2} \mathbf{u} \right)$$

$$= \mathbf{u} \cdot \mathbf{v} - \dfrac{\mathbf{v} \cdot \mathbf{u}}{|\mathbf{u}|^2} (\mathbf{u} \cdot \mathbf{u})$$

$$= \mathbf{u} \cdot \mathbf{v} - \dfrac{\mathbf{u} \cdot \mathbf{u}}{|\mathbf{u}|^2} (\mathbf{v} \cdot \mathbf{u})$$

$$= \mathbf{u} \cdot \mathbf{v} - \dfrac{|\mathbf{u}|^2}{|\mathbf{u}|^2} (\mathbf{u} \cdot \mathbf{v}) = 0.$$

Thus, the vectors $\mathbf{u}$ and $\mathbf{w}$ are orthogonal.

78. $\text{comp}_{\mathbf{u}} \mathbf{v} = \mathbf{v} \cdot \dfrac{\mathbf{u}}{|\mathbf{u}|} = \langle 2, 6 \rangle \cdot \dfrac{\langle 1, -1 \rangle}{\sqrt{1 + 1}} = \dfrac{2 - 6}{\sqrt{2}} = -\dfrac{4}{\sqrt{2}} = -2\sqrt{2}$

81. (a) $\text{comp}_{\mathbf{v}} \mathbf{u} = \mathbf{u} \cdot \dfrac{\mathbf{v}}{|\mathbf{v}|} = (-5\mathbf{i} + 5\mathbf{j}) \cdot \dfrac{-3\mathbf{i} + 4\mathbf{j}}{5} = 7$

$\text{proj}_{\mathbf{v}} \mathbf{u} = (\text{comp}_{\mathbf{v}} \mathbf{u}) \dfrac{\mathbf{v}}{|\mathbf{v}|} = 7 \left(\dfrac{-3\mathbf{i} + 4\mathbf{j}}{5} \right) = -\dfrac{21}{5}\mathbf{i} + \dfrac{28}{5}\mathbf{j}$

(b) $\text{comp}_{\mathbf{u}} \mathbf{v} = \mathbf{v} \cdot \dfrac{\mathbf{u}}{|\mathbf{u}|} = (-3\mathbf{i} + 4\mathbf{j}) \cdot \dfrac{-5\mathbf{i} + 5\mathbf{j}}{5\sqrt{2}} = \dfrac{15 + 20}{5\sqrt{2}} = \dfrac{7}{\sqrt{2}}$

$\text{proj}_{\mathbf{u}} \mathbf{v} = \text{comp}_{\mathbf{u}} \mathbf{v} \left(\dfrac{\mathbf{u}}{|\mathbf{u}|} \right) = \dfrac{7}{\sqrt{2}} \left(\dfrac{-5\mathbf{i} + 5\mathbf{j}}{5\sqrt{2}} \right) = \dfrac{7}{2}(-\mathbf{i} + \mathbf{j}) = -\dfrac{7}{2}\mathbf{i} + \dfrac{7}{2}\mathbf{j}$

84. $\mathbf{u} - \mathbf{v} = 5\mathbf{i} + 2\mathbf{j}; \quad |\mathbf{u} - \mathbf{v}| = \sqrt{29}; \quad \text{comp}_{(\mathbf{u} - \mathbf{v})} \mathbf{v} = \mathbf{v} \cdot \dfrac{\mathbf{u} - \mathbf{v}}{|\mathbf{u} - \mathbf{v}|} = \dfrac{-\mathbf{i} + \mathbf{j}}{\sqrt{29}} = \dfrac{-3}{\sqrt{29}};$

$\text{proj}_{(\mathbf{u} - \mathbf{v})} \mathbf{v} = (\text{comp}_{(\mathbf{u} - \mathbf{v})}) \dfrac{\mathbf{u} - \mathbf{v}}{|\mathbf{u} - \mathbf{v}|} = -\dfrac{3}{\sqrt{29}} \left(\dfrac{5\mathbf{i} + 2\mathbf{j}}{\sqrt{29}} \right) = -\dfrac{15}{29}\mathbf{i} - \dfrac{6}{29}\mathbf{j}$

87. See Figure 5.5.27 in the text. Using $\mathbf{d} = 6\mathbf{i} + 2\mathbf{j}$ and $\mathbf{F} = 3 \left(\dfrac{3}{5}\mathbf{i} + \dfrac{4}{5}\mathbf{j} \right)$, $W = \mathbf{F} \cdot \mathbf{d} = \left\langle \dfrac{9}{5}, \dfrac{12}{5} \right\rangle \cdot \langle 6, 2 \rangle = \dfrac{78}{5}$ ft-lb.

Miscellaneous Applications

90. The mail sack will travel at an angle of

$$\theta = \tan^{-1} \frac{10}{15} \approx 33.69°$$

from the direction in which it is thrown from the train.

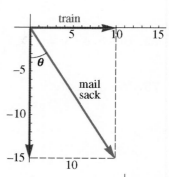

93. The wind is represented by the vector **v** in the figure. The speed of the wind is the magnitude $|\mathbf{v}|$ of **v**. Since the triangle has a right angle where **v** and the vector of magnitude 300 along the positive y-axis meet, we have

$$\tan 10° = \frac{|\mathbf{v}|}{300} \qquad \text{or} \qquad \mathbf{v} = 300 \tan 10° \approx 52.90 \text{ mi/h}.$$

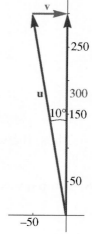

Chapter 5 Review Exercises

A. Fill in the Blanks

3. When two sides and the included angle are known you would first use the Law of Cosines.

6. $\sin\theta = \dfrac{1}{3} \Rightarrow \cos^2\theta = 1 - \dfrac{1}{9} = \dfrac{8}{9} \Rightarrow \cos\theta = \dfrac{2\sqrt{2}}{3} \Rightarrow \tan\theta = 2\sqrt{2}.$

9. A unit vector in the opposite direction of **v** is

$$\mathbf{u} = -\frac{1}{\sqrt{12^2 + 5^2}} \langle 12, -5 \rangle = \left\langle \tfrac{12}{13}, -\tfrac{5}{13} \right\rangle$$

12. By Theorem 5.5.3

$$\mathbf{u} \cdot \mathbf{v} = |\mathbf{u}||\mathbf{v}| \cos\theta = 4(3) \cos \frac{2\pi}{3} = 12\left(-\frac{1}{2}\right) = -6.$$

B. True/False

3. This is false because in a right triangle $\csc\theta = \dfrac{\text{hyp}}{\text{opp}}$.

6. True; if α and β are the acute angles in a right triangle, then $\sin\alpha = \cos\beta$.

9. False; if θ is the angle between ground and ladder, then

$$\cos\theta = \tfrac{5.2}{20} \Rightarrow \theta = \cos^{-1}\tfrac{5.2}{20} \approx 75°$$

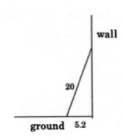

12. True; since if $|\mathbf{u}| = |\mathbf{v}| = 1$ then

$$(\mathbf{u}+\mathbf{v})\cdot(\mathbf{u}-\mathbf{v}) = \mathbf{u}\cdot\mathbf{u} - \mathbf{v}\cdot\mathbf{v} = |\mathbf{u}|^2 - |\mathbf{v}|^2 = 1 - 1 = 0.$$

C. Review Exercises

3. With $\alpha = 51°$, $b = 20$, and $c = 10$, and using the Law of Cosines, we have

$$a^2 = b^2 + c^2 - 2bc\cos\alpha = 20^2 + 10^2 - 2(20)(10)\cos 51° \approx 248.27,$$

so $a = 15.76$. We use the Law of Sines to find γ:

$$\frac{\sin\alpha}{a} = \frac{\sin\gamma}{c} \qquad \text{or} \qquad \sin\gamma = \frac{10\sin 51°}{15.76} \approx 0.49,$$

so

$$\gamma = \sin^{-1}(0.49) \approx 29.55°.$$

Finally,

$$\beta = 180° - 51° - 29.55° = 99.45°.$$

6. Let a denote the horizontal distance and c the straight-line distance as shown in the figure. Then, $\tan 43° = 20000/a$ and $\sin 43° = 20000/c$, so

$$a = \frac{20000}{\tan 43°} = 21447.37 \text{ ft}$$

and

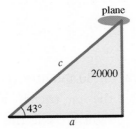

$$c = \frac{20000}{\sin 43°} = 29325.58 \text{ ft}.$$

9. (a) Refer to the figure for this problem in the text. Using the Law of Sines, we have

$$\frac{\sin\theta}{R} = \frac{\sin\rho}{H+R} \qquad \text{or} \qquad \sin\alpha = \frac{(36000 + 6370)\sin 6.5°}{6370} \approx 0.75,$$

where α is the related angle to ρ and ρ is the obtuse angle at point P. Then

$$\alpha = \sin^{-1} 0.75 \approx 48.85° \qquad \text{and} \qquad \rho = 180° - \alpha = 180° - 48.85° = 131.15°.$$

Since the sum of the angles in a triangle is $180°$, we have

$$\phi = 180° - 6.5° - 131.15° = 42.35°.$$

(b) Let S be the distance from the satellite to the point P. Then, using the Law of Sines, we have

$$\frac{\sin\theta}{R} = \frac{\sin\phi}{S} \qquad \text{or} \qquad \sin\theta = \frac{R\sin\phi}{S}.$$

Using the Law of Cosines, we have

$$R^2 = S^2 + (H+R)^2 - 2S(H+R)\cos\theta$$

or

$$\cos\theta = \frac{S^2 + (H+R)^2 - R^2}{2S(H+R)}.$$

Dividing the above two results, we have

$$\tan\theta = \frac{\sin\theta}{\cos\theta} = \frac{R\sin\phi}{S} \bigg/ \frac{s^2 + (H+R)^2 - R^2}{2S(H+R)} = \frac{2(H+R)R\sin\phi}{S^2 + (H+R)^2 - R^2}.$$

To remove S^2 from this expression we again use the Law of Cosines:

$$S^2 = (H+R)^2 + R^2 - 2(H+R)R\cos\phi.$$

Substituting into the expression for $\tan\theta$, we have

$$\tan\theta = \frac{2(H+R)R\sin\phi}{(H+R)^2 + R^2 - 2(H+R)R\cos\phi + (H+R)^2 - R^2}$$

$$= \frac{2(H+R)R\sin\phi}{2(H+R)^2 - 2(H+R)R\cos\phi} = \frac{R\sin\phi}{H+R-R\cos\phi}$$

$$= \frac{R\sin\phi}{H + R(1 - \cos\phi)}.$$

12. In 30 minutes the airplane travels 200 miles and the car travels 30 miles. (See the figure to the right.) Using the right triangle with angle $6°$ and hypotenuse 200, we have

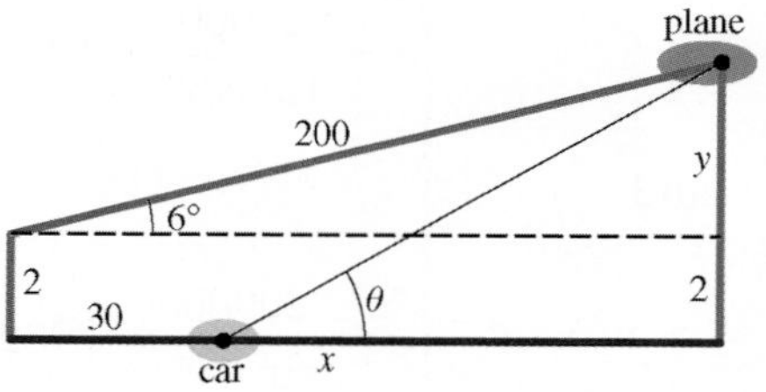

$$\cos 6° = \frac{x}{200} \qquad \text{or} \qquad x = 200\cos 6° \approx 198.90.$$

The altitude of the airplane after 30 minutes is $y + 2$, where

$$\sin 6° = \frac{y}{200} \qquad \text{or} \qquad y = 200\sin 6° = 20.91.$$

Thus, the altitude of the plane is $2 + 20.91 = 22.91$. The angle of elevation from the car to the airplane is θ, where

$$\tan\theta = \frac{22.91}{198.90} \approx 0.12 \qquad \text{so} \qquad \theta = \tan^{-1}(0.12) \approx 6.57°.$$

15. Let d_1 and d_2 be, respectively, the distances from towers A and B to the swimmer. From the Law of Sines:

$$\frac{\sin 44°}{d_2} = \frac{\sin 63°}{d_1} = \frac{\sin 73°}{250} \Rightarrow d_1 = 250\frac{\sin 63°}{\sin 73°} \text{ and } d_2 = 250\frac{\sin 44°}{\sin 73°} \Rightarrow$$

$d_1 \approx 233$ ft and $d_2 \approx 182$ ft

18. Let θ_1 denote the angle of elevation to the top of the freeway sign. Then

$$\tan\theta_1 = \frac{18+4}{x} \qquad \text{so} \qquad \theta_1 = \arctan\frac{22}{x}.$$

Let θ_2 denote the angle of elevation to the bottom of the freeway sign. Then

$$\tan\theta_2 = \frac{18}{x} \qquad \text{so} \qquad \theta_2 = \arctan\frac{18}{x}.$$

Thus, we see that

$$\theta = \theta_1 - \theta_2 = \arctan\frac{22}{x} - \arctan\frac{18}{x}.$$

21. To find the volume of the box, we begin by finding the area of the side of the box shown in the figure at the right. We see that $\tan\theta = 5/x$, so $x = 5\cot\theta$. Then the area of the figure is given by

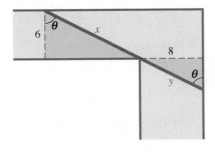

$$A = (12)(5) + \frac{1}{2}(5\cot\theta)(5) = 60 + \frac{25}{2}\cot\theta.$$

The volume of the box is

$$V = 6\left(60 + \frac{25}{2}\cot\theta\right) = 360 + 75\cot\theta.$$

24. Let x denote the length of the section of pipe that stretches across the 6-foot-wide hallway to the corner, and let y denote the length of the section of pipe that stretches from the corner across the 8-foot-wide hallway. From the figure, we see that

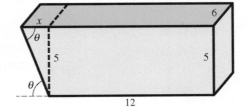

$$\cos\theta = \frac{6}{x} \qquad \text{or} \qquad x = \frac{6}{\cos\theta} = 6\sec\theta,$$

and

$$\sin\theta = \frac{8}{y} \qquad \text{or} \qquad y = \frac{8}{\sin\theta} = 8\csc\theta.$$

Thus, we have

$$L = x + y = 6\sec\theta + 8\csc\theta.$$

27. $-5\mathbf{u} + 3\mathbf{v} = -5(-2\mathbf{i} + 3\mathbf{j}) + 3(\mathbf{i} + \mathbf{j}) = (10\mathbf{i} - 15\mathbf{j}) + (3\mathbf{i} + 3\mathbf{j}) = 13\mathbf{i} - 12\mathbf{j}$

30. $4\mathbf{u} - (3\mathbf{v} + \mathbf{w}) = 4(-2\mathbf{i} + 3\mathbf{j}) - [3(\mathbf{i} + \mathbf{j}) + (\mathbf{i} - 4\mathbf{j})] = (-8\mathbf{i} + 12\mathbf{j}) - [(3\mathbf{i} + 3\mathbf{j}) + (\mathbf{i} - 4\mathbf{j})] =$
$(-8\mathbf{i} + 12\mathbf{j}) - [4\mathbf{i} - \mathbf{j}] = -12\mathbf{i} + 13\mathbf{j}$

33. $\text{comp}_{\mathbf{w}}\,\mathbf{v} = \mathbf{v} \cdot \left(\dfrac{\mathbf{w}}{|\mathbf{w}|}\right) = (\mathbf{i} + \mathbf{j}) \cdot \left(\dfrac{\mathbf{i} - 4\mathbf{j}}{\sqrt{1+16}}\right) = (\mathbf{i} + \mathbf{j}) \cdot \left(\dfrac{1}{\sqrt{17}}\mathbf{i} - \dfrac{4}{\sqrt{17}}\mathbf{j}\right)$

$= \dfrac{1}{\sqrt{17}} - \dfrac{4}{\sqrt{17}} = -\dfrac{3}{\sqrt{17}}$

36. $\text{proj}_{\mathbf{w}}(\mathbf{u} + \mathbf{v}) = \left(\dfrac{(\mathbf{u} + \mathbf{v}) \cdot \mathbf{w}}{\mathbf{w} \cdot \mathbf{w}}\right)\mathbf{v} = \left(\dfrac{(-\mathbf{i} + 4\mathbf{j}) \cdot (\mathbf{i} - 4\mathbf{j})}{(\mathbf{i} - 4\mathbf{j}) \cdot (\mathbf{i} - 4\mathbf{j})}\right)(\mathbf{i} - 4\mathbf{j})$

$= \dfrac{-1 - 16}{1 + 16}(\mathbf{i} - 4\mathbf{j}) = -\mathbf{i} + 4\mathbf{j}$

39. To find the trigonometric form of $2\mathbf{v}$ we must find the magnitude $|2\mathbf{v}|$ and the direction angle θ. First,

$$|2\mathbf{v}| = |2(\mathbf{i} + \mathbf{j})| = |2\mathbf{i} + 2\mathbf{j}| = \sqrt{4 + 4} = 2\sqrt{2},$$

and then

$$\tan\theta = \frac{2}{2} = 1 \qquad \text{so} \qquad \theta = \frac{\pi}{4}.$$

Thus,

$$2\mathbf{v} = 2\sqrt{2}\left(\cos\frac{\pi}{4}\,\mathbf{i} + \sin\frac{\pi}{4}\,\mathbf{j}\right).$$

42. To find the angle θ between $\mathbf{v}$ and $\mathbf{w}$ we use

$$\cos\theta = \frac{\mathbf{v} \cdot \mathbf{w}}{|\mathbf{v}||\mathbf{w}|} = \frac{(\mathbf{i} + \mathbf{j}) \cdot (\mathbf{i} - 4\mathbf{j})}{|\mathbf{i} + \mathbf{j}||\mathbf{i} - 4\mathbf{j}|} = \frac{1 - 4}{\sqrt{2}\sqrt{17}} = -\frac{3}{\sqrt{34}}.$$

Then

$$\cos^{-1}\left(-\frac{3}{\sqrt{34}}\right) \approx 2.11122 \text{ radians}$$

and the angle between $\mathbf{v}$ and $\mathbf{w}$ is 2.11122 radians $\approx 120.964°$.

Chapter 6

Exponential and Logarithmic Functions

6.1 | Exponential Functions

3. Since $f(0) = -2^0 = -1$, the y-intercept is $(0, -1)$. The x-axis is a horizontal asymptote and the graph of $f(x) = -2^x$ is the graph of $y = 2^x$ reflected in the x-axis. The function is decreasing.

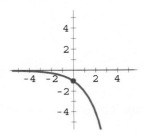

6. Since $f(0) = 2^{2-0} = 4$, the y-intercept is $(0, 4)$. The x-axis is a horizontal asymptote and the graph of $f(x) = 2^{2-x} = 2^{-(x-2)}$ is the graph of $y = 2^{-x}$ shifted 2 units to the right. The function is decreasing.

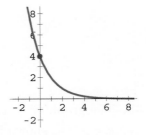

9. Since $f(0) = 3 - (\frac{1}{5})^0 = 3 - 1 = 2$, the y-intercept is $(0, 2)$. The line $y = 3$ is a horizontal asymptote and the graph of $f(x) = 3 - (\frac{1}{5})^x$ is the graph of $y = -(\frac{1}{5})^x = -5^{-x}$ shifted up 3 units. The function is increasing.

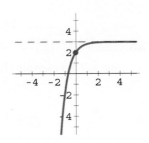

12. Since $f(0) = -3 - e^{0+5} = -3 - e^5$ the y-intercept is $(0, -3 - e^5)$. The line $y = -3$ is a horizontal asymptote and the graph of $f(x) = -3 - e^{x+5}$ is the graph of $y = -e^{-x}$ shifted down 3 units and to the left 5 units. The function is decreasing.

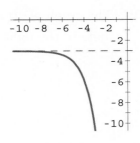

15. Letting $x = -1$ and $f(-1) = e^2$, we have $f(-1) = e^2 = b^{-1}$, so $b = e^{-2}$ and $f(x) = (e^{-2})^x = e^{-2x}$.

18. $\left(\frac{1}{2},\ 6\right) \Rightarrow f\left(\frac{1}{2}\right) = b^{1/2} = 6 \Rightarrow b = 36 \Rightarrow f(x) = 36^x$ x

21. $f(x) = 3^x - 2 \Rightarrow$ range is $(-2,\ \infty)$

24. Since $f(0) = -3^0 + 9 = -1 + 9 = 8$, the y-intercept is $(0, 8)$. To find any x-intercepts, we solve $-3^{2x} + 9 = 0$. Then $3^{2x} = 9$, so $x = 1$, and the x-intercept is $(1, 0)$.

27. Since $f(0) = 0$, the y-intercept is $(0, 0)$. To find any x-intercepts we solve

$$x^3 8^x + 5x^2 8^x + 6x 8^x = 0 \qquad \text{or} \qquad 8^x x(x^2 + 5x + 6) = 0.$$

Since $8^x > 0$ for all x, this implies

$$x(x^2 + 5x + 6) = x(x + 2)(x + 3) = 0,$$

so $x = 0$, $x = -2$, and $x = -3$, and the x-intercepts are $(0, 0)$, $(-2, 0)$, and $(-3, 0)$.

30. Graphing $y = e^x$ and $y = 1$, we see that $e^x \le 1$ for $x \le 0$.

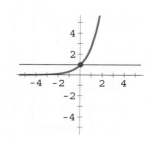

33. The graph of $f(x) = e^{-(x-3)^2}$, shown in blue, is the graph of $y = e^{-x^2}$, shown in brown, shifted right 3 units.

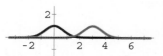

36. Since $f(-x) = e^{-|-x|} = e^{-|x|} = f(x)$, we see that $f(x)$ is even. To sketch the graph, we note that

$$e^{-|x|} = \begin{cases} e^{-x}, & x \ge 0 \\ e^x, & x < 0. \end{cases}$$

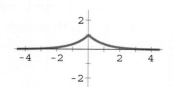

39. The graph of $f(x) = -e^{|x-3|}$ is the graph of $y = e^{|x|}$ shifted 3 units to the right and then reflected in the x-axis.

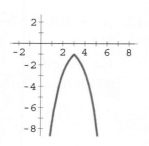

42. Since

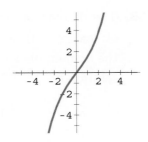

$$f(-x) = 2^{-x} - 2^{-(-x)} = 2^{-x} - 2^{x} = -(2^{x} - 2^{-x}) = -f(x),$$

we see that $f(x)$ is odd.

45. The line passes through $(-1, 1/3)$ and $(2, 9)$. The slope of this line is $(9 - 1/3)/(2 + 1) = 26/9$. The equation is

$$y - 9 = \frac{26}{9}(x - 2) \quad \text{or} \quad y = \frac{26}{9}x + \frac{29}{9}.$$

6.2   Logarithmic Functions

3. $10^4 = 10{,}000$ is equivalent to $\log_{10} 10{,}000 = 4$.

6. $(a + b)^2 = a^2 + 2ab + b^2$ is equivalent to $\log_{a+b}(a^2 + 2ab + b^2) = 2$.

9. $\log_{\sqrt{3}} 81 = 8$ is equivalent to $(\sqrt{3})^8 = 81$.

12. $\log_b b^2 = 2$ is equivalent to $b^2 = b^2$.

15. $\log_2(2^2 + 2^2) = \log_2(4 + 4) = \log_2 8 = \log_2 2^3 = 3$

18. $\ln(e^4 e^9) = \ln e^{4+9} = \ln e^{13} = 13$

21. $e^{-\ln 7} = e^{\ln 7^{-1}} = 7^{-1} = \frac{1}{7}$

24. We solve $\frac{1}{3} = \log_b 4$ or $b^{1/3} = 4$. Cubing both sides, we have $b = 4^3 = 64$ and $f(x) = \log_{64} x$.

27. The domain of $\log_2(x)$ is determined by $x > 0$, so the domain of $\log_2(-x)$ is determined by $-x > 0$ or $x < 0$. That is, the domain of $\log_2(-x)$ is $(-\infty, 0)$. The x-intercept is the solution of $\log_2(-x) = 0$. This is equivalent to $2^0 = -x$ or $x = -2^0 = -1$, so the x-intercept is $(-1, 0)$. The vertical asymptote is $-x = 0$ or $x = 0$, which is the y-axis.

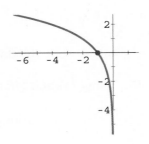

30. The domain of $\log_4 x$ is determined by $x > 0$, so the domain of $\log_4(x - 4)$ is determined by $x - 4 > 0$ or $x > 4$. Thus, the domain of $1 - 2\log_4(x - 4)$ is $(4, \infty)$. The x-intercept is the solution of

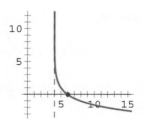

$$1 - 2\log_4(x - 4) = 0$$
$$2\log_4(x - 4) = 1$$
$$\log_4(x - 4) = \frac{1}{2}$$
$$4^{1/2} = x - 4$$
$$x = 4 + 2 = 6,$$

so the x–intercept is $(6, 0)$. The vertical asymptote is $x - 4 = 0$ or $x = 4$.

33. From the graph, we see that $\ln(x + 1) < 0$ for $-1 < x < 0$.

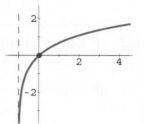

36. The graph of $y = \ln|x - 2|$, shown in dark blue, is the graph of $y = \ln|x|$, shown in light blue, shifted 2 units to the right. The x-intercepts occur where $\ln|x - 2| = 0$ or $|x - 2| = 1$. They are $(1, 0)$ and $(3, 0)$. The vertical asymptote is $y = 2$.

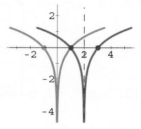

39. The domain of $\ln(2x - 3)$ is determined by $2x - 3 > 0$ or $x > \frac{3}{2}$. Thus, the domain is $\left(\frac{3}{2}, \infty\right)$.

42. The domain of $\ln(x^2 - 2x)$ is determined by $x^2 - 2x > 0$. This is equivalent to $x(x - 2) > 0$. From the sign chart below

x	$-$	0	$+$	$+$	$+$
$x - 2$	$-$		$-$	0	$+$
$x(x - 2)$	$+$	0	$-$	0	$+$

$$\xleftarrow{\hspace{3cm}} \quad 0 \quad \xrightarrow{\hspace{3cm}} 2 \quad \xrightarrow{\hspace{2cm}}$$

we see that the domain is $(-\infty, 0) \cup (2, \infty)$.

45. Using (i) and (iii) of the laws of logarithms in the text, we have

$$\log_{10} 2 + 2\log_{10} 5 = \log_{10} 2 + \log_{10} 5^2 = \log_{10}(2 \cdot 5^2) = \log_{10} 50.$$

48. Using (i), (ii), and (iii) of the laws of logarithms in the text, we have

$$\ln\left(\frac{x}{y}\right) - 2\ln x^3 - 4\ln y = \ln\left(\frac{x}{y}\right) - \ln x^6 - \ln y^4 = \ln\left(\frac{x}{y}\right) - \ln(x^6 y^4)$$

$$= \ln\left(\frac{x}{y} \cdot \frac{1}{x^6 y^4}\right) = \ln\left(\frac{1}{x^5 y^5}\right) = -\ln(x^5 y^5).$$

51. $\log_b 2 = \log_b 4^{1/2} = \dfrac{1}{2}\log_b 4 = \dfrac{1}{2}(0.6021) = 0.3011$

54. $\log_b 625 = \log_b 5^4 = 4\log_b 5 = 4(0.6690) = 2.7960$

57. $\log_b \sqrt[3]{4} = \dfrac{1}{3}\log_b 4 = \dfrac{1}{3}(0.6021) = 0.2007$

60. $\log_b 3.2 = \log_b \dfrac{16}{5} = \log_b 4^2 - \log_b 5 = 2(0.6021) - 0.6990 = 0.5052$

63. Using the laws of logarithms in the text, we have

$$\ln y = \ln \frac{x^{10}\sqrt{x^2+5}}{\sqrt[3]{8x^3+2}}$$

$$= \ln x^{10} + \ln(x^2+5)^{1/2} - \ln(8x^3+2)^{1/3}$$

$$= 10\ln x + \frac{1}{2}\ln(x^2+5) - \frac{1}{3}\ln(8x^3+2).$$

66. Using the laws of logarithms in the text, we have

$$\ln y = \ln\left(64x^6\sqrt{x+1}\sqrt[3]{x^2+2}\right)$$

$$= \ln 64 + \ln x^6 + \ln(x+1)^{1/2} + \ln(x^2+2)^{1/3}$$

$$= 6\ln 2 + 6\ln x + \frac{1}{2}\ln(x+1) + \frac{1}{3}\ln(x^2+2).$$

69. $\ln|\sec x - \tan x| = \ln\left|\sec x - \tan x \dfrac{\sec x + \tan x}{\sec x + \tan x}\right|$

$$= \ln\left|\sec^2 x - \tan^2 x\right| - \ln|\sec x + \tan x| == \ln|1| - \ln|\sec x + \tan x| = -\ln|\sec x + \tan x|$$

6.3 | Exponential and Logarithmic Equations

3. In logarithmic form the equation is

$$-2x = \log_{10} 10^{-4} = -4.$$

Thus $x = 2$.

6. Writing $(1/e)^x = e^{-x} = e^3$, we see that $-x = 3$, so $x = -3$.

9. Writing the equation in the form

$$2^{x^2} = \left(2^3\right)^{2x-3} = 2^{6x-9},$$

and using the fact that 2^x is a one-to-one function we have

$$x^2 = 6x - 9$$
$$x^2 - 6x + 9 = (x-3)^2 = 0.$$

Thus $x = 3$.

12. Taking the base 10 logarithm of both sides of the equation we have

$$\log_{10} 7^{-x} = \log_{10} 9 \qquad \text{or} \qquad -x = \frac{\log_{10} 9}{\log_{10} 7} \approx \frac{0.9542}{0.8451} \approx 1.1292.$$

Thus $x \approx 1.1292$.

15. Raising both sides of the equation to -1 we obtain

$$3 = 2^{|x|-2} - 1$$
$$2^{|x|-2} = 4 = 2^2.$$

Since 2^x is a one-to-one function

$$|x| - 2 = 2 \qquad \text{so} \qquad |x| = 4,$$

and $x = -4$ or $x = 4$.

18. Writing the equation in the form $e^{2x^2} = e^{-(5x+3)}$ we have

$$2x^2 = -(5x + 3)$$
$$2x^2 + 5x + 3 = 0$$
$$(2x + 3)(x + 1) = 0.$$

Thus, $x = -\frac{3}{2}$ or $x = -1$.

21. Since $\log_3 x$ is a one-to-one function, $5x = 160$ and $x = 32$.

24. First, write the equation as $\log_8 x^3 = \log_8 \dfrac{36(12)}{2} = \log_8 216$. Since $\log_8 x$ is a one-to-one function, $x^3 = 216$ and $x = 6$.

27. Rewriting the equation in exponential form, we have

$$\log_3 x = 2^2 = 4.$$

Rewriting the resulting equation in exponential form, we have

$$x = 3^4 = 81.$$

30. Using properties of logarithms, we have

$$\log_2 8^x = \frac{1}{2}\log_2\frac{1}{4}$$

$$x\log_2 8 = -\frac{1}{2}\log_2 4$$

$$3x = -\frac{1}{2}(2)$$

$$x = -\frac{1}{3}.$$

33. Combining the base 2 logarithms we have

$$\log_2 x(10-x) = 4$$

$$x(10-x) = 2^4 = 16$$

$$x^2 - 10x + 16 = (x-2)(x-8) = 0.$$

Thus $x = 2$ or $x = 8$.

36. Combining the base 10 logarithms we have $\log_{10}\dfrac{54}{2} = \log_{10}\dfrac{x^2}{\sqrt{x}}$. Then, since $\log_{10} x$ is a one-to-one function,

$$\frac{54}{2} = \frac{x^2}{\sqrt{x}}$$

$$27 = x^{3/2}$$

$$x = 27^{2/3} = 9.$$

39. We use the laws of logarithms and the fact that $\ln x$ is one-to-one:

$$\ln 3 + \ln(2x-1) = \ln 4 + \ln(x+1)$$

$$\ln[3(2x-1)] = \ln[4(x+1)]$$

$$3(2x-1) = 4(x+1)$$

$$6x - 3 = 4x + 4$$

$$2x = 7$$

$$x = \frac{7}{2}.$$

42. We write the equation as

$$(8^2)^x - 10(8^x) + 16 = (8^x)^2 - 10(8^x) + 16 = (8^x - 2)(8^x - 8) = 0.$$

Then

$$8^x = 2, \quad (2^3)^x = 2, \quad 2^{3x} = 2, \quad \text{or} \quad 3x = 1$$

and $x = \frac{1}{3}$. Also, $8^x = 8$ or $x = 1$. The solutions are $x = \frac{1}{3}$ and $x = 1$.

45. Since the equation does not factor with integer coefficients, we apply the quadratic formula to obtain

$$5^x = \frac{-(-2) \pm \sqrt{(-2)^2 - 4(1)(-1)}}{2(1)} = \frac{2 \pm \sqrt{4+4}}{2} = \frac{2 \pm 2\sqrt{2}}{2} = 1 \pm \sqrt{2}.$$

Since $5^x > 0$, we disregard $1 - \sqrt{2}$ and we solve $5^x = 1 + \sqrt{2}$. Taking the logarithm base of 5 of both sides, we have $x = \log_5(1 + \sqrt{2})$.

48. Writing

$$(\log_{10} 2x)^2 - \log_{10}(2x)^2 = (\log_{10} 2x)^2 - 2\log_{10} 2x = 0$$

and then factoring, we have

$$\log_{10} 2x (\log_{10} 2x - 2) = 0.$$

Thus, $\log_{10} 2x = 0$ and $\log_{10} 2x = 2$. Writing these solutions in exponential form, we obtain $10^0 = 1 = 2x$ and $10^2 = 100 = 2x$, so $x = \frac{1}{2}$ and $x = 50$.

51. The x-intercept is the solution of $e^{x+4} - e = 0$ or $e^{x+4} = e^1$. Thus, $x + 4 = 1$ and $x = -3$, so the x-intercept is $(-3, 0)$.

54. The x-intercept is the solution of $-3^{2x} + 5 = 0$. From this equation we see that $3^{2x} = 5$ or $2x \log_{10} 3 = \log_{10} 5$. Thus $x = \frac{1}{2}\left(\frac{\log_{10} 5}{\log_{10} 3}\right) \approx 0.7325$, and the x-intercept is $(0.7325, 0)$.

57. $f(x) = 2^x - 5 \Rightarrow f(0) = 1 - 5 = -4 \Rightarrow y$-intercept is $(0, -4)$. The x-coordinate of the x-intercept is given by:

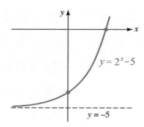

$$2^x - 5 = 0 \Rightarrow x = \log_2 5 = \frac{\ln 5}{\ln 2} \approx 2.3219 \Rightarrow x\text{-intercept } (2.3219, 0).$$

60. The graph of $f(x)$ is shown in blue and the graph of $g(x)$ is shown in red. The approximate x-intercept of the point of intersection is $x = 0.5850$.

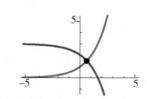

63. The graph of $f(x)$ is shown in blue and the graph of $g(x)$ is shown in red. The approximate x-intercept of the point of intersection $x = 3.1623$.

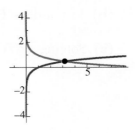

66. Taking $\log_{10}$ of both sides we have

$$\log_{10} x^{\log_{10} x} = \log_{10} \frac{1000}{x^2} = 3\log_{10} 10 - 2\log_{10} x$$

$$(\log_{10} x)(\log_{10} x) = 3 - 2\log_{10} x \qquad \leftarrow 1000 = 10^3$$

$$(\log_{10} x)^2 + 2\log_{10} x - 3 = 0$$

$$(\log_{10} x + 3)(\log_{10} x - 1) = 0.$$

Thus $\log_{10} x = -3$ and $\log_{10} x = 1$. Writing these equation in exponential form we have $x = 10^{-3} = 0.001$ and $x = 10$.

69. We want to solve $6^x = 51$. This is equivalent to

$$\ln 6^x = \ln 51$$

$$x \ln 6 = \ln 51$$

$$x = \frac{\ln 51}{\ln 6} \approx 2.1944.$$

6.4   Exponential and Logarithmic Models

3. The initial population is 1500, so an increase of 25% is $0.25(1500) = 375$. Thus, $P(10) = 1500e^{10k} = 1875$. Solving for k gives

$$e^{10k} = \frac{1875}{1500}$$

$$10k = \ln \frac{1875}{1500} \approx 0.2231$$

$$k = 0.0223.$$

Thus, $P(t) = 1500e^{0.0223t}$ and $P(20) = 1500e^{0.0223(2)} \approx 2344$.

6. (a) We are given $P(2) = 200$ and $P(5) = 400$. That is,

$$200 = P_0 e^{2k} \qquad \text{and} \qquad 400 = P_0 e^{5k}.$$

Dividing, we have $e^{5k}/e^{2k} = 400/200 = 2$ or $e^{3k} = 2$. Thus, $3k = \ln 2$ and $k = \frac{1}{3}\ln 2 \approx 0.2310$. Now, solving $P(2) = P_0 E^{0.2310(2)} = 200$ for P_0, we find $P_0 = 200e^{-0.4621} \approx 126$. Thus, the model is $P(t) = 126e^{0.2310t}$.

(b) In ten days $P(10) \approx 1270$.

(c) Solving $P(t) = 126e^{0.2310t} = 5000$ for t, we have

$$e^{0.2310t} = \frac{5000}{126}$$

$$0.2310t = \ln \frac{5000}{126}$$

$$t = \frac{1}{0.2310} \ln \frac{5000}{126} \approx 15.9 \text{ days}.$$

9. We are given $A_0 = 200$, so $A(t) = 200e^{kt}$. Since 3% of 200 is 6, $A(6) = 200e^{6k} = 194$ and

$$e^{6k} = \frac{194}{200}$$

$$6k = \ln \frac{194}{200}$$

$$k = \frac{1}{6} \ln \frac{194}{200} \approx -0.0051.$$

Thus, $A(t) = 200e^{-0.0051t}$ and $A(24) \approx 177.1$ mg.

12. We are given $A_0 = 400$, so $A(t) = 400e^{kt}$. After 8 hours, one-half, or 200 grams, remains. Solving $A(8) = 400e^{8k} = 200$ for k we have

$$e^{8k} = \frac{200}{400} = \frac{1}{2}$$

$$8k = \ln \frac{1}{2}$$

$$k = \frac{1}{8} \ln \frac{1}{2} \approx -0.0866.$$

Thus, $A(t) = 400e^{-0.0866t}$ so $A(17) \approx 91.7$, $A(23) \approx 54.5$, and $A(33) \approx 22.9$.

15. The initial amount is not given so we represent it using A_0 and express the remaining amounts in terms of A_0. Since the half-life is 140 days, $A(140) = A_0 e^{140k} = \frac{1}{2}A_0$, which we solve for k:

$$e^{140k} = \frac{1}{2}$$

$$140k = \ln \frac{1}{2}$$

$$k = \frac{1}{140} \ln \frac{1}{2} \approx -0.0050.$$

Thus, $A(t) = A_0 e^{-0.0050t}$ and

$$A(80) = A_0 e^{-0.0050(80)} \approx 0.6730 A_0$$

$$A(300) = A_0 e^{-0.0050(300)} \approx 0.2264 A_0.$$

18. From Example 4 in the text, we see that the amount of C-14 remaining at t years is given by $A(t) = A_0 e^{-0.00012097t}$, where A_0 is the initial amount of C-14 in the organism. We want to solve this equation for t when $A(t) = 0.10 A_0$, and again when $A(t) = 0.05 A_0$. (When 90% is lost, it means that 10% is remaining.) This gives

$$A_0 e^{-0.00012097t} = 0.10 A_0$$

$$-0.00012097t = \ln 0.1$$

$$t = -\frac{1}{0.00012097} \ln 0.1 \approx 19{,}034$$

and

$$A_0 e^{-0.00012097t} = 0.05 A_0$$

$$-0.00012097t = \ln 0.05$$

$$t = -\frac{1}{0.00012097} \ln 0.05 \approx 24{,}764,$$

respectively, thus, the bone is between 19,034 and 24,764 years old.

21. We identify $T_0 = 400$, $T_m = 80$, and $T(3) = 275$, where $T(t) = T_m + (T_0 - T_m)e^{kt}$. Using this information, we solve for k:

$$T(3) = 80 + (400 - 80)e^{3k} = 275$$

$$320e^{3k} = 195$$

$$e^{3k} = \frac{195}{320}$$

$$3k = \ln \frac{195}{320}$$

$$k = \frac{1}{3} \ln \frac{195}{320} \approx -0.1651.$$

Thus, $T(t) = 80 + 320e^{-0.1651t}$.

(a) $T(5) \approx 220.2°\text{F}$

(b) We solve $T(t) = 150$ for t:

$$80 + 320e^{-0.1651t} = 150$$

$$320e^{-0.1651t} = 70$$

$$e^{-0.1651t} = \frac{7}{32}$$

$$-0.1651t = \ln \frac{7}{32}$$

$$t = -\frac{1}{0.1651} \ln \frac{7}{32} \approx 9.2 \text{ min.}$$

(c) After a very long time, $e^{-0.1651t}$ approaches 0, so T approaches $80°$ F.

24. We identify $T_m = 5$, $T(1) = 59$, and $T(5) = 32$. Then

$$5 + (T_0 - 5)e^k = 59 \qquad \text{and} \qquad 5 + (T_0 - 5)e^{5k} = 32$$

or

$$(T_0 - 5)e^k = 54 \qquad \text{and} \qquad (T_0 - 5)e^{5k} = 27.$$

Dividing, we have $e^{4k} = \frac{1}{2}$, so $e^k = \left(\frac{1}{2}\right)^{1/4}$. Then

$$(T_0 - 5)\left(\frac{1}{2}\right)^{1/4} = 54 \qquad \text{or} \qquad T_0 = 54\left(\frac{1}{2}\right)^{-1/4} + 5 \approx 69.2° \text{ F}.$$

Thus, the temperature inside the house is $69.2°$ F.

27. Using $S(t) = Pe^{rt}$ we identify $P = 0.01$ and $r = 0.01$. Then

$$S(2000) = 0.01e^{0.01(2000)} \approx \$4{,}851{,}651.95.$$

If $r = 0.02$, then

$$S(2000) = 0.01e^{0.02(2000)} \approx \$2{,}353{,}852{,}668{,}370{,}199.85 \approx \$2.35 \times 10^{15}.$$

30. Using $P(t) = Se^{-rt}$ we identify $S = 100{,}000$, $r = 0.03$, and $t = 30$. then

$$P(30) = 100{,}000e^{-0.03(30)} \approx \$40{,}656.97.$$

33. From

$$I = \frac{E}{R}\left(1 - e^{-(R/L)t}\right)$$

we have

$$\frac{IR}{E} = 1 - e^{-(R/L)t}$$

$$e^{-(R/L)t} = 1 - \frac{IR}{E}$$

$$-\frac{R}{L}t = \ln\left(1 - \frac{IR}{E}\right)$$

$$t = -\frac{L}{R}\ln\left(1 - \frac{IR}{E}\right).$$

36. The magnitude of an earthquake is defined by $M = \log_{10}(A/A_0)$, where A and A_0 are amplitudes of seismic waves. Using this formula, we have

$$9.3 = \log_{10}\left(\frac{A}{A_0}\right)_{\text{Sumatra}} \quad \text{and} \quad 8.9 = \log_{10}\left(\frac{A}{A_0}\right)_{\text{Alaska}}.$$

Then

$$\left(\frac{A}{A_0}\right)_{\text{Sumatra}} = 10^{9.3} \quad \text{and} \quad \left(\frac{A}{A_0}\right)_{\text{Alaska}} = 10^{8.9}.$$

Since $9.3 = 8.9 + 0.4$, it follows from the laws of exponents that

$$\left(\frac{A}{A_0}\right)_{\text{Sumatra}} = 10^{9.3} = 10^{0.4}10^{8.9} = 10^{0.4}\left(\frac{A}{A_0}\right)_{\text{Alaska}} \approx 2.51\left(\frac{A}{A_0}\right)_{\text{Alaska}}.$$

Thus, the Sumatra earthquake was about $2\frac{1}{2}$ times as intense as the Alaska earthquake.

39. Since $\text{pH} = -\log_{10}[\text{H}^+]$ where $[\text{H}^+]$ is the hydrogen-ion concentration, we have

$$\text{pH}(10^{-6}) = -\log_{10}10^{-6} = 6.$$

42. Since $\text{pH} = -\log_{10}[\text{H}^+]$ where $[\text{H}^+]$ is the hydrogen-ion concentration, we have

$$\text{pH}(5.1 \times 10^{-5}) = -\log_{10}(5.1 \times 10^{-5}) = -\log_{10}5.1 - \log_{10}10^{-5}$$

$$= 5 - \log_{10}5.1 \approx 5 - 0.71 = 4.29.$$

45. We need to solve $6.6 = -\log_{10}[\text{H}^+]$:

$$-6.6 = \log_{10}[\text{H}^+]$$

$$10^{-6.6} = [\text{H}^+]$$

$$[\text{H}^+] \approx 2.5 \times 10^{-7}.$$

48. From the formula for pH we have

$$1 = -\log_{10}[\text{H}^+]_{\text{battery acid}} \quad \text{and} \quad 13 = -\log_{10}[\text{H}^+]_{\text{lye}}.$$

Then

$$[\text{H}^+]_{\text{battery acid}} = 10^{-1} \quad \text{and} \quad [\text{H}^+]_{\text{lye}} = 10^{-13}.$$

Since $-1 = 12 - 13$, it follows from the laws of exponents that

$$[\text{H}^+]_{\text{battery acid}} = 10^{-1} = 10^{12}10^{-13} = 10^{12}[\text{H}^+]_{\text{lye}}.$$

Thus, lye is 10^{12} times more acidic than battery acid.

51. (a) We need to solve $M = \frac{2}{3}[\log_{10} E - 11.8]$ for E given that $M = 9.3$:

$$\frac{2}{3}[\log_{10} E - 11.8] = 9.3$$

$$\log_{10} E - 11.8 = \frac{3}{2}(9.3) = 13.95$$

$$\log_{10} E = 25.7$$

$$E = 10^{25.75} \approx 5.62 \times 10^{25} \text{ ergs.}$$

(b) Solving

$$M = \frac{2}{3}[\log_{10} E - 11.8]$$

for E we have

$$\frac{3}{2}M = \log_{10} E - 11.8$$

$$\log_{10} E = \frac{3}{2}M + 11.8$$

$$E = 10^{\frac{3}{2}M + 11.8}.$$

Then

$$f_{\Delta E} = \frac{E_1}{E_2} = \frac{10^{\frac{3}{2}M_1 + 11.8}}{10^{\frac{3}{2}M_2 + 11.8}} = 10^{\frac{3}{2}[M_1 - M_2]}.$$

(c) Letting $M_1 = 1 + M_2$ we have from part **(b)** that

$$E_1 = 10^{\frac{3}{2}M_1 + 11.8} = 10^{\frac{3}{2}(1 + M_2) + 11.8} = 10^{\frac{3}{2}}10^{\frac{3}{2}M_2 + 11.8} = 10^{\frac{3}{2}}E_2.$$

Since $10^{\frac{3}{2}} \approx 31.62$ we see that E_1 is about $32\,E_2$.

If $M_1 = 2 + M_2$, we have from part **(b)** that

$$E_1 = 10^{\frac{3}{2}M_1 + 11.8} = 10^{\frac{3}{2}(2 + M_2) + 11.8} = 10^3 10^{\frac{3}{2}M_2 + 11.8} = 10^3 E_2.$$

Since $10^3 = 1000$ we see that E_1 is $1000\,E_2$ in this case.

54. From (10) in the text, we have

$$b_1 = 10 \log_{10} \frac{I_1}{I_0} \quad \text{and} \quad b_2 = 10 \log_{10} \frac{I_2}{I_0}.$$

Then

$$b_2 - b_1 = 10 \log_{10} \frac{I_2}{I_0} - 10 \log_{10} \frac{I_1}{I_0} = 10 \log_{10} \frac{I_2/I_0}{I_1/I_0} = 10 \log_{10} \frac{I_2}{I_1}.$$

From (11) in the text, we have

$$I_1 = \frac{k}{d_1^2} \quad \text{and} \quad I_2 = \frac{k}{d_2^2},$$

so

$$b_2 - b_1 = 10 \log_{10} \frac{I_2}{I_1} = 10 \log_{10} \frac{k/d_2^2}{k/d_1^2} = 10 \log_{10} \left(\frac{d_1}{d_2}\right)^2 = 20 \log_{10} \frac{d_1}{d_2}.$$

Thus,

$$b_2 = b_1 + 2 \log_{10} \frac{d_1}{d_2}.$$

57. (a) When d is the pupil diameter and $B = 255$, we have

$$\log_{10} d = 0.8558 - 0.000401(8.1 + \log_{10} 255)^3 \approx 0.3907,$$

so

$$d = 10^{0.3907} \approx 2.46 \text{ mm}.$$

(b) From $\log_{10} d = -0.10446$ when $B = 190{,}000$ mL we have $d = 10^{-0.10446} \approx 0.79$ mm, and from $\log_{10} d = -0.728144$ when $B = 51{,}000{,}000$ mL we have $d = 10^{-0.728144} \approx 0.19$ mm.

(c) We want to solve

$$\log_{10} 7 = 0.8558 - 0.000401(8.1 + \log_{10} B)^3$$

for B:

$$0.8451 = 0.8558 - 0.000401(8.1 + \log_{10} B)^3$$

$$0.000401(8.1 + \log_{10} B)^3 = 0.0107$$

$$(8.1 + \log_{10} B)^3 = \frac{0.0107}{0.000401} \approx 26.6882$$

$$8.1 + \log_{10} B = 26.6882^{1/3} \approx 2.9884$$

$$\log_{10} B = 2.9884 - 8.1 = -5.1116$$

$$B = 10^{-5.1116} \approx 7.73 \times 10^{-6} \text{ mL}.$$

6.5 $\int$ **Calculus PREVIEW**   **The Hyperbolic Functions**

3. From Problem 2, we have that the derivative of $f(x) = \log_b x$ is $f'(x) = (1/x) \log_b e$. Thus, when $f(x) = \log_{10} x$, we have $f'(x) = (1/x) \log_{10} e$. Using (7) from Section 6.3 in the text, we can also write

$$f'(x) = \frac{1}{x} \log_{10} e = \frac{1}{x} \frac{\ln e}{\ln 10} = \frac{1}{x \ln 10}.$$

6. We first write $f(x) = \log_{10} 6x = \log_{10} 6 + \log_{10} x$. We then compute the difference quotient:

$$\frac{f(x+h) - f(x)}{h} = \frac{\log_{10} 6 + \log_{10}(x+h) - [\log_{10} 6 + \log_{10} x]}{h}$$

$$= \frac{\log_{10}(x+h) - \log_{10} x}{h}.$$

Thus,

$$\lim_{h \to 0} \frac{f(x+h) - f(x)}{h} = \lim_{h \to 0} \frac{\log_{10}(x+h) - \log_{10} x}{h},$$

which we recognize as the derivative of $\log_{10} x$. By Problem 3, this is $1/(x \ln 10)$, so the derivative of $f(x) = \log_{10} 6x$ is $f'(x) = 1/(x \ln 10)$.

9. Using $\cosh x = \frac{1}{2}(e^x + e^{-x})$ and $\sinh x = \frac{1}{2}(e^x - e^{-x})$, we have

$$\cosh^2 x - \sinh^2 x = \frac{1}{4}(e^x + e^{-x})^2 - \frac{1}{4}(e^x - e^{-x})^2$$

$$= \frac{1}{4}(e^{2x} + 2 + e^{-2x}) - \frac{1}{4}(e^{2x} - 2 + e^{-2x})$$

$$= \frac{2}{4} + \frac{2}{4} = 1.$$

12. Using $\sinh x = \frac{1}{2}(e^x - e^{-x})$, we have

$$\sinh(-x) = \frac{1}{2}\left(e^{-x} - e^{-(-x)}\right) = -\frac{1}{2}(e^x - e^{-x}) = -\sinh x.$$

15. (a) If $\sinh x = -\frac{3}{2}$, use the identity given in Problem 9 to find the value of $\cosh x$.

$$\cosh^2 x - \left(-\frac{3}{2}\right)^2 = 1 \Rightarrow \cosh^2 x = \frac{13}{4} \Rightarrow \cosh x = \frac{\sqrt{13}}{2} \Rightarrow$$

(b) Use the result of part **(a)** to find the numerical values of $\tanh x$, $\coth x$, $\mathrm{sech}x$, and $\mathrm{csch}x$.

$$\tanh x = \frac{-3/2}{\sqrt{13}/2} = -\frac{3}{\sqrt{13}} \Rightarrow \coth x = -\frac{\sqrt{13}}{3},$$

$$\mathrm{sech}x = \frac{2}{\sqrt{13}}, \quad \mathrm{csch}x = -\frac{2}{3}$$

18. (a)

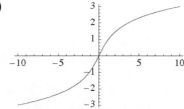

(b) Since the domain and range of $y = \sinh x$ are both $(-\infty, \infty)$, this is also the domain and range of $y = \sinh^{-1} x$.

21. The domain of $y = \tanh^{-1} x$ is $(-1, 1)$ and its range is $(-\infty, \infty)$.

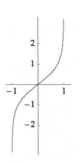

Chapter 6 Review Exercises

A. Fill in the Blanks

3. Since $y = \ln x$ has x-intercept at 1 and vertical asymptote $x = -4$, we use the fact that the graph of $y = \ln(x + 4)$ is the graph of $y = \ln x$ shifted 4 units to the left to see that $y = \ln(x + 4)$ has x-intercept $(-3, 0)$ and vertical asymptote $x = -4$.

6. $6 \ln e + 3 \ln \dfrac{1}{e} = 6(1) + 3 \ln e^{-1} = 6 - 3 \ln e = 6 - 3 = 3$

9. $\log_4(4 \cdot 4^2 \cdot 4^3) = \log_4 4^{1+2+3} = 1 + 2 + 3 = 6$

12. If $\log_b 6 = \frac{1}{2}$, then $b^{1/2} = 6$ and $b = 36$.

15. If $-1 + \ln(x - 3) = 0$, then

$$\ln(x - 3) = 1$$

$$x - 3 = e$$

$$x = 3 + e.$$

18. If $3^x = 5$, then

$$3^{-2x} = (3^x)^{-2} = 5^{-2} = \frac{1}{5^2} = \frac{1}{25}.$$

21. We let $x = 2$ and $y = 9$ and solve for C:

$$9 = e^{2-2} + c = 1 + c$$

$$c = 8.$$

B. True/False

3. True; since $y = 10^{-x} = (10^{-1})^x = \left(\frac{1}{10}\right)^x = (0.1)^x$.

6. False; since $\dfrac{2^{x^2}}{2^x} = 2^{x^2-x} = 2^{x(x-1)}$.

9. True; since $-\ln 2 = \ln(2^{-1}) = \ln(\tfrac{1}{2})$.

12. True; since $\ln\sqrt{43} = \ln(43^{1/2}) = \dfrac{1}{2}\ln 43 = \dfrac{\ln 43}{2}$.

15. True. $\ln\dfrac{1}{e} = -1$ **16.** False. $\dfrac{\ln 10}{\ln 2} = \ln 5$

18. True. If $f(x) = b^x$, then $f(nx) = [f(x)]^n$.

21. True. $\ln e \cdot \ln e^2 \cdots \ln e^n = n!$

C. Review Exercises

3. $\log_9 27 = 1.5$ is equivalent to $9^{1.5} = 27$.

6. Since $3^{2x} = 81 = 3^4$, $2x = 4$ and $x = 2$.

9. Taking the logarithm base 2 of both sides:

$$\log_2 2^{1-x} = \log_2 7$$

$$1 - x = \log_2 7$$

$$x = 1 - \log_2 7.$$

Alternatively, taking the natural logarithm of both sides:

$$\ln 2^{1-x} = \ln 7$$

$$(1 - x)\ln 2 = \ln 7$$

$$x = 1 - \frac{\ln 7}{\ln 2}.$$

12. Dividing both sides of the equation by e^x:

$$3 = 4\frac{e^{-3x}}{e^x} = 4e^{-4x}$$

$$e^{-4x} = \frac{3}{4}$$

$$-4x = \ln\frac{3}{4}$$

$$x = -\frac{1}{4}\ln\frac{3}{4} \approx 0.0719.$$

15. The graph of $y = 4^x$ is shown in blue and the graph of $y = \log_4 x$ is shown in red.

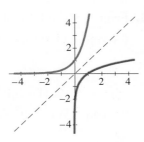

18. For $y = \log_2 x$, when $x = 3$, $y = \log_2 3$. The corresponding point on the graph of $y = 2^x$ is $(\log_2 3, 3)$. Solving $\log_2 2 = \log_2 x$, we see that $x = 2$. The corresponding point on the graph of $y = 2^x$ is $(\log_2 2, 2)$. Solving $-2 \log_2 2 = \log_2 x$, we see that $x = \frac{1}{4}$. The corresponding point on the graph of $y = 2^x$ is $(-2 \log_2 2, \frac{1}{4})$. Finally, solving $2^x = 1$, we see that $x = 0$, and the corresponding point on the graph of $y = \log_2 x$ is $(1, 0)$.

21. This looks like the graph of $y = \ln x$ revolved around the x-axis (giving $-\ln x$) and shifted up 2 units. Thus, this is the graph of (ii) $y = 2 - \ln x$.

24. This looks like the graph of $y = \ln x$ revolved around the x-axis (giving $-\ln x$). But then the graph should pass through $(-1, 0)$. Instead, it appears to pass through $(\frac{1}{2}, 0)$. Thus, this is the graph of (v) $y = -\ln(2x)$.

27. Since $\ln e = \ln e + \ln x = 1 + \ln x$ the graph of $f(x)$ is the graph of $\ln x$ shifted up by 1 unit.

30. Since $f(0) = \frac{1}{2}$, we see that $\frac{1}{2} = A10^{k \cdot 0} = A$ and the function is $f(x) = \frac{1}{2}(10^{kx})$. Since $f(3) = \frac{1}{2}(10^{3k}) = 8$, we have $10^{3k} = 16$. Taking the logarithm base 10 of both sides, we have

$$\log_{10} 10^{3k} = \log_{10} 16$$

$$3k = \log_{10} 16$$

$$k = \frac{\log_{10} 16}{3}$$

and

$$f(x) = \frac{1}{2}\left(10^{\left[(\log_{10} 16)/3\right]x}\right) \approx \frac{1}{2}\left(10^{0.4014x}\right).$$

33. After doubling, it will take another 9 hours to double again. Thus, it will take a total of 18 hours for the population to grow to 4 times the initial population.

36. From Example 4 of Section 6.4 in the text, we see that the amount of C-14 remaining at t years is given by

$$A(t) = A_0 e^{-0.00012097t},$$

where A_0 is the initial amount of C-14. We want to solve this equation for t when $A(t) = 0.03A_0$. (When 97% is lost, it means that 3% is remaining.) This gives

$$A_0 e^{-0.00012097t} = 0.03A_0$$

$$-0.00012097t = \ln 0.03$$

$$t = \frac{\ln 0.03}{-0.00012097} \approx 28{,}987.$$

Thus, the skeleton is approximately 28,987 years old.

39. Solving $y = ae^{-be^{-ct}}$ we find

$$\frac{y}{a} = e^{-be^{-ct}}$$

$$\ln y - \ln a = -be^{-ct}$$

$$\ln a - \ln y = be^{-ct}$$

$$\frac{1}{b}(\ln a - \ln y) = e^{-ct}$$

$$\ln \frac{1}{b} + \ln(\ln a - \ln y) = -ct$$

$$\ln(\ln a - \ln y) - \ln b = -ct$$

$$t = -\frac{1}{c}\left[\ln(\ln a - \ln y) - \ln b\right]$$

$$t = \frac{1}{c}\left[\ln b - \ln(\ln a - \ln y)\right]$$

42. Writing $x = \log_{a^2} b^2$ as an exponential equation we have

$$\left(a^2\right)^x = b^2$$

$$a^{2x} = b^2$$

$$\left(a^x\right)^2 = b^2 \qquad \leftarrow \text{since both } a \text{ and } b \text{ are positive}$$

$$a^x = b.$$

Writing the last equation in logarithmic form we have $x = \log_a b$. Thus,

$$\log_{a^2} b^2 = x = \log_a b.$$

45. $\ln y = \ln x + 3\ln 2 = \ln 8x \Rightarrow y = 8x$

48. $\ln y - \ln(x - 1) = \ln(x + 4) - \ln 2 \Leftrightarrow \ln y - \ln 2 = \ln(x + 4) - \ln(x - 1)$

$\Leftrightarrow \ln(y/2) = \ln[(x + 4)/(x - 1)] \Rightarrow y = \dfrac{2x + 8}{x - 1}$

51. $\ln y = -3x + \ln 5 \Leftrightarrow \ln(y/5) = -3x \Rightarrow y/5 = e^{-3x} \Rightarrow y = 5e^{-3x}$

Chapter 7

Conic Sections

7.1 | The Parabola

*In Problems 3–24 the directrix of the parabola is shown as a dashed line (unless it coincides with one of the coordinate axes), the vertex is indicated with a **V**, and the focus with an **F**.*

3. The form of the parabola is $y^2 = 4cx$, $c < 0$, so the parabola opens to the left, and $4c = -\frac{4}{3}$ so $c = -\frac{1}{3}$. The vertex is at the origin, the focus is at $F(c, 0)$ or $(-\frac{1}{3}, 0)$, the directrix is $x = -c = \frac{1}{3}$, and the axis is $y = 0$. The ends of the focal chord are $(-\frac{1}{3}, \pm\frac{2}{3})$. The graph of the parabola is shown, along with its focus F and directrix.

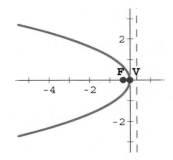

6. The form of the parabola is $x^2 = 4cy$, $c > 0$, so the parabola opens up, and $4c = \frac{1}{10}$ so $c = \frac{1}{40}$. The vertex is at the origin, the focus is at $F(0, c)$ or $(0, \frac{1}{40})$, the directrix is $y = -c = -\frac{1}{40}$, and the axis is $x = 0$. The ends of the focal chord are $(\pm\frac{1}{20}, \frac{1}{40})$. The graph of the parabola is shown, along with its focus F and directrix.

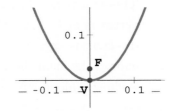

9. The form of the parabola is $(y - k)^2 = 4c(x - h)$, $c > 0$, so the parabola opens to the right, and $4c = 16$ so $c = 4$. The vertex is at $(0, 1)$, the focus is at $F(h + c, k)$ or $(4, 1)$, the directrix is $x = -c = -4$, and the axis is $y = 1$. The ends of the focal chord are $(4, \pm 8)$. The graph of the parabola is shown, along with its focus F and directrix.

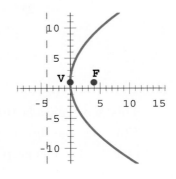

12. The form of the parabola is $(x-h)^2 = 4c(y-k)$, $c < 0$, so the parabola opens down, and $4c = -1$ so $c = -\frac{1}{4}$. The vertex is at $(2,0)$, the focus is at $F(2,c)$ or $(2, -\frac{1}{4})$, the directrix is $y = -c = \frac{1}{4}$, and the axis is $x = 2$. The ends of the focal chord are $(2 \pm \frac{1}{2}, -\frac{1}{4})$. The graph of the parabola is shown, along with its focus F and directrix.

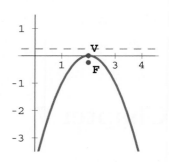

15. We rewrite the equation in standard form (3) in the text:

$$x^2 + 5x \qquad = \frac{1}{4}y - 6$$

$$x^2 + 5x + \frac{25}{4} = \frac{1}{4}y - 6 + \frac{25}{4}$$

$$\left(x + \frac{5}{2}\right)^2 = \frac{1}{4}y + \frac{1}{4} = 4\left(\frac{1}{16}\right)(y+1).$$

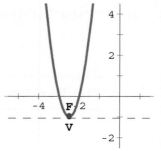

This is a parabola that opens up, with $c = \frac{1}{16}$. The vertex is at $(-\frac{5}{2}, -1)$, the focus is at $F(-\frac{5}{2}, -1 + \frac{1}{16})$ or $(-\frac{5}{2}, -\frac{15}{16})$, the directrix is $y = -1 - \frac{1}{16} = -\frac{17}{16}$, and the axis is $x = -\frac{5}{2}$. The ends of the focal chord are $(-\frac{5}{2} \pm \frac{1}{8}, -\frac{15}{16})$. The graph of the parabola is shown, along with its focus F and directrix.

18. We rewrite the equation in standard form (4) in the text:

$$y^2 - 4y \qquad = 4x - 3$$

$$y^2 - 4y + 4 = 4x - 3 + 4$$

$$(y-2)^2 = 4x + 1 = 4\left(x + \frac{1}{4}\right).$$

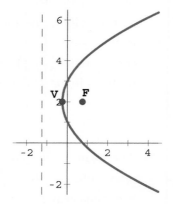

This is a parabola that opens to the right, with $c = 1$. The vertex is at $(-\frac{1}{4}, 2)$, the focus is at $F(-\frac{1}{4} + 1, 2)$ or $(\frac{3}{4}, 2)$, the directrix is $x = -\frac{1}{4} - 1 = -\frac{5}{4}$, and the axis is $y = 2$. The ends of the focal chord are $(\frac{3}{4}, 2 \pm 2)$. The graph of the parabola is shown, along with its focus F and directrix.

21. We rewrite the equation in standard form (3) in the text:

$$2x^2 - 12x \qquad = -8y - 18$$

$$x^2 - 6x \qquad = -4y - 9$$

$$x^2 - 6x + 9 = -4y - 9 + 9$$

$$(x-3)^2 = -4y.$$

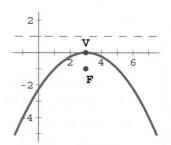

This is a parabola that opens down, with $c = -1$. The vertex is at $(3,0)$, the focus is at $F(3,-1)$, the directrix is $y = 1$, and the axis is $x = 3$. The ends of the focal chord are $(3 \pm 2, -1)$. The graph of the parabola is shown, along with its focus F and directrix.

24. We rewrite the equation in standard form (3) in the text:

$$3x^2 + 30x = 8y - 75$$

$$x^2 + 10x = \frac{8}{3}y - 25$$

$$x^2 + 10x + 25 = \frac{8}{3}y - 25 + 25$$

$$(x+5)^2 = 4\left(\frac{2}{3}\right)y.$$

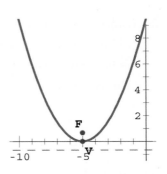

This is a parabola that opens up, with $c = \frac{2}{3}$. The vertex is at $(-5,0)$, the focus is at $F(-5,\frac{2}{3})$, the directrix is $y = -\frac{2}{3}$, and the axis is $x = -5$. The ends of the focal chord are $(-5 \pm \frac{4}{3}, \frac{2}{3})$. The graph of the parabola is shown, along with its focus F and directrix.

27. Since the directrix is a vertical line and the focus is to the left of the directrix, the parabola opens to the left. Its equation, then, has the form $(y - k)^2 = 4c(x - h)$, where $c < 0$. The vertex is halfway between the focus and the directrix, so $h = 0$ and $k = 0$. Since the distance from the focus to the vertex is $|c|$, we have $c = -4$. Therefore, the equation of the parabola is

$$(y - 0)^2 = 4(-4)(x - 0) \qquad \text{or} \qquad y^2 = -16x.$$

30. Since the axis is the line through the focus and the vertex, and the vertex is above the focus, the parabola opens down. Its equation, then, has the form $(x - h)^2 = 4c(y - k)$, where $c < 0$. The vertex is $(0,0)$ and, since the distance from the vertex to the focus is $|c|$, we have $c = -10$. Therefore, the equation of the parabola is

$$(x - 0)^2 = 4(-10)(y - 0) \qquad \text{or} \qquad x^2 = -40y.$$

33. Since the directrix is a vertical line and the focus is to the left of the directrix, the parabola opens to the left. Its equation, then, has the form $(y - k)^2 = 4c(x - h)$, where $c < 0$. The vertex is halfway between the focus and the directrix, so $h = 2$ and $k = 4$. Since the distance from the focus to the vertex is $|c|$, we have $c = -3$. Therefore, the equation of the parabola is

$$(y - 4)^2 = 4(-3)(x - 2) \qquad \text{or} \qquad (y - 4)^2 = -12(x - 2).$$

36. Since the axis is the line through the focus and the vertex, and the vertex is above the focus, the parabola opens down. Its equation, then, has the form $(x - h)^2 = 4c(y - k)$, where $c < 0$. The vertex is $(-2,5)$ and, since the distance from the vertex to the focus is $|c|$, we have $c = -2$. Therefore, the equation of the parabola is

$$(x + 2)^2 = 4(-2)(y - 5) \qquad \text{or} \qquad (x + 2)^2 = -8(y - 5).$$

39. Since the directrix is a horizontal line and the vertex is above the directrix, the parabola opens up. Its equation, then, has the form $(x - h)^2 = 4c(y - k)$, where $c > 0$. The vertex is halfway between the focus and the directrix, so the focus is $(0, \frac{7}{4})$. Since the distance from the vertex to the directrix is $|c|$, we have $c = \frac{7}{4}$. Therefore, the equation of the parabola is

$$(x - 0)^2 = 4\left(\frac{7}{4}\right)(y - 0) \qquad \text{or} \qquad x^2 = 7y.$$

42. Since the directrix is a vertical line and the focus is to the left of the directrix, the parabola opens to the left. Its equation, then, has the form $(y - k)^2 = 4c(x - h)$, where $c < 0$. The vertex is halfway between the focus and the directrix, so the focus is $(-2, 4)$. Since the distance from the focus to the vertex is $|c|$, we have $c = -1$. Therefore, the equation of the parabola is

$$(y - 4)^2 = 4(-1)(x + 1) \qquad \text{or} \qquad (y - 4)^2 = -4(x + 1).$$

45. To find the x-intercepts, we set $y = 0$ and solve for x:

$$(0 + 4)^2 = 4(x + 1)$$
$$4 = x + 1$$
$$x = 3.$$

The x-intercept is $(3, 0)$. To find the y-intercepts, we set $x = 0$ and solve for y:

$$(y + 4)^2 = 4(0 + 1) = 4$$
$$y + 4 = \pm 2$$
$$y = \pm 2 - 4.$$

The y-intercepts are $(0, -6)$ and $(0, -2)$.

48. To find the x-intercepts, we set $y = 0$ and solve for x:

$$0 - 8(0) - x + 15 = 0$$
$$x = 15.$$

The x-intercept is $(15, 0)$. To find the y-intercepts, we set $x = 0$ and solve for y:

$$y^2 - 8y - 0 + 15 = 0$$
$$y^2 - 8y + 15 = 0$$
$$(y - 3)(y - 5) = 0.$$

The y-intercepts are $(0, 3)$ and $(0, 5)$.

Miscellaneous Applications

51. Since light emanating from the focus of a parabola is reflected along a line parallel to the axis of the parabola (the x-axis in this case), the equation of the reflected ray is $y = -2$, where $x \geq 1$.

54. Let the ground be the x-axis and suppose the vertex of the parabola is on the y-axis at $(0, 20)$. Then the equation of the parabola is $x^2 = 4c(y - 20)$, where $c < 0$. Since the point $(4, 18)$ is on the parabola, we have

$$4^2 = 4c(18 - 20)$$

$$4 = c(-2)$$

$$c = -2.$$

Thus, the equation of the parabola is $x^2 = -8(y - 20)$. The water hits the ground at an x-intercept of the parabola. Setting $y = 0$ in the equation of the parabola, we have $x^2 = 160$, so $x = \pm\sqrt{160} = \pm 4\sqrt{10}$. The water hits the ground $4\sqrt{10} \approx 12.65$ m from the point on the ground directly beneath the end of the pipe.

57. (a) For the parabola $x^2 = 8y = 4(2)y$, we identify $c = 2$. Since the vertex is at the origin, the parabola opens up around the y-axis. The ends of the focal chord are at $(-2c, c)$ and $(2c, c)$, so the focal width of the parabola is $2(2) - (-2)(2) = 8$.

(b) The parabolas $x^2 = 4cy$ and $y^2 = 4cx$ have the same size and shape, so their focal widths are the same, and we consider the parabola $x^2 = 4cy$. This is a parabola with vertex at the origin and axis on the y-axis. The focus is at $(0, c)$, so the ends of the focal chord are at $(-2c, c)$ and $(2c, c)$. The focal width of the parabola is $|2c - (-2c)| = 4|c|$.

7.2 | The Ellipse

3. We identify $h = 0$, $k = 0$, $a = \sqrt{16} = 4$, and $b = \sqrt{1} = 1$. Since the form of the equation is

$$\frac{x^2}{b^2} + \frac{y^2}{a^2} = 1,$$

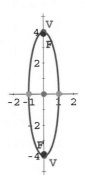

the major axis is vertical. From $b^2 = a^2 - c^2$ we have $c^2 = a^2 - b^2 = 16 - 1 = 15$, so $c = \sqrt{15}$. We use $e = c/a$ to find the eccentricity.

Center: $(0, 0)$;   *Foci*: $(0, -\sqrt{15}), (0, \sqrt{15})$;   *Vertices*: $(0, -4), (0, 4)$;

Endpoints of Minor Axis: $(-1, 0), (1, 0)$;   *Eccentricity*: $\dfrac{\sqrt{15}}{4}$

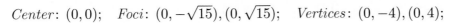

6. Dividing by 4, we write the equation as

$$\frac{x^2}{2} + \frac{y^2}{4} = 1.$$

We identify $h = 0$, $k = 0$, $a = \sqrt{4} = 2$, and $b = \sqrt{2}$. Since the form of the equation is

$$\frac{x^2}{b^2} + \frac{y^2}{a^2} = 1,$$

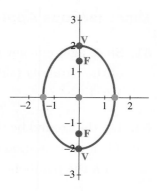

the major axis is vertical. From $b^2 = a^2 - c^2$ we have $c^2 = a^2 - b^2 = 4 - 2 = 2$, so $c = \sqrt{2}$. We use $e = c/a$ to find the eccentricity.

Center: $(0,0)$; *Foci:* $(0, -\sqrt{2}), (0, \sqrt{2})$; *Vertices:* $(0, -2), (0, 2)$;

Endpoints of Minor Axis: $(-\sqrt{2}, 0), (\sqrt{2}, 0)$; *Eccentricity:* $\dfrac{\sqrt{2}}{2}$

9. We identify $h = 1$, $k = 3$, $a = \sqrt{49} = 7$, and $b = \sqrt{36} = 6$. Since the form of the equation is

$$\frac{(x - h)^2}{a^2} + \frac{(y - k)^2}{b^2} = 1,$$

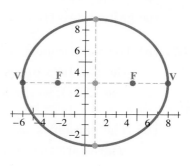

the major axis is vertical. From $b^2 = a^2 - c^2$ we have $c^2 = a^2 - b^2 = 49 - 36 = 13$, so $c = \sqrt{13}$. We use $e = c/a$ to find the eccentricity.

Center: $(1, 3)$; *Foci:* $(1 - \sqrt{13}, 3), (1 + \sqrt{13}, 3)$; *Vertices:* $(-6, 3), (8, 3)$;

Endpoints of Minor Axis: $(1, -3), (1, 9)$; *Eccentricity:* $\dfrac{\sqrt{13}}{7}$

12. We identify $h = 3$, $k = -4$, $a = \sqrt{81} = 9$, and $b = \sqrt{64} = 8$. Since the form of the equation is

$$\frac{(x - h)^2}{b^2} + \frac{(y - k)^2}{a^2} = 1,$$

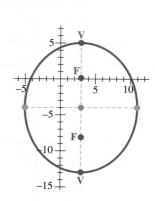

the major axis is vertical. From $b^2 = a^2 - c^2$ we have $c^2 = a^2 - b^2 = 81 - 64 = 17$, so $c = \sqrt{17}$. We use $e = c/a$ to find the eccentricity.

Center: $(3, -4)$; *Foci:* $(3, -4 - \sqrt{17}), (3, -4 + \sqrt{17})$;

Vertices: $(3, -13), (3, 5)$; *Endpoints of Minor Axis:* $(-5, -4), (11, -4)$; *Eccentricity:* $\dfrac{\sqrt{17}}{9}$

15. Dividing by 45, we write the equation as

$$\frac{(x-1)^2}{9} + \frac{(y+2)^2}{15} = 1.$$

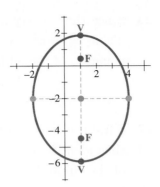

We identify $h = 1$, $k = -2$, $a = \sqrt{15}$, and $b = \sqrt{9} = 3$. Since the form of the equation is

$$\frac{(x-h)^2}{b^2} + \frac{(y-k)^2}{a^2} = 1,$$

the major axis is vertical. From $b^2 = a^2 - c^2$ we have $c^2 = a^2 - b^2 = 15 - 9 = 6$, so $c = \sqrt{6}$. We use $e = c/a$ to find the eccentricity.

Center: $(1, -2)$;   *Foci:* $(1, -2 - \sqrt{6}), (1, -2 + \sqrt{6})$;   *Vertices:* $(1, -2 - \sqrt{15}), (1, -2 + \sqrt{15})$;

Endpoints of Minor Axis: $(-2, -2), (4, -2)$;   *Eccentricity:* $\dfrac{\sqrt{6}}{\sqrt{15}} = \sqrt{\dfrac{2}{5}}$

18. Factoring 9 from the x^2 and x terms and 5 from the y^2 and y terms, and then completing the square, we have

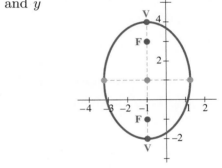

$$9(x^2 + 2x \quad) + 5(y^2 - 2y \quad) = 31$$

$$9(x^2 + 2x + 1) + 5(y^2 - 2y + 1) = 31 + 9 + 5$$

$$9(x + 1)^2 + 5(y - 1)^2 = 45$$

$$\frac{(x+1)^2}{5} + \frac{(y-1)^2}{9} = 1.$$

We identify $h = -1$, $k = 1$, $a = \sqrt{9} = 3$, and $b = \sqrt{5}$. Since the form of the equation is

$$\frac{(x-h)^2}{b^2} + \frac{(y-k)^2}{a^2} = 1,$$

the major axis is vertical. From $b^2 = a^2 - c^2$ we have $c^2 = a^2 - b^2 = 9 - 5 = 4$, so $c = 2$. We use $e = c/a$ to find the eccentricity.

Center: $(-1, 1)$;   *Foci:* $(-1, -1), (-1, 3)$;   *Vertices:* $(-1, -2), (-1, 4)$;

Endpoints of Minor Axis: $(-1 - \sqrt{5}, 1), (-1 + \sqrt{5}, 1)$;   *Eccentricity:* $\dfrac{2}{3}$

21. Since the vertices are at $(\pm 5, 0)$ and the foci are at $(\pm 3, 0)$, we identify $a = 5$ and $c = 3$. Then $b^2 = a^2 - c^2 = 25 - 9 = 16$ and $b = 4$. The major axis is the x-axis, so the equation of the ellipse is

$$\frac{x^2}{25} + \frac{y^2}{16} = 1.$$

24. Since the vertices are at $(0, \pm 7)$ and the foci are at $(0, \pm 3)$, we identify $a = 7$ and $c = 3$. Then $b^2 = a^2 - c^2 = 49 - 9 = 40$ and $b = 2\sqrt{10}$. The major axis is the y-axis, so the equation of the ellipse is

$$\frac{x^2}{40} + \frac{y^2}{49} = 1.$$

27. Since the center of the ellipse is the midpoint of the major axis, the center is at $(1, -3)$. We then identify $h = 1$, $k = -3$, $a = \frac{1}{2}[5 - (-3)] = 4$, and $b = \frac{1}{2}[-1 - (-5)] = 2$. The equation of the ellipse is

$$\frac{(x-1)^2}{16} + \frac{(y+3)^2}{4} = 1.$$

30. Since the center is at the origin and a focus is at $(1, 0)$, we have $c = 1$. Using $a = 3$ we see that $b^2 = a^2 - c^2 = 3^2 - 1^2 = 8$. The equation of the ellipse is

$$\frac{x^2}{9} + \frac{y^2}{8} = 1.$$

33. The equation is centered at the origin with its major axis along the y-axis, so the form of its equation is

$$\frac{x^2}{b^2} + \frac{y^2}{a^2} = 1 \qquad \text{or} \qquad \frac{x^2}{a^2 - c^2} + \frac{y^2}{a^2} = 1.$$

Since the distance from the center to a focus is $c = 3$ and the ellipse passes through $P(-1, 2\sqrt{2})$, we have

$$\frac{(-1)^2}{a^2 - 9} + \frac{(2\sqrt{2})^2}{a^2} = 1$$
$$a^2 + 8(a^2 - 9) = a^2(a^2 - 9)$$
$$9a^2 - 72 = a^4 - 9a^2$$
$$a^4 - 18a^2 + 72 = 0$$
$$(a^2 - 6)(a^2 - 12) = 0.$$

Since $a^2 = 6$ implies that $a < 3$, we cannot have $a^2 - 6 = 0$. From $a^2 = 12$ we see that the equation of the ellipse is

$$\frac{x^2}{3} + \frac{y^2}{12} = 1.$$

36. Since the center is at $(1, -1)$ and a focus is at $(1, 1)$, the major axis is vertical and $c = 1 - (-1) = 2$. The form of the equation is

$$\frac{(x-1)^2}{b^2} + \frac{(y+1)^2}{a^2} = 1.$$

From $a = 5$ and $c = 2$ we have $b^2 = a^2 - c^2 = 25 - 4 = 21$. Thus, the equation of the ellipse is

$$\frac{(x-1)^2}{21} + \frac{(y+1)^2}{25} = 1.$$

39. Since the endpoints of the minor axis are at $E_1(0, -1)$ and $E_2(0, 5)$, the center is at $(0, 2)$ and $b = 5 - 2 = 3$. Because the focus at $(6, 2)$ is $c = 6$ units from the center, we have $a^2 = b^2 + c^2 = 3^2 + 6^2 = 45$. Thus, the equation of the ellipse is

$$\frac{x^2}{a^2} + \frac{(y-2)^2}{b^2} = 1 \qquad \text{or} \qquad \frac{x^2}{45} + \frac{(y-2)^2}{9} = 1.$$

42. $x^2 + \dfrac{y^2}{16} = 1$; upper half-ellipse $\Rightarrow y^2 = 16(1 - x^2) \Rightarrow$

$y = 4\sqrt{1 - x^2} \Rightarrow f(x) = 4\sqrt{1 - x^2} \Rightarrow$ domain is $[-1, \ 1]$.

45. We identify $a = \frac{1}{2}(72) = 36$ million and $b = \frac{1}{2}(70.4) = 35.2$ million. Then

$$c^2 = a^2 - b^2 = 36^2 - 35.2^2 = 1296 - 12309.04 = 56.96,$$

so $c \approx 7.5$ million. Thus, the least distance between Mercury and the Sun is $36 - 7.5 = 28.5$ million miles, and the greatest distance is $36 + 7.5 = 43.5$ million miles.

48. We assume that the earth is spherical and superimpose a coordinate system whose origin lies at the center of the earth. We then assume that the satellite has an elliptical orbit with major axis lying along the x-axis and maximum altitude of 1000 miles to the right of the earth and minimum altitude of 200 miles to the left of the earth, as shown in the figure (which is not to scale). Since the length of the major axis is $200 + 4000 + 4000 + 1000 = 9200$, the center of the ellipse is $a = 9200/2 = 4600$ miles from the

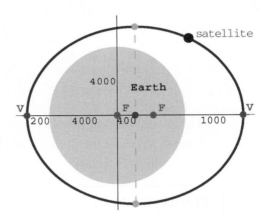

left vertex. That is the center of the ellipse is at $(400, 0)$. Thus, $c = 400$ and

$$b^2 = a^2 - c^2 = 4600^2 - 400^2 = 21{,}000{,}000.$$

Using $a^2 = 4600^2 = 21{,}160{,}000$, we see that the equation of the satellite's orbit is

$$\frac{(x - 400)^2}{21{,}160{,}000} + \frac{y^2}{21{,}000{,}000} = 1.$$

51. Since $a = \frac{1}{2}(4) = 2$ and $b = \frac{1}{2}(3) = \frac{3}{2}$, the length of the string should be $2a = 4$ feet long. Since

$$c^2 = a^2 - b^2 = 4 - \frac{9}{4} = \frac{7}{4} \qquad \text{or} \qquad c = \frac{\sqrt{7}}{2},$$

we see that the tacks should be placed $\sqrt{7}/2$ feet from the center of the rectangle on the major axis of the ellipse.

54. (a) Identifying $a^2 = 9$ and $b^2 = 4$, we have $c^2 = 9 - 4 = 5$, so $c = \sqrt{5}$. Letting $x = \sqrt{5}$, we obtain

$$\frac{(\sqrt{5})^2}{9} + \frac{y^2}{4} = 1$$

$$\frac{y^2}{4} = \frac{4}{9}$$

$$y^2 = \frac{16}{9}.$$

Thus, $y = \frac{4}{3}$ and the focal width is $2(\frac{4}{3}) = \frac{8}{3}$.

(b) The x-coordinate of a point on a focal chord is $\pm c$. We substitute $x = c$ into the equation of the ellipse and solve for y:

$$\frac{c^2}{a^2} + \frac{y^2}{b^2} = 1$$

$$y^2 = b^2 \left(1 - \frac{c^2}{a^2}\right) = \frac{b^2}{a^2}(a^2 - c^2) = \frac{b^2}{a^2}(b^2) = \frac{b^4}{a^2}$$

$$y = \pm\frac{b^2}{a}.$$

Therefore, the endpoints of the right-hand focal chord are $(c, b^2/a)$ and $(c, -b^2/a)$. It follows that the focal width of the ellipse is $2b^2/a$.

7.3   The Hyperbola

3. From the equation of the hyperbola, we identify $a = \sqrt{64} = 8$ and $b = \sqrt{9} = 3$. Thus, the vertices are $(0, -8)$ and $(0, 8)$. The center of the hyperbola is the origin. From $c^2 = a^2 + b^2 = 64 + 9 = 73$, we see that $c = \sqrt{73}$ and the foci are at $(0, -\sqrt{73})$ and $(0, \sqrt{73})$. Factoring

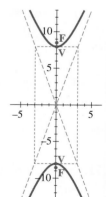

$$\frac{y^2}{64} - \frac{x^2}{9} = 0$$

we get

$$\left(\frac{y}{8} - \frac{x}{3}\right)\left(\frac{y}{8} + \frac{x}{3}\right) = 0,$$

so the asymtotes are $y = 8x/3$ and $y = -8x/3$. The eccentricity is $e = c/a = \sqrt{73}/8$.

6. Dividing both sides of the equation by 25, we have

$$\frac{x^2}{5} - \frac{y^2}{5} = 1.$$

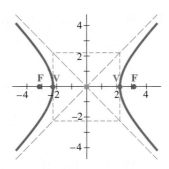

From this, we identify $a = \sqrt{5}$ and $b = \sqrt{5}$. The center of the hyperbola is the origin, and the vertices are $(-\sqrt{5}, 0)$ and $(\sqrt{5}, 0)$. From $c^2 = a^2 + b^2 = 5 + 5 = 10$, we see that $c = \sqrt{10}$ and the foci are at $(0, -\sqrt{10})$ and $(0, \sqrt{10})$. Factoring

$$\frac{x^2}{5} - \frac{y^2}{5} = 0$$

we get

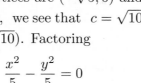

$$\left(\frac{x}{\sqrt{5}} - \frac{y}{\sqrt{5}}\right)\left(\frac{x}{\sqrt{5}} + \frac{y}{\sqrt{5}}\right) = 0,$$

so the asymptotes are $y = x$ and $y = -x$. The eccentricity is $e = c/a = \sqrt{10}/\sqrt{5} = \sqrt{2}$.

9. From the equation of the hyperbola, we see that the center of the hyperbola is $(5, -1)$. We identify $a = \sqrt{4} = 2$ and $b = \sqrt{49} = 7$. Thus, the vertices are $(3, -1)$ and $(7, -1)$. From $c^2 = a^2 + b^2 = 4 + 49 = 53$, we see that $c = \sqrt{53}$ and the foci are at $(5 - \sqrt{53}, -1)$ and $(5 + \sqrt{53}, -1)$. Factoring

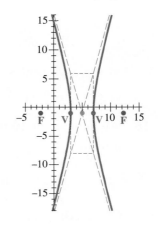

$$\frac{(x-5)^2}{4} - \frac{(y+1)^2}{49} = 0$$

we get

$$\left(\frac{x-5}{2} - \frac{y+1}{7}\right)\left(\frac{x-5}{2} + \frac{y+1}{7}\right) = 0,$$

so the asymptotes are $y = \frac{7}{2}x - \frac{37}{2}$ and $y = -\frac{7}{2}x + \frac{33}{2}$. The eccentricity is $e = c/a = \sqrt{53}/2$.

12. From the equation of the hyperbola, we see that the center of the hyperbola is $(-3, \frac{1}{4})$. We identify $a = \sqrt{4} = 2$ and $b = \sqrt{9} = 3$. Thus, the vertices are $(-3, -\frac{7}{4})$ and $(-3, \frac{9}{4})$. From $c^2 = a^2 + b^2 = 4 + 9 = 13$, we see that $c = \sqrt{13}$ and the foci are at $(-3, \frac{1}{4} - \sqrt{13})$ and $(-3, \frac{1}{4} + \sqrt{13})$. Factoring

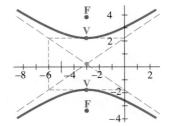

$$\frac{\left(y - \frac{1}{4}\right)^2}{4} - \frac{(x+3)^2}{9} = 0$$

we get

$$\left(\frac{y - \frac{1}{4}}{2} - \frac{x+3}{3}\right)\left(\frac{y - \frac{1}{4}}{2} + \frac{x+3}{3}\right) = 0,$$

so the asymptotes are $y = \frac{2}{3}x + \frac{9}{4}$ and $y = -\frac{2}{3}x - \frac{7}{4}$. The eccentricity is $e = c/a = \sqrt{13}/2$.

15. We first write the equation in the form

$$5(y - 7)^2 - 8(x + 4)^2 = 40,$$

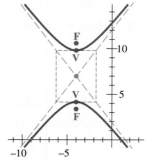

and then divide both sides by 40. This gives

$$\frac{(y - 7)^2}{8} - \frac{(x + 4)^2}{5} = 1.$$

We see that the center is $(-4, 7)$, and we identify $a = \sqrt{8}$ $= 2\sqrt{2}$ and $b = \sqrt{5}$. Thus, the vertices are $(-4, 7 - 2\sqrt{2})$ and $(-4, 7 + 2\sqrt{2})$. From $c^2 = a^2 + b^2 = 8 + 5 = 13$, we see that $c = \sqrt{13}$ and the foci are at $(-4, 7 - \sqrt{13})$ and $(-4, 7 + \sqrt{13})$. Factoring

$$\frac{(y - 7)^2}{8} - \frac{(x + 4)^2}{5} = 0$$

we get

$$\left(\frac{y - 7}{\sqrt{8}} - \frac{x + 4}{\sqrt{5}} \right) \left(\frac{y - 7}{\sqrt{8}} + \frac{x + 4}{\sqrt{5}} \right) = 0,$$

so the asymptotes are $y = \sqrt{8/5}\,(x + 4) + 7$ and $y = -\sqrt{8/5}\,(x + 4) + 7$. The eccentricity is $e = c/a = \sqrt{13/8}$.

18. We first factor 16 from the two x-terms and -25 from the two y-terms:

$$16(x^2 - 16x \quad\;) - 25(y^2 + 6y \quad\;) = -399.$$

Completing the square in x and y, we get

$$16(x^2 - 16x + 64) - 25(y^2 + 6y + 9) = -399 + 1024 - 225 = 400.$$

Dividing both sides by 400, we obtain

$$\frac{(x - 8)^2}{25} - \frac{(y + 3)^2}{16} = 1.$$

We see that the center is $(8, -3)$, and we identify $a = \sqrt{25} = 5$ and $b = \sqrt{16} = 4$. Thus, the vertices are $(3, -3)$ and $(13, -3)$. From $c^2 = a^2 + b^2 = 25 + 16 = 41$, we see that $c = \sqrt{41}$ and the foci are at $(8 - \sqrt{41}, -3)$ and $(8 + \sqrt{41}, -3)$. Factoring

$$\frac{(x - 8)^2}{25} - \frac{(y + 3)^2}{16} = 0$$

we get

$$\left(\frac{x - 8}{5} - \frac{y + 3}{4} \right) \left(\frac{x - 8}{5} + \frac{y + 3}{4} \right) = 0,$$

so the asymptotes are $y = \frac{4}{5}x - \frac{47}{5}$ and $y = -\frac{4}{5}x + \frac{17}{5}$. The eccentricity is $e = c/a = \sqrt{41}/5$.

21. We identify $c = 5$ and note that $h = 0$ and $k = 0$. Using $a = 3$, we have $a^2 + b^2 = c^2$ or $3^2 + b^2 = 5^2$, so that $b^2 = 25 - 9 = 16$ and $b = 4$. The equation of the hyperbola is

$$\frac{x^2}{9} - \frac{y^2}{16} = 1.$$

24. We identify $c = 3$ and note that $h = 0$ and $k = 0$. Using $a = \frac{3}{2}$, we have $a^2 + b^2 = c^2$ or $\left(\frac{3}{2}\right)^2 + b^2 = 3^2$, so that $b^2 = 9 - \frac{9}{4} = \frac{27}{4}$ and $b = 3\sqrt{3}/2$. The equation of the hyperbola is

$$\frac{y^2}{\frac{9}{4}} - \frac{x^2}{\frac{27}{4}} = 1.$$

27. We are given $h = 0$, $k = 0$, $a = \frac{5}{2}$, and $c = 3$. Then, using $a^2 + b^2 = c^2$ we have $\left(\frac{5}{2}\right)^2 + b^2 = 3^2$, so $b^2 = 9 - \frac{25}{4} = \frac{11}{4}$. The equation of the hyperbola is

$$\frac{y^2}{\frac{25}{4}} - \frac{x^2}{\frac{11}{4}} = 1.$$

30. We are given $h = 0$, $k = 0$, $a = 1$, and $c = 5$. Then, using $a^2 + b^2 = c^2$ we have $1^2 + b^2 = 5^2$, so $b^2 = 25 - 1 = 24$. The equation of the hyperbola is

$$\frac{x^2}{1} - \frac{y^2}{24} = 1.$$

33. The vertices are on the x-axis equidistant from the origin, so the x-axis is the transverse axis and the center of the hyperbola is the origin. Thus, $h = 0$, $k = 0$, and we identify $a = 2$. The form of the equation of the hyperbola is $y = \pm(b/a)x$, so $b/a = 4/3$, and $b = \frac{4}{3}a = \frac{4}{3}(2) = \frac{8}{3}$. The equation of the hyperbola is

$$\frac{x^2}{4} - \frac{y^2}{\frac{64}{9}} = 1.$$

36. We identify $h = 2$, $k = 3$, $c = 2 - 0 = 2$, and $a = 3 - 2 = 1$. The transverse axis is on the horizontal line $y = 3$. Using $a^2 + b^2 = c^2$, we have $1^2 + b^2 = 2^2$, so $b^2 = 4 - 1 = 3$. The equation of the hyperbola is

$$\frac{(x-2)^2}{1} - \frac{(y-3)^2}{3} = 1.$$

39. The center is at the origin and the transverse axis is on the x-axis, so the form of the equation of the hyperbola is

$$\frac{x^2}{a^2} - \frac{y^2}{b^2} = 1.$$

Since $a = 2$ and $P(2\sqrt{3}, 4)$ is on the hyperbola,

$$\frac{(2\sqrt{3})^2}{2^2} - \frac{4^2}{b^2} = 1$$

$$3 - 1 = \frac{16}{b^2}$$

$$b^2 = \frac{16}{2} = 8.$$

Thus, the equation of the hyperbola is

$$\frac{x^2}{4} - \frac{y^2}{8} = 1.$$

42. We identify $h = 3$, $k = -5$, and $a = -2 - (-5) = 3$. Since the transverse axis is vertical, the equation of the hyperbola has the form

$$\frac{(y + 5)^2}{3^2} - \frac{(x - 3)^2}{b^2} = 1.$$

Since the hyperbola passes through $(1, 1)$,

$$\frac{(-1 + 5)^2}{9} - \frac{(1 - 3)^2}{b^2} = 1$$

$$\frac{16}{9} - 1 = \frac{4}{b^2}$$

$$b^2 = \frac{4}{\frac{7}{9}} = \frac{36}{7}.$$

Thus, the equation of the hyperbola is

$$\frac{(y + 5)^2}{9} - \frac{(x - 3)^2}{\frac{36}{7}} = 1.$$

45. $\dfrac{x^2}{16} - \dfrac{y^2}{25} = 1 \Rightarrow \dfrac{y^2}{25} = \dfrac{x^2}{16} - 1 \Rightarrow y = 5\sqrt{\dfrac{x^2}{16} - 1} \Rightarrow$

$f(x) = \frac{5}{4}\sqrt{x^2 - 16} \Rightarrow$ domain is $(-\infty, \ -4] \cup [4, \ \infty)$.

The graph is the upper two half-branches of the hyperbola for $y \geq 0$.

48. $9(x - 1)^2 - 81(y - 2)^2 = 9 \Rightarrow (x - 1)^2 - 9(y - 2)^2 = 1 \Rightarrow$

$9(y - 2)^2 = (x - 1)^2 - 1 \Rightarrow y - 2 = \frac{1}{3}\sqrt{(x - 1)^2 - 1} \Rightarrow$

$f(x) = 2 + \frac{1}{3}\sqrt{(x - 1)^2 - 1} \Rightarrow$ domain is $(-\infty, \ 0] \cup [2, \ \infty)$

The graph is the upper two half-branches of the hyperbola for $y \geq 2$.

51. If (x, y) is a point on the hyperbola, then $d_1 + d_2 = 8$ where

$$d_1 = \sqrt{(x - 0)^2 + (y + 2)^2} \qquad \text{and} \qquad d_2 = \sqrt{(x - 8)^2 + (y - 4)^2}.$$

That is,

$$\sqrt{x^2 + (y+2)^2} + \sqrt{(x-8)^2 + (y-4)^2} = 8$$

$$\sqrt{(x-8)^2 + (y-4)^2} = 8 - \sqrt{x^2 + (y+2)^2}$$

$$(x-8)^2 + (y-4)^2 = 64 - 16\sqrt{x^2 + (y+2)^2} + x^2 + (y+2)^2$$

$$x^2 - 16x + 64 + y^2 - 8y + 16 = x^2 + y^2 + 4y + 68 - 16\sqrt{x^2 + (y+2)^2}$$

$$\sqrt{x^2 + (y+2)^2} = 16x + 12y - 12$$

$$4\sqrt{x^2 + (y+2)^2} = 4x + 3y - 3$$

$$16\left[x^2 + (y+2)^2\right] = 16x^2 + 24xy + 9y^2 - 24x - 18y + 9$$

$$16x^2 + 16y^2 + 64y + 64 = 16x^2 + 24xy + 9y^2 - 24x - 18y + 9$$

Simplifying, we see that the equation of the hyperbola is

$$7y^2 - 24xy + 24x + 82y + 55 = 0.$$

7.4   Rotation of Axes

3. Using the equations in (4) in the text we obtain

$$x' = (-1)\cos 60° + (-1)\sin 60° = -\frac{1}{2} - \frac{\sqrt{3}}{2}$$

$$y' = -(-1)\sin 60° + (-1)\cos 60° = \frac{\sqrt{3}}{2} - \frac{1}{2}.$$

6. Using the equations in (5) in the text we obtain

$$x = -5\cos 45° - 7\sin 45° = -\frac{5}{\sqrt{2}} - \frac{7}{\sqrt{2}} = -\frac{12}{\sqrt{2}} = -6\sqrt{2}$$

$$y = -5\sin 45° + 7\cos 45° = -\frac{5}{\sqrt{2}} + \frac{7}{\sqrt{2}} = \frac{2}{\sqrt{2}} = \sqrt{2}.$$

9. Using the equations in (5) in the text we obtain

$$x = 4\cos 15° - 6\sin 15° = 4\left(\frac{1+\sqrt{3}}{2\sqrt{2}}\right) - 6\left(\frac{-1+\sqrt{3}}{2\sqrt{2}}\right) = \frac{5-\sqrt{3}}{\sqrt{2}} \approx 2.31$$

$$y = 4\sin 15° + 6\cos 15° = 4\left(\frac{-1+\sqrt{3}}{2\sqrt{2}}\right) + 6\left(\frac{1+\sqrt{3}}{2\sqrt{2}}\right) = \frac{1+5\sqrt{3}}{\sqrt{2}} \approx 6.83.$$

The half-angle formula was used to find $\sin 15°$ and $\cos 15°$.

12. From $2x^2 - 3xy - 2y^2 = 5$ we identify $A = 2$, $B = -3$, and $C = -2$. Since $B^2 - 4AC = (-3)^2 - 4(2)(-2) = 25 > 0$, the graph is a hyperbola and

$$\cot 2\theta = \frac{A - C}{B} = \frac{2 - (-2)}{-3} = -\frac{4}{3}.$$

Using the note in this manual preceding the solution of Problem 11, we have

$$\csc 2\theta = \sqrt{1 + \left(-\frac{4}{3}\right)^2} = \sqrt{1 + \frac{16}{9}} = \sqrt{\frac{25}{9}} = \frac{5}{3}.$$

so that

$$\sin 2\theta = \frac{3}{5} \qquad \text{and} \qquad \cos 2\theta = \cot 2\theta \sin 2\theta = -\frac{4}{3}\left(\frac{3}{5}\right) = -\frac{4}{5}.$$

Using the half-angle formulas we find

$$\sin\theta = \sqrt{\frac{1 - \cos 2\theta}{2}} = \sqrt{\frac{1 + \frac{4}{5}}{2}} = \frac{3}{\sqrt{10}} \qquad \text{and} \qquad \cos\theta = \sqrt{\frac{1 + \cos 2\theta}{2}} = \sqrt{\frac{1 - \frac{4}{5}}{2}} = \frac{1}{\sqrt{10}}.$$

Then, using (5) in the text,

$$x = \frac{1}{\sqrt{10}}x' - \frac{3}{\sqrt{10}}y' = \frac{1}{\sqrt{10}}(x' - 3y')$$

$$y = \frac{3}{\sqrt{10}}x' + \frac{1}{\sqrt{10}}y' = \frac{1}{\sqrt{10}}(3x' + y').$$

Substituting into the original equation we have

$$2\left(\frac{1}{10}\right)(x' - 3y')^2 - 3\frac{1}{\sqrt{10}}(x' - 3y')\frac{1}{\sqrt{10}}(3x' + y') - 2\left(\frac{1}{10}\right)(3x' + y')^2 = 5$$

$$\frac{1}{5}(x'^2 - 6x'y' + 9y'^2) - \frac{3}{10}(3x'^2 - 8x'y' - 3y'^2) - \frac{1}{5}(9x'^2 + 6x'y' + y'^2) = 5$$

$$-25x'^2 + 25y'^2 = 50$$

$$y'^2 - x'^2 = 2.$$

The angle of rotation is $\theta = \sin^{-1}(3/\sqrt{10}) \approx 71.57°$. The axes of the $x'y'$-plane are shown below in the first figure in red. The graph of the hyperbola in the $x'y'$-plane is shown in the second figure. In the third figure the graph is shown in the xy-plane.

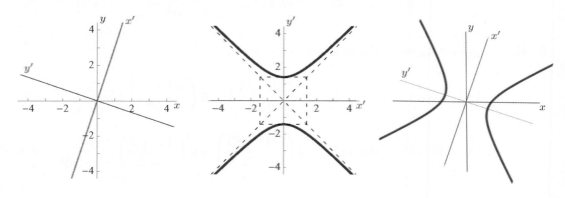

15. From $x^2 + 4xy - 2y^2 = 6$ we identify $A = 1$, $B = 4$, and $C = -2$. Since $B^2 - 4AC = 4^2 - 4(1)(-2) = 24 > 0$, the graph is a hyperbola and

$$\cot 2\theta = \frac{A - C}{B} = \frac{1 - (-2)}{4} = \frac{3}{4}.$$

Using the note in this manual preceding the solution of Problem 11, we have

$$\csc 2\theta = \sqrt{1 + \left(\frac{3}{4}\right)^2} = \sqrt{1 + \frac{9}{16}} = \sqrt{\frac{25}{16}} = \frac{5}{4},$$

so that

$$\sin 2\theta = \frac{4}{5} \quad \text{and} \quad \cos 2\theta = \cot 2\theta \sin 2\theta = \frac{3}{4}\left(\frac{4}{5}\right) = \frac{3}{5}.$$

Using the half-angle formulas we find

$$\sin \theta = \sqrt{\frac{1 - \cos 2\theta}{2}} = \sqrt{\frac{1 - \frac{3}{5}}{2}} = \frac{1}{\sqrt{5}} \quad \text{and} \quad \cos \theta = \sqrt{\frac{1 + \cos 2\theta}{2}} = \sqrt{\frac{1 + \frac{3}{5}}{2}} = \frac{2}{\sqrt{5}}.$$

Then, using (5) in the text,

$$x = \frac{2}{\sqrt{5}} x' - \frac{1}{\sqrt{5}} y' = \frac{1}{\sqrt{5}} (2x' - y')$$

$$y = \frac{1}{\sqrt{5}} x' + \frac{2}{\sqrt{5}} y' = \frac{1}{\sqrt{5}} (x' + 2y').$$

Substituting into the original equation we have

$$\left(\frac{1}{5}\right)(2x' - y')^2 + 4\frac{1}{\sqrt{5}}(2x' - y')\frac{1}{\sqrt{5}}(x' + 2y') - 2\left(\frac{1}{5}\right)(x' + 2y')^2 = 6$$

$$\frac{1}{5}(4x'^2 - 4x'y' + y'^2) + \frac{4}{5}(2x'^2 + 3x'y' - 2y'^2) - \frac{2}{5}(x'^2 + 4x'y' + 4y'^2) = 6$$

$$10x'^2 - 15y'^2 = 30$$

$$\frac{x'^2}{3} - \frac{y'^2}{2} = 1.$$

The angle of rotation of $\theta = \sin^{-1}(1/\sqrt{5}) \approx 26.57°$. The axes of the $x'y'$-plane are shown below in the first figure as dashed lines. The graph of the hyperbola in the $x'y$-plane is shown in the second figure.

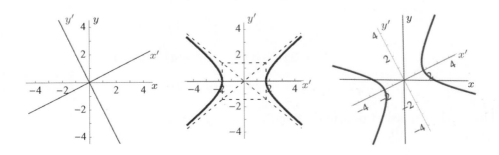

18. From $-x^2 + 6\sqrt{3}\,xy + 5y^2 - 8\sqrt{3}\,x + 8y = 12$ we identify $A = -1$, $B = 6\sqrt{3}$, and $C = 5$, and we see that

$$\cot 2\theta = \frac{A - C}{B} = \frac{-1 - 5}{6\sqrt{3}} = -\frac{1}{\sqrt{3}}.$$

Using the note in this manual preceding the solution of Problem 12, we have $45° < \theta < 90°$, and

$$\csc 2\theta = \sqrt{1 + \left(-\frac{1}{\sqrt{3}}\right)^2} = \sqrt{1 + \frac{1}{\sqrt{3}}} = \frac{2}{\sqrt{3}}.$$

Then

$$\sin 2\theta = \frac{\sqrt{3}}{2} \qquad \text{and} \qquad \cos 2\theta = \cot 2\theta \sin 2\theta = -\frac{1}{\sqrt{3}}\left(\frac{\sqrt{3}}{2}\right) = -\frac{1}{2}.$$

Using the half-angle formulas we find

$$\sin \theta = \sqrt{\frac{1 - \cos 2\theta}{2}} = \sqrt{\frac{1 - \left(-\frac{1}{2}\right)}{2}} = \frac{\sqrt{3}}{2} \qquad \text{and} \qquad \cos \theta = \sqrt{\frac{1 + \cos 2\theta}{2}} = \sqrt{\frac{1 + \left(-\frac{1}{2}\right)}{2}} = \frac{1}{2}.$$

Then, using (5) in the text,

$$x = \frac{1}{2}x' - \frac{\sqrt{3}}{2}y' = \frac{1}{2}\left(x' - \sqrt{3}\,y'\right)$$

$$y = \frac{\sqrt{3}}{2}x' + \frac{1}{2}y' = \frac{1}{2}\left(\sqrt{3}\,x' + y'\right).$$

Substituting into the given equation we have

$$-\frac{1}{4}\left(x' - \sqrt{3}\,y'\right)^2 + 6\sqrt{3}\left(\frac{1}{2}\right)\left(x' - \sqrt{3}\,y'\right)\frac{1}{2}\left(\sqrt{3}\,x' + y'\right) + 5\left(\frac{1}{4}\right)\left(\sqrt{3}\,x' + y'\right)^2$$

$$-8\sqrt{3}\left(\frac{1}{2}\right)\left(x' - \sqrt{3}y'\right) + 8\left(\frac{1}{2}\right)\left(\sqrt{3}\,x' + y'\right) = 12$$

$$-\left(x'^2 - 2\sqrt{3}\,x'y' + 3y'^2\right) + 6\sqrt{3}\left(\sqrt{3}\,x'^2 - 2x'y' - \sqrt{3}\,y'^2\right)$$

$$+ 5\left(3x'^2 + 2\sqrt{3}\,x'y' + y'^2\right) - 16\sqrt{3}\,x' + 48y' + 16\sqrt{3}\,x' + 16y' = 48$$

$$32x'^2 - 16y'^2 + 64y' = 48$$

$$2x'^2 - y'^2 + 4y' = 3$$

$$2x'^2 - \left(y'^2 - 4y' + 4\right) = 3 - 4$$

$$2x'^2 - (y' - 2)^2 = 7.$$

This is the equation of a hyperbola.

21. From $3x^2 + 2\sqrt{3}\,xy + y^2 + 2x - 2\sqrt{3}\,y = 0$ we identify $A = 3$, $B = 2\sqrt{3}$, and $C = 1$

(a) We see that

$$\cot 2\theta = \frac{A - C}{B} = \frac{3 - 1}{2\sqrt{3}} = \frac{1}{\sqrt{3}}.$$

Using the note in this manual preceding the solution of Problem 12, we have that $0° < \theta < 45°$, and

$$\csc 2\theta = \sqrt{1 + \left(\frac{1}{\sqrt{3}}\right)^2} = \sqrt{1 + \frac{1}{3}} = \frac{2}{\sqrt{3}}.$$

Then

$$\sin 2\theta = \frac{\sqrt{3}}{2} \quad \text{and} \quad \cos 2\theta = \cot 2\theta \sin 2\theta = \frac{1}{\sqrt{3}}\left(\frac{\sqrt{3}}{2}\right) = \frac{1}{2}.$$

Using the half-angle formulas we find

$$\sin \theta = \sqrt{\frac{1 - \cos 2\theta}{2}} = \sqrt{\frac{1 - \frac{1}{2}}{2}} = \frac{1}{2} \quad \text{and} \quad \cos \theta = \sqrt{\frac{1 + \cos 2\theta}{2}} = \sqrt{\frac{1 + \frac{1}{2}}{2}} = \frac{\sqrt{3}}{2}.$$

Then, using (5) in the text,

$$x = \frac{\sqrt{3}}{2}\,x' - \frac{1}{2}\,y' = \frac{1}{2}\left(\sqrt{3}\,x' - y'\right)$$

$$y = \frac{1}{2}\,x' + \frac{\sqrt{3}}{2}\,y' = \frac{1}{2}\left(x' + \sqrt{3}\,y'\right).$$

Substituting into the original equation we have

$$3\left(\frac{1}{4}\right)\left(\sqrt{3}\,x' - y'\right)^2 + 2\sqrt{3}\left(\frac{1}{2}\right)\left(\sqrt{3}\,x' - y'\right)\left(\frac{1}{2}\right)\left(x' + \sqrt{3}\,y'\right) + \frac{1}{4}\left(x' + \sqrt{3}\,y'\right)^2$$

$$+ 2\left(\frac{1}{2}\right)\left(\sqrt{3}\,x' - y'\right) - 2\sqrt{3}\left(\frac{1}{2}\right)\left(x' + \sqrt{3}\,y'\right) = 0$$

(multiply both sides by 4)

$$3\left(3x'^2 - 2\sqrt{3}\,x'y' + y'^2\right) + 2\sqrt{3}\left(\sqrt{3}\,x'^2 + 2x'y' - \sqrt{3}\,y'^2\right) + \left(x'^2 + 2\sqrt{3}\,x'y' + 3y'^2\right)$$

$$+ 4\left(\sqrt{3}\,x' - y'\right) - 4\sqrt{3}\left(x' + \sqrt{3}\,y'\right) = 0$$

$$16x'^2 - 16y' = 0$$

$$y' = x'^2.$$

(b) From Section 7.1 in the text, the $x'y'$-equation of a parabola with focus at $(0, c)$ is $x'^2 = 4cy'$. Thus $c = \frac{1}{4}$, and in this case the $x'y'$-coordinates of the focus are $(0, \frac{1}{4})$. Using equation (5) in the text with the values for $\sin\theta$ and $\cos\theta$ from part **(a)** of this problem, we have

$$x = 0\left(\frac{\sqrt{3}}{2}\right) - \frac{1}{4}\left(\frac{1}{2}\right) = -\frac{1}{8} \quad \text{and} \quad y = 0\left(\frac{1}{2}\right) + \frac{1}{4}\left(\frac{\sqrt{3}}{2}\right) = \frac{\sqrt{3}}{8}.$$

The xy-coordinates of the focus are $\left(-\frac{1}{8}, \frac{\sqrt{3}}{8}\right)$.

(c) Again, using Section 7.1 in the text, the $x'y'$-equation of the directrix is $y' = -\frac{1}{4}$. By the second equation in (4) in this section of the text,

$$y' = -x\sin\theta + y\cos\theta = -\frac{1}{2}x + \frac{\sqrt{3}}{2}y.$$

Setting this equal to $-\frac{1}{4}$, we have

$$-\frac{1}{2}x + \frac{\sqrt{3}}{2}y = -\frac{1}{4} \quad \text{or} \quad 2x - 2\sqrt{3}\,y = 1.$$

Thus, the xy-equation of the directrix is

$$y = \frac{1}{\sqrt{3}}x - \frac{1}{2\sqrt{3}}.$$

24. Identifying $A = 2$, $B = -2$, and $C = 2$, the discriminant is $B^2 - 4AC = (-2)^2 - 4(2)(2) = 4 - 16 = -12 < 0$, so the graph is an ellipse.

27. Identifying $A = 1$, $B = 1$, and $C = 1$, the discriminant is $B^2 - 4AC = 1^2 - 4(1)(1) = 1 - 4 = -3 < 0$, so the graph is an ellipse.

7.5 $\boxed{\displaystyle\int \begin{array}{l}\textbf{Calculus}\\ \text{PREVIEW}\end{array} \ \textbf{3-Space}}$

3 and 6.

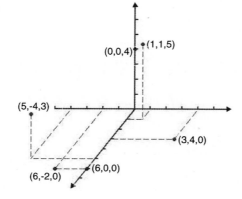

9. A line perpendicular to the xy-plane at $(2, 3, 0)$.

12.

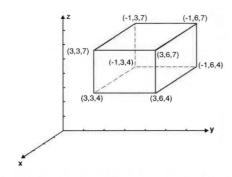

15. The union of the planes $x = 0$, $y = 0$, and $z = 0$.

18. The union of the planes $x = 2$ and $z = 8$.

21. $d = \sqrt{(3-6)^2 + (-1-4)^2 + (2-8)^2} = \sqrt{70}$

24. **(a)** 2; **(b)** $d = \sqrt{(-6)^2 + 2^2 + (-3)^2} = 7$

27. $d(P_1, P_2) = \sqrt{((4-1)^2 + (1-2)^2 + (3-3)^2} = \sqrt{10}$
$d(P_1, P_3) = \sqrt{(4-1)^2 + (6-2)^2 + (4-3)^2} = \sqrt{26}$
$d(P_2, P_3) = \sqrt{((4-4)^2 + (6-1)^2 + (4-3)^2} = \sqrt{26}$; The triangle is an isosceles triangle.

30. $d(P_1, P_2) = \sqrt{((0-1)^2 + (3-2)^2 + (2-(-1))^2} = \sqrt{11}$
$d(P_1, P_3) = \sqrt{(1-1)^2 + (1-2)^2 + ((-3)-(-1))^2} = \sqrt{5}$
$d(P_2, P_3) = \sqrt{((1-0)^2 + (1-3)^2 + ((-3)-2)^2} = \sqrt{30}$
Since adding any two of the above distances will not result in the third, the points cannot be collinear.

33. $\sqrt{(2-x)^2 + (1-2)^2 + (1-3)^2} = \sqrt{21} \longrightarrow x^2 - 4x + 9 = 21 \longrightarrow x^2 - 4x + 4 = 16 \longrightarrow$
$(x-2)^2 = 16 \longrightarrow x = 2 + -4$ or $x = 6, -2$

36. $\left(\dfrac{0+4}{2}, \dfrac{5+1}{2}, \dfrac{-8+(-6)}{2}\right) = (2, 3, -7)$

39.

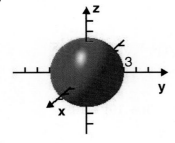

42.

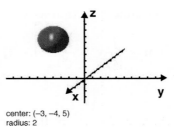

center: $(-3, -4, 5)$
radius: 2

45. $x^2 + y^2 + (z^2 - 16z = 64) = 64$; center: $(0, 0, 8)$; radius: 8

48. $x^2 + (y-3)^2 + z^2 = 25/16$

51. There are two solutions: one sphere is inside the given sphere and the other is outside.
$x^2 + (y-8)^2 + z^2 = 4$ or $x^2 + (y-4)^2 + z^2 = 4$.

54.

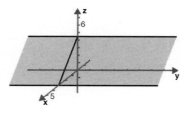

57.

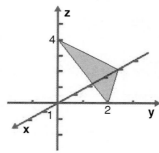

60. From the points $(3,5,2)$ and $(2,3,1)$ we obtain the vector $\mathbf{u} = \mathbf{i} + 2\mathbf{j} + \mathbf{k}$. From the points $(2,3,1)$ and $(-1,-1,4)$ we obtain the vector $\mathbf{v} = 3\mathbf{i} + 4\mathbf{j} - 3\mathbf{k}$. From the points $(-1,-1,4)$ and (x,y,z) we obtain the vector $\mathbf{w} = (x+1)\mathbf{i} + (y+1)\mathbf{j} + (z-4)\mathbf{k}$. Then, a normal vector is

$$\mathbf{u} \times \mathbf{v} = \begin{vmatrix} \mathbf{i} & \mathbf{j} & \mathbf{k} \\ 1 & 2 & 1 \\ 3 & 4 & -3 \end{vmatrix} = -10\mathbf{i} + 6\mathbf{j} - 2\mathbf{k}$$

A vector equation of the plane is $-10(x+1) + 6(y+1) - 2(z-4) = 0$ or $5x - 3y + z = 2$.

63.
$$A + 2B - C + D = 0$$
$$4A + 3B + C + D = 0 \Rightarrow$$
$$7A + 4B + C + D = 0$$
$A = \tfrac{1}{5}D, \ B = -\tfrac{3}{5}D, C = 0 \Rightarrow -x + 3y = 5$ when $D = -5$

66. $2\mathbf{u} - (\mathbf{v} - \mathbf{w}) = \langle 2, -6, 4 \rangle - \langle -3, -5, -8 \rangle = \langle 5, -1, 12 \rangle$

69. $|\mathbf{u} + \mathbf{w}| = |\langle 3, \ 3, \ 11 \rangle| = \sqrt{9 + 9 + 121} = \sqrt{139}$

72. $|\mathbf{v}| \, \mathbf{u} + |\mathbf{u}| \, \mathbf{v} = \sqrt{3} \langle 1, -3, 2 \rangle + \sqrt{14} \langle -1, 1, 1 \rangle = \langle \sqrt{3} - \sqrt{14}, -3\sqrt{3} + \sqrt{14}, 2\sqrt{3} + \sqrt{14} \rangle$

75. $(\mathbf{u} + \mathbf{v}) \cdot \mathbf{w} = \langle 0, \ -2, \ 3 \rangle \cdot \langle 2, \ 6, \ 9 \rangle = 0 - 12 + 27 = 15$

78. A unit vector in the same direction as $\mathbf{v} = \mathbf{i} - 3\mathbf{j} + 2\mathbf{k}$ is

$$\mathbf{u} = \frac{1}{|\mathbf{v}|}\mathbf{v} = \frac{1}{\sqrt{14}} (\mathbf{i} - 3\mathbf{j} + 2\mathbf{k}) = \tfrac{1}{\sqrt{14}}\mathbf{i} - \tfrac{3}{\sqrt{14}}\mathbf{j} + \tfrac{2}{\sqrt{14}}\mathbf{k}.$$

81. $\mathbf{u} = \langle 4, \ -2, \ 5 \rangle, \ \mathbf{v} = \langle 3, \ 1, \ -1 \rangle \Rightarrow \mathbf{u} \times \mathbf{v} = \begin{vmatrix} \mathbf{i} & \mathbf{j} & \mathbf{k} \\ 4 & -2 & 5 \\ 3 & 1 & -1 \end{vmatrix} = -3\mathbf{i} + 19\mathbf{j} + 10\mathbf{k}$

84. $\mathbf{u} = 8\mathbf{i} + \mathbf{j} - 6\mathbf{k}, \ \mathbf{v} = \mathbf{i} - 2\mathbf{j} + 10\mathbf{k} \Rightarrow \mathbf{u} \times \mathbf{v} = \begin{vmatrix} \mathbf{i} & \mathbf{j} & \mathbf{k} \\ 8 & 1 & -6 \\ 1 & -2 & 10 \end{vmatrix} = -2\mathbf{i} - 86\mathbf{j} - 17\mathbf{k}$

Chapter 7 Review Exercises

A. Fill in the Blanks

3. Since the directrix is a horizontal line and the vertex is halfway between the focus and the directrix, the vertex is at $(1, -5)$. The focus is above the directrix, so the parabola opens up. Because c is the distance from the vertex to the focus, $c = 2$. Thus, the equation of the parabola is $(x - 1)^2 = 8(y + 5)$.

6. We write the equation of the parabola in the form $(x + 4)^2 = \frac{1}{8}(y - 2) = 4(\frac{1}{32})(y - 2)$. The vertex of the parabola is $(-4, 2)$. The parabola opens up and $c = \frac{1}{32}$, so the focus is $(-4, \frac{65}{32})$.

9. The transverse axis of the hyperbola is vertical and the center is $(-3, 0)$. Since $a = 1$, the vertices are $(-3, -1)$ and $(-3, 1)$.

12. $9x^2 + y^2 = 1 \Rightarrow \dfrac{x^2}{\frac{1}{9}} + \dfrac{y^2}{1} = 1 \Rightarrow c^2 = 1 - \frac{1}{9} = \frac{8}{9} \Rightarrow e = \dfrac{c}{a} = \dfrac{\frac{2\sqrt{2}}{3}}{1} = \frac{2\sqrt{2}}{3}.$

15. $4x^2 + 9y^2 = 25 \Rightarrow \dfrac{x^2}{\frac{25}{4}} + \dfrac{y^2}{\frac{25}{9}} = 1 \Rightarrow b^2 = \frac{25}{9} \Rightarrow$ minor axis endpoints are

$(0, -\frac{5}{3})$, $(0, \frac{5}{3}) \Rightarrow$ length is $\frac{10}{3}$.

18. Because $B^2 - 4AC > 0$ (Fill in with < 0, $= 0$ or > 0), the conic $3x^2 - xy - y^2 + 1 = 0$ is a hyperbola.

B. True/False

3. The graphs of $y = \sqrt{5 + x^2}$ and $x = \sqrt{y^2 - 5} \geq 0$ are shown below. In rough terms, their graphs are one-half the graph of the equation $y^2 - x^2 = 5$. So, false.

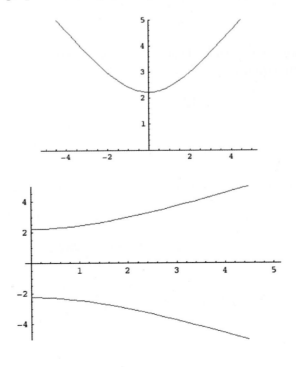

6. The first equation is a parabola centered at the origin, opening upward. The second equation is a hyperbola, also centered at the origin, and opening upward and downward. Since the parabola lies entirely above the x-axis, it can intersect the hyperbola in at most two points. The statement is true.

9. Since the axis is defined as the line through the focus perpendicular to the directrix, and the vertex is the point of intersection of the axis and the parabola, the statement is true.

12. Since the hyperbola is of the form $x^2/a^2 - y^2/b^2 = 1$, the axis is horizontal, so the statement is false.

15. The perpendicular distance between a point P on the parabola and the directrix equals the distance between P and the focus, so the statement is false.

18. If $y = 3x + 8$ is an asymptote of a hyperbola, then the slope of the other asymptote is $m = -3$. So, true.

C. Review Exercises

3. Completing the square

$$x^2 - 2x \quad = -4y - 1$$
$$x^2 - 2x + 1 = -4y - 1 + 1$$
$$(x - 1)^2 = -4y$$

we identify $p = -1$. Then

$$\text{Vertex}: (1, 0) \qquad \text{Focus}: (1, -1)$$
$$\text{Directrix}: y = 1 \qquad \text{Axis}: x = 1$$

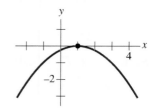

6. From the figure we deduce that the parabola opens to the right and $c = 3$. Its equation is

$$(y + 1)^2 = 12x.$$

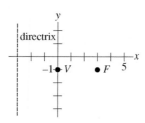

9. Identifying $a = 5$ and $b = \sqrt{3}$ we compute $c = \sqrt{25 - 3} = \sqrt{22}$. Then

$$\text{Center}: (0, -5) \qquad \text{Vertices}: (0, -10),\ (0, 0)$$
$$\text{Foci}: (0, -5 - \sqrt{22}),\ (0, -5 + \sqrt{22})$$

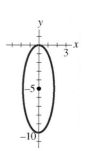

12. Completing the square

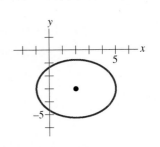

$$5(x^2 - 4x \quad) + 9(y^2 + 6y \quad) = -56$$

$$5(x^2 - 4x + 4) + 9(y^2 + 6y + 9) = -56 + 20 + 81$$

$$\frac{(x-2)^2}{9} + \frac{(y+3)^2}{5} = 1.$$

Identifying $a = 3$ and $b = \sqrt{5}$ we compute $c = \sqrt{9 - 5} = 2$. Then

$$Center: (2, -3) \qquad Vertices: (-1, -3),\ (5, -3) \qquad Foci: (0, -3),\ (4, -3)$$

15. The center is $(0, -2)$ and the form of the equation is

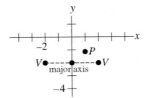

$$\frac{x^2}{4} + \frac{(y+2)^2}{s^2} = 1.$$

Letting $x = 1$ and $y = -2 + \sqrt{3}/2$ we have

$$\frac{1}{4} + \frac{(\sqrt{3}/2)^2}{s^2} = 1$$

$$\frac{3}{4s^2} = \frac{3}{4},$$

so $s^2 = 1$. The equation is

$$\frac{x^2}{4} + (y+2)^2 = 1.$$

18. Identifying $a = 1$ and $b = 2$ we compute $c = \sqrt{1+4} = \sqrt{5}$. To find the asymptotes we solve

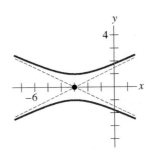

$$y^2 - \frac{(x+3)^2}{4} = 0$$

$$y^2 = \frac{1}{4}(x+3)^2.$$

Then

$$Center: (-3,\ 0) \qquad Vertices: (-3,\ -1),\ (-3,\ 1)$$

$$Foci: (-3,\ -\sqrt{5}),\ (-3,\ \sqrt{5}) \qquad Asymptotes: y = \pm\frac{1}{2}(x+3)$$

21. The transverse axis is on the x-axis, $a = 6$, and $c = 8$, so $b^2 = 64 - 36 = 28$. The equation is

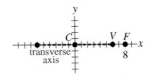

$$\frac{x^2}{36} - \frac{y^2}{28} = 1.$$

24. The transverse axis is on the line $x = -3$, the center is at $(-3, 3)$, $a = 1$, and $c = \sqrt{2}$, so $b^2 = 2 - 1 = 1$. The equation is

$$(y - 3)^2 - (x + 3)^2 = 1.$$

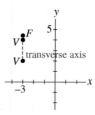

27. To obtain the translated equation we replace x with $x + 5$ and y with $y - 2$. This gives

$$4(x + 5)^2 + (y - 2)^2 = 4 \qquad \text{or} \qquad (x + 5)^2 + \frac{(y - 2)^2}{4} = 1.$$

30. $\dfrac{x^2}{a^2} + \dfrac{y^2}{b^2} = 1$

33. The center of the sphere is

$$\left(\frac{0 + 2}{2}, \frac{-4 + 12}{2}, \frac{7 - 3}{2} \right) = (1,\ 4,\ 2),$$

its radius is

$$r = \sqrt{(2 - 1)^2 + (12 - 4)^2 + (-3 - 2)^2} = \sqrt{1^2 + 8^2 + (-5)^2} = \sqrt{1 + 64 + 25} = \sqrt{90},$$

so its equation is

$$(x - 1)^2 + (y - 4)^2 + (z - 2)^2 = 90.$$

Chapter 8

Polar Coordinates

8.1 | **The Polar Coordinate System**

3. We measure backwards from the pole $\frac{1}{2}$ unit along the extension onto the negative y-axis of the ray $\pi/2$.

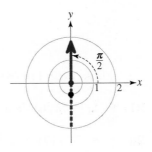

6. We measure $\frac{2}{3}$ unit along the ray $7\pi/4$.

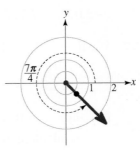

9. (a) We use $\theta = -2\pi + \pi/3 = -5\pi/3$, so the point is $(4, -5\pi/3)$.

(b) We use $\theta = 2\pi + \pi/3 = 7\pi/3$, so the point is $(4, 7\pi/3)$.

(c) We use $\theta = \pi + \pi/3 = 4\pi/3$, so the point is $(-4, 4\pi/3)$.

(d) We use $\theta = \pi/3 - \pi = -2\pi/3$, so the point is $(-4, -2\pi/3)$.

12. (a) We use $\theta = -2\pi + 7\pi/6 = -5\pi/6$, so the point is $(3, -5\pi/6)$.

(b) We use $\theta = 2\pi + 7\pi/6 = 19\pi/6$, so the point is $(3, 19\pi/6)$.

(c) We use $\theta = 7\pi/6 - \pi = \pi/6$, so the point is $(-3, \pi/6)$.

(d) We want to express the ray given by $\pi/6$ in part **(c)** as a negative angle. We use $\theta = -2\pi + \pi/6 = -11\pi/6$, so the point is $(-3, -11\pi/6)$.

15. With $r = \frac{1}{2}$ and $\theta = 2\pi/3$ we find

$$x = \frac{1}{2}\cos\frac{2\pi}{3} = \frac{1}{2}\left(-\frac{1}{2}\right) = -\frac{1}{4}$$

$$y = \frac{1}{2}\sin\frac{2\pi}{3} = \frac{1}{2}\left(\frac{\sqrt{3}}{2}\right) = \frac{\sqrt{3}}{4}.$$

Thus, $(\frac{1}{2}, 2\pi/3)$ is equivalent to $(-\frac{1}{4}, \sqrt{3}/4)$ in rectangular coordinates.

18. With $r = \sqrt{2}$ and $\theta = 11\pi/6$, we find

$$x = \sqrt{2}\cos\frac{11\pi}{6} = \sqrt{2}\left(\frac{\sqrt{3}}{2}\right) = \frac{\sqrt{6}}{2}$$

$$y = \sqrt{2}\sin\frac{11\pi}{6} = \sqrt{2}\left(-\frac{1}{2}\right) = -\frac{\sqrt{2}}{2}.$$

Thus, $(\sqrt{2}, 11\pi/6)$ is equivalent to $(\sqrt{6}/2, -\sqrt{2}/2)$ in rectangular coordinates.

21. $(-1, -5\pi/6) \Rightarrow x = -\cos(-5\pi/6) = \sqrt{3}/2,\ y = -\sin(-5\pi/6) = \frac{1}{2} \Rightarrow (\frac{\sqrt{3}}{2}, \frac{1}{2})$

24. $(-8, 5\pi/12) \Rightarrow x = -8\cos(5\pi/12) \approx -2.071,\ y = -8\sin(5\pi/12) \approx -7.727 \Rightarrow$ $(-2.071, -7.727)$

27. With $x = 1$ and $y = -\sqrt{3}$, we have, from (2) in the text,

$$r^2 = 1^2 + (-\sqrt{3})^2 = 4 \qquad \text{and} \qquad \tan\theta = -\sqrt{3}.$$

(a) For $r > 0$ we take $r = 2$, and for $-\pi < \theta < \pi$ we take $\theta = -\pi/3$. Thus, $(1, -\sqrt{3})$ is equivalent to $(2, -\pi/3)$ in polar coordinates.

(b) For $r < 0$ we take $r = -2$. Then we use $\theta = -\pi/3 + \pi = 2\pi/3$. Thus, $(1, -\sqrt{3})$ is also equivalent to $(-2, 2\pi/3)$ in polar coordinates.

30. With $x = 1$ and $y = 2$, we have, from (2) in the text,

$$r^2 = 1^2 + 2^2 = 5 \qquad \text{and} \qquad \tan\theta = \frac{2}{1} = 2.$$

(a) For $r > 0$ we take $r = \sqrt{5}$. Then we use $\theta = \tan^{-1}2$, which is a first quadrant angle. Thus, $(1, 2)$ is equivalent to $(\sqrt{5}, \tan^{-1}2)$ in polar coordinates.

(b) For $r < 0$ we take $r = -\sqrt{5}$ and use the third quadrant angle $\pi + \tan^{-1}2$ for θ. In this case, $(1, 2)$ is equivalent to $(-\sqrt{5}, \pi + \tan^{-1}2)$ in polar coordinates.

33.

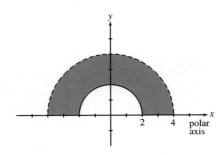

36.

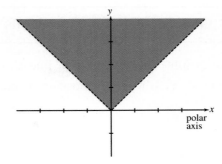

39. Substituting $y = r \sin \theta$ into the given equation, we find

$$r \sin \theta = 5$$

$$r = \frac{5}{\sin \theta} = 5 \csc \theta.$$

42. Substituting $x = r \cos \theta$ and $y = r \sin \theta$ into the given equation, we find

$$3r \cos \theta + 8r \sin \theta + 6 = 0$$

$$r(3 \cos \theta + 8 \sin \theta) = -6$$

$$r = -\frac{6}{3 \cos \theta + 8 \sin \theta}.$$

45. We use $r^2 = x^2 + y^2$ and we obtain $r^2 = 36$ or $r = \pm 6$. These two equations are equivalent and we take the polar equation to be $r = 6$.

48. Substituting $x = r \cos \theta$ and $y = r \sin \theta$ into the given equation, we find

$$r^3 \cos^3 \theta + r^3 \sin^3 \theta - (r \cos \theta)(r \sin \theta) = 0$$

$$r^3 \cos^3 \theta + r^3 \sin^3 \theta - r^2 \cos \theta \sin \theta = 0$$

$$r \cos^3 \theta + r \sin^3 \theta - \cos \theta \sin \theta = 0$$

$$r \cos^3 \theta + r \sin^3 \theta = \cos \theta \sin \theta$$

$$r(\cos^3 \theta + \sin^3 \theta) = \cos \theta \sin \theta$$

$$r = \frac{\cos \theta \sin \theta}{\cos^3 \theta + \sin^3 \theta}.$$

51. $r \cos \theta = 2;\quad x = 2$

54. $2(x^2 + y^2)^{1/2} = y/x;\quad 4x^2(x^2 + y^2) = y^2$

57. $r^2 + 5r \sin \theta = 0;\quad x^2 + y^2 + 5y = 0$

60. $4r - r \sin \theta = 10;\ 4\sqrt{x^2 + y^2} - y = 10;\ 4\sqrt{x^2 + y^2} = y + 10;\ 16(x^2 + y^2) = y^2 + 20y + 100;$
$16x^2 + 15y^2 - 20y - 100 = 0$

8.2 | Graphs of Polar Equations

3. The graph of $\theta = \pi/3$ is a line through the origin that makes an angle of $\pi/3$ with the positive x-axis.

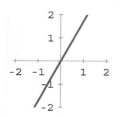

6. The graph of $r = 3\theta$, $\theta \geq 0$, is a spiral.

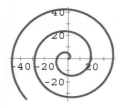

9. The graph of $r = 2(1 + \sin\theta)$ is a cardioid.

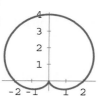

12. Identifying $a = 2$ and $b = 4$ in $r = 2 + 4\sin\theta$, we have $a/b = \frac{1}{2} < 1$, so the graph is a limaçon with an interior loop.

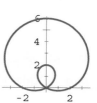

15. Identifying $a = 4$ and $b = 1$ in $r = 4 + \cos\theta$, we have $a/b = 4 > 2$, so the graph is a convex limaçon.

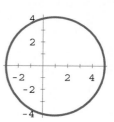

18. The graph of $r = 3\sin 4\theta$ is a rose curve. Since 4 is an even number, there are $2(4) = 8$ petals.

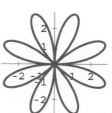

21. The graph of $r = \cos 5\theta$ is a rose curve. Since 5 is an odd number, there are 5 petals.

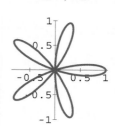

24. The graph of $r = -2\cos\theta$ is a circle centered at $(-1, 0)$.

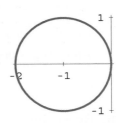

27. The graph of $r^2 = 4\sin 2\theta$ is a lemniscate.

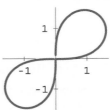

30. Since r^2 is nonnegative, $\sin 2\theta \le 0$, so $\pi < 2\theta < 2\pi$ and $\pi/2 < \theta < \pi$. In this case, $r = \pm 3\sqrt{-\sin 2\theta}$.

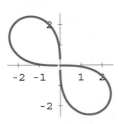

33. $r = \dfrac{5}{2}$

36. $r = 2 + 3\sin\theta$

39. The graph of $r = 2$ is shown in blue, and the graph of $r = 4\sin\theta$ is shown in red. The two equations imply $4\sin\theta = 2$, so $\sin\theta = \frac{1}{2}$ and $\theta = \pi/6,\ 5\pi/6$. The points of intersection are $(2, \pi/6)$ and $(2, 5\pi/6)$.

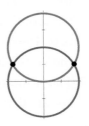

42. The graph of $r = 3 - 3\cos\theta$ is shown in blue, and the graph of $r = 3\cos\theta$ is shown in red. Adding the two equations, we obtain $2r = 3$, so $r = \frac{3}{2}$. This implies $\cos\theta = \frac{1}{2}$ or $\theta = \pi/3,\ 5\pi/3$. The points of intersection are $(3/2, \pi/3)$ and $(3/2, 5\pi/3)$. From the figure, we see that $(0, 0)$ is also a point of intersection.

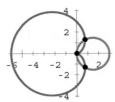

8.3   Conic Sections in Polar Coordinates

3. By writing the equation as

$$r = \frac{4}{1 + \frac{1}{4}\cos\theta},$$

we see that the eccentricity is $e = \frac{1}{4}$, and so the conic is an ellipse. Since the equation has the form given in (5) in the text, the axis of the ellipse lies along the x-axis. In polar form, the vertices

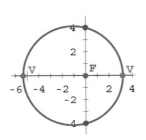

are where $\theta = 0$ and $\theta = \pi$, so they are at $(\frac{16}{5}, 0)$ and $(\frac{16}{3}, \pi)$ in polar coordinates. The y-intercepts occur where $\theta = \pi/2$ and $\theta = 3\pi/2$, so they are at $(4, \pi/2)$ and $(4, 3\pi/2)$.

6. By writing the equation as

$$r = \frac{4}{1 - \cos\theta},$$

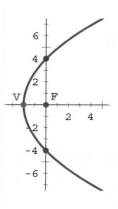

we see that the eccentricity is $e = 1$, and so the conic is a parabola. Since the equation has the form given in (5) in the text, the axis of the parabola lies along the x-axis. In polar form, the vertex is where $\theta = \pi$, so it is at $(2, \pi)$ in polar coordinates. The y-intercepts occur where $\theta = \pi/2$ and $\theta = 3\pi/2$, so they are at $(4, \pi/2)$ and $(4, 3\pi/2)$.

9. From

$$r = \frac{6}{1 - \cos\theta},$$

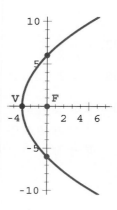

we see that the eccentricity is $e = 1$, and so the conic is a parabola. Since the equation has the form given in (5) in the text, the axis of the parabola lies along the x-axis. In polar form, the vertex is where $\theta = \pi$, so it is at $(3, \pi)$ in polar coordinates. The y-intercepts occur where $\theta = \pi/2$ and $\theta = 3\pi/2$, so they are at $(6, \pi/2)$ and $(6, 3\pi/2)$.

12. By writing the equation as

$$r = \frac{5}{1 - \frac{3}{2}\cos\theta},$$

we see that the eccentricity is $e = \frac{3}{2}$, and so the conic is a hyperbola. Now, write the equation in the form

$$r - \frac{3}{2}r\cos\theta = 5 \qquad \text{or} \qquad r = 5 + \frac{3}{2}r\cos\theta.$$

Substituting $r = \pm\sqrt{x^2 + y^2}$ and $x = r\cos\theta$ and simplifying, we have

$$\pm\sqrt{x^2 + y^2} = 5 + \frac{3}{2}x$$

$$x^2 + y^2 = 25 + 15x + \frac{9}{4}x^2$$

$$y^2 - \frac{5}{4}x^2 - 15x = 25$$

$$y^2 - \frac{5}{4}(x^2 + 12x) = 25$$

$$y^2 - \frac{5}{4}(x^2 + 12x + 36) = 25 - \frac{5}{4}(36) = -20$$

$$\frac{(x+6)^2}{16} - \frac{y^2}{20} = 1.$$

$c^2 = a^2 + b^2 = 36$. Thus, $c = 6$ and $e = c/a = \frac{6}{4} = \frac{3}{2}$.

15. Since the directrix is a vertical line to the right of the origin, the equation of the conic has the form $r = ep/(1 + e\cos\theta)$. Since $e = 1$, the conic is a parabola, and since the directrix is 3 units from the origin, $p = 3$. Thus, the equation of the parabola is

$$r = \frac{1(3)}{1 + 1\cos\theta} = \frac{3}{1 + \cos\theta}.$$

18. Since the directrix is a vertical line to the right of the origin, the equation of the conic has the form $r = ep/(1 + e\cos\theta)$. Since $e = \frac{1}{2}$, the conic is an ellipse, and since the directrix is 4 units from the origin, $p = 4$. Thus, the equation of the ellipse is

$$r = \frac{\frac{1}{2}(4)}{1 + \frac{1}{2}\cos\theta} = \frac{4}{2 + \cos\theta}.$$

21. Because the conic is a parabola, the eccentricity is $e = 1$. Since the vertex lies on the y-axis below the origin, the equation of the parabola has the form $r = p/(1 - \sin\theta)$. The distance from the focus to the directrix is p and the vertex is halfway between the two, so $p = 2(\frac{3}{2}) = 3$ and

$$r = \frac{3}{1 - \sin\theta}.$$

24. Because the conic is a parabola, the eccentricity is $e = 1$. Since the vertex lies on the x-axis to the right of the origin, the equation of the parabola has the form $r = p/(1 + \cos\theta)$. The distance from the focus to the directrix is p and the vertex is halfway between the two, so $p = 2(2) = 4$ and

$$r = \frac{4}{1 + \cos\theta}.$$

27. Because the conic is a parabola, the eccentricity is $e = 1$. Since the vertex lies on the y-axis below the origin, the equation of the parabola has the form $r = p/(1 - \sin\theta)$. The distance from the focus to the directrix is p and the vertex is halfway between the two, so $p = 2(\frac{1}{4}) = \frac{1}{2}$ and

$$r = \frac{\frac{1}{2}}{1 - \sin\theta} = \frac{1}{2 - 2\sin\theta}.$$

30. This is the ellipse $r = \dfrac{5}{3 + 2\cos\theta} = \dfrac{5/3}{1 + 2/3\cos\theta}$ rotated counterclockwise by $\pi/3$. The original ellipse had vertices at $\theta = 0$ and $\theta = \pi$. The polar coordinates of the vertices were $(1, 0)$ and $(5, \pi)$. After rotation, the vertices are located at $(1, \pi/3)$ and $(5, 4\pi/3)$.

33. We are given $r_a = 12{,}000$ and $e = 0.2$. We let r_p be the perigee distance and solve equation (7) in the text for r_p:

$$0.2 = \frac{r_a - r_p}{r_a + r_p}$$

$$0.2 r_a + 0.2 r_p = r_a - r_p$$

$$1.2 r_p = r_a - 0.2 r_a$$

$$r_p = \frac{0.8 r_a}{1.2} = \frac{0.8(12{,}000)}{1.2} \approx 8{,}000 \text{ km.}$$

36. (a) We are given $2a = 3.34 \times 10^9$, so $a = 1.67 \times 10^9$. Also, we are given $e = 0.97$, so

$$c = ae = (1.67 \times 10^9)(0.97) \approx 1.62 \times 10^9.$$

Now, the distance from the focus at $(0,0)$ to the vertex on the negative x-axis is $a - c$. Thus,

$$r(\pi) = \frac{0.97p}{1 - 0.97(-1)} = \frac{0.97p}{1.97} = a - c = 1.67 \times 10^9 - 1.62 \times 10^9 = 0.05 \times 10^9,$$

so $0.97p = 1.97 \times 0.05 \times 10^9 = 0.0985 \times 10^9$ and

$$p = \frac{0.0985 \times 10^9}{0.97} \approx 0.1 \times 10^9 = 10^8.$$

The equation of the orbit is

$$r = \frac{0.97 \times 10^8}{1 - 0.97\cos\theta}.$$

(b) $r_a = r(0) = \dfrac{0.97 \times 10^8}{1 - 0.97} = \dfrac{0.97 \times 10^8}{0.03} \approx 32.3 \times 10^8 = 3.23 \times 10^9$ mi

$r_p = r(\pi) = 0.05 \times 10^9 = 5 \times 10^7$ mi

8.4   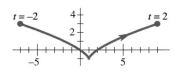   **Parametric Equations**

3.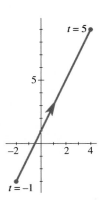

6.

9.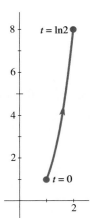

12. Substituting $t^3 + t = x - 4$ into y, we obtain

$$y = -2(x - 4) = -2x + 8.$$

15. From $y = 3 \ln t$ we have $t = e^{y/3}$. Substituting this into x, we obtain

$$x = t^3 = (e^{y/3})^3 = e^y \qquad \text{or} \qquad y = \ln x,$$

where $x > 0$.

18. Squaring both sides of $x + 1 = \cos t$ and $y - 2 = \sin t$, we obtain $(x + 1)^2 = \cos^2 t$ and $(y - 2)^2 = \sin^2 t$. Adding the equations, we get

$$(x + 1)^2 + (y - 2)^2 = 1.$$

21. The rectangular graph is shown in blue and the parametric curve in red. We see that the parameterized curve is only a portion of the rectangular graph.

24. The rectangular graph is shown in blue and the parametric curve in red. We see that the parameterized curve is a ray lying on the line $y = 2x - 2$, starting at $t = 0$ or $(-1, -4)$.

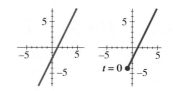

27.

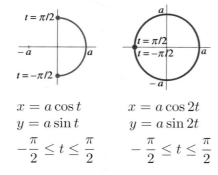

$$x = a \cos t \qquad x = a \cos 2t$$
$$y = a \sin t \qquad y = a \sin 2t$$
$$-\frac{\pi}{2} \le t \le \frac{\pi}{2} \qquad -\frac{\pi}{2} \le t \le \frac{\pi}{2}$$

30. We find the x-intercepts by solving

$$y = t^2 + t - 6 = 0 \qquad \text{or} \qquad y = (t + 3)(t - 2) = 0.$$

Thus, the x-intercepts occur for $t = -3$ and $t = 2$. When $t = -3$, $x = 6$, and when $t = 2$, $x = 6$. The only x-intercept is $(6, 0)$. We find the y-intercepts by solving

$$x = t^2 + t = 0 \qquad \text{or} \qquad x = t(t + 1) = 0.$$

Thus, the y-intercepts occur for $t = 0$ and $t = -1$. When $t = 0$, $y = -6$, and when $t = -1$, $y = -6$. The only y-intercept is $(0, -6)$.

33. Setting $v_0 = 190$, $\theta_0 = 45°$, and $g = 32$, by (1) in the text, parametric equations of the ball's path are

$$x = (190 \cos 45°)t = 95\sqrt{2}\, t \approx 134.35t$$
$$y = -\frac{1}{2}(32)t^2 + (190 \sin 45°)t = -16t^2 + 95\sqrt{2}\, t \approx -16t^2 + 134.35t, \quad t \ge 0.$$

When $t = 2$, $x = 268.7$ and $y = -64 + 268.7 = 204.7$, so the ball is at the point $(268.7, 204.7)$.

36. Refer to the figure at the right. Using triangle BCO, we have $\cot\theta = x/2a$ or $x = 2a \cot\theta$. Since angle OAC subtends an arc of $180°$, OAC is a right triangle and $\sin\theta = h/2a$ or $h = 2a \sin\theta$. Now, using triangle ODA, $\sin\theta = y/h$. Thus, $y = h \sin\theta = 2a \sin^2\theta$. The parametric equations are $x = 2a \cot\theta$ and $y = 2a \sin^2\theta$.

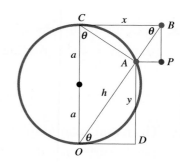

Chapter 8 Review Exercises

A. Fill in the Blanks

3. Identifying $x = 0$ and $y = -10$, we see that the point is 10 units down the negative y-axis. Thus, polar coordinates of the point are $(10, 3\pi/2)$.

6. This is the equation of a circle with a diameter on the x-axis and radius $|-2a|/2 = a$. Since $-2a$ is negative, the circle lies in the second and third quadrants and the center is at $(-a, 0)$.

9. Identifying $a = 2$ and $b = 1$, we have $a/b = 2 > 1$, so the curve is a convex limaçon.

12. The graph of the polar equation $r = 2 + 2\sin(\theta - \pi/4)$ is a cardioid rotated $\pi/4$ radians counterclockwise about the polar axis.

15. Solving $x = t + 2$ for t, we obtain $t = x - 2$. Substituting into $y = 3 + \frac{1}{2}t$, we get $y = 3 + \frac{1}{2}(x - 2) = \frac{1}{2}x + 2$. This is a line, and since $-\infty < t < \infty$, the parametric equations represent the entire straight line.

B. True/False

3. By convention, (r, θ) and $(-r, \theta + \pi)$ represent the same point. Since $\left(-(-3), (-5\pi/6) + \pi\right)$ is the same as $(3, \pi/6)$, the statement is true.

6. For $r = 2 + 4\sin\theta$, we identify $a = 2$ and $b = 4$. Since $a/b = \frac{1}{2} < 1$, the graph is a limaçon with an interior loop and the statement is true.

9. False; the same point can be expressed as $(-4, \pi/2)$, which does satisfy the equation.

12. False; the eccentricity of a parabola is 1.

15. True; both in rectangular form are $x^2 + y^2 = 25$.

18. True; since $x = 1 + \cos t$, $y = 1 + \sin t$ has rectangular form $(x - 1)^2 + (y - 1)^2 = 1$.

C. Review Exercises

3. Since $r^2 = x^2 + y^2$ and $y = r\sin\theta$, a polar equation having the same graph as $x^2 + y^2 - 4y = 0$ is $r^2 - 4r\sin\theta = 0$ or $r(r - 4\sin\theta) = 0$. From $r = 4\sin\theta$, we see that when $\theta = 0$, $r = 0$, so we do not need to use $r = 0$ from $r(r - 4\sin\theta) = 0$. The polar equation is $r = 4\sin\theta$.

6. Since the vertices lie on the y-axis, the polar form of the equation is $r = ep/(1 \pm e \sin \theta)$. We are given $e = 2$ and both vertices are below the origin, so $r = 2p/(1 - 2\sin \theta)$, and we need only find p. From $r(\pi/2) = -2p$ and $r(3\pi/2) = 2p/3$ we have $-2p = -4$ so $p = 2$, and $2p/3 = 4/3$, so $p = 2$. The equation of the hyperbola is $r = 4/(1 - 2\sin\theta)$.

9.

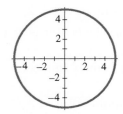

The graph is a circle centered at the origin with radius 5.

12.

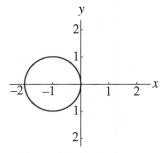

The graph is a circle with radius 1, centered on the x-axis at $(-1, 0)$.

15.

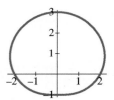

This is the equation of a limaçon with $a = 2$ and $b = 1$, so that $a/b = 2 > 1$ and the curve is a convex limaçon.

18.

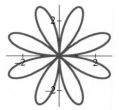

This graph is a rose curve with $8 = 4 \cdot 2$ petals.

21. An equation is $r = 3\sin(10\theta)$.

24. (a) A polar equation of the rotated graph is $r = \dfrac{1}{1 + \cos(\theta + \pi/6)}$.

(b) Using the identity for the cosine of the sum of two angles, we have

$$r = \frac{1}{1 + \cos\theta\cos\pi/6 - \sin\theta\sin\pi/6}$$

$$r = \frac{1}{1 + \frac{\sqrt{3}}{2}\cos\theta - \frac{1}{2}\sin\theta}$$

$$r\left(1 + \frac{\sqrt{3}}{2}\cos\theta - \frac{1}{2}\sin\theta\right) = 1$$

$$r + \frac{\sqrt{3}}{2}r\cos\theta - \frac{1}{2}r\sin\theta = 1$$

$$\sqrt{x^2 + y^2} + \frac{\sqrt{3}}{2}x - \frac{1}{2}y = 1$$

$$\sqrt{x^2 + y^2} = 1 - \frac{\sqrt{3}}{2}x + \frac{1}{2}y$$

$$x^2 + y^2 = \frac{3}{4}x^2 - \frac{\sqrt{3}xy}{2} - \sqrt{3}x + \frac{y^2}{4} + y + 1$$

$$\frac{1}{4}x^2 + \frac{\sqrt{3}xy}{2} + \sqrt{3}x + \frac{3}{4}y^2 - y = 1.$$

Chapter 9

Systems of Equations and Inequalities

9.1 | **Systems of Linear Equations**

In Problems 3–9 we will generally use the method of substitution when one of the equations contains an unknown with coefficient either −1 or 1. Otherwise, we will use the method of elimination.

3. We use the method of substitutuion. Solving the first equation for y gives $y = 4x + 1$. We substitute this expression into the second equation and solve for x:

$$x + 3(4x + 1) + 9 = 0 \quad \text{or} \quad 13x + 12 = 0 \quad \text{or} \quad x = -\frac{12}{13}.$$

Substituting this value back into $y = 4x + 1$ gives

$$y = 4\left(-\frac{12}{13}\right) + 1 = -\frac{35}{13}.$$

Thus, the only solution of the system is $\left(-\frac{12}{13}, -\frac{35}{13}\right)$, the system is consistent, and the equations are independent.

6. We use the method of elimination.

$$\left. \begin{array}{r} 6x - 4y = 9 \\ -3x + 2y = -4.5 \end{array} \right\} \xrightarrow{\frac{1}{2}E_1 + E_2} \left\{ \begin{array}{r} 6x - 4y = 9 \\ 0 = 0 \end{array} \right.$$

Thus, the system is equivalent to one equation in two variables. Letting $y = \alpha$ in the first equation we have

$$6x - 4\alpha = 9 \quad \text{or} \quad 6x = 4\alpha + 9 \quad \text{or} \quad x = \frac{2}{3}\alpha + \frac{3}{2}.$$

The solution of the system is all ordered pairs of the form $\left(\frac{2}{3}\alpha + \frac{1}{2}, \alpha\right)$, where α is any real number. The system is consistent with dependent equations.

9. We use the method of elimination.

$$\left. \begin{array}{r} -x - 2y = -4 \\ 5x + 10y = 20 \end{array} \right\} \xrightarrow{5E_1 + E_2} \left\{ \begin{array}{r} -x - 2y = -4 \\ 0 = 0 \end{array} \right.$$

Thus, the system is equivalent to one equation in two variables. Letting $y = \alpha$ in the first equation we have

$$-x - 2y = -4 \quad \text{or} \quad x = -2y + 4.$$

The solution is all ordered pairs of the form $(-2\alpha + 4,\ \alpha)$, where α is any real number. The system is consistent with dependent equations.

12. By elimination

$$\left.\begin{array}{r} x + y + z = 8 \\ x - 2y + z = 4 \\ x + y - z = -4 \end{array}\right\} \xrightarrow[-E_1+E_3]{-E_1+E_2} \left\{\begin{array}{r} x + y + z = 8 \\ -3y = -4 \\ -2z = -12 \end{array}\right.$$

From the third equation we see that $z = 6$, and from the second equation $y = \frac{4}{3}$. Substituting these values into the first equation we have

$$x + \frac{4}{3} + 6 = 8 \quad \text{or} \quad x = \frac{2}{3}.$$

Thus, the system is consistent with independent equations, and the solution is $\left(\frac{2}{3},\ \frac{4}{3},\ 6\right)$.

15. We begin by interchanging the first and second equations.

$$\left.\begin{array}{r} 2x + y + z = 1 \\ x - y + 2z = 5 \\ 3x + 4y - z = -2 \end{array}\right\} \xrightarrow{E_1 \leftrightarrow E_2} \left\{\begin{array}{r} x - y + 2z = 5 \\ 2x + y + z = 1 \\ 3x + 4y - z = -2 \end{array}\right.$$

$$\xrightarrow[-3E_1+E_3]{-2E_1+E_2} \left\{\begin{array}{r} x - y + 2z = 5 \\ y - z = -9 \\ y - z = -7 \end{array}\right.$$

Since the second and third equations imply that

$$-9 = y - z = -7,$$

we see that the system is inconsistent and thus has no solutions.

18. Since the system is homogeneous, it has $(0,\ 0,\ 0)$, at least, as a solution. To see if there are any nontrivial solutions we use elimination.

$$\left.\begin{array}{r} -5x + y + z = 0 \\ 4x - y = 0 \\ 2x - y + 2z = 0 \end{array}\right\} \xrightarrow{E_1 \leftrightarrow E_3} \left\{\begin{array}{r} 2x - y + 2z = 0 \\ 4x - y = 0 \\ -5x + y + z = 0 \end{array}\right.$$

$$\xrightarrow[\frac{5}{2}E_1+E_3]{-2E_1+E_2} \left\{\begin{array}{r} 2x - y + 2z = 0 \\ y - 4z = 0 \\ -\frac{3}{2}y + 6z = 0 \end{array}\right.$$

$$\xrightarrow{\frac{3}{2}E_2+E_3} \left\{\begin{array}{r} 2x - y + 2z = 0 \\ y - 4z = 0 \\ 0 = 0 \end{array}\right.$$

Letting $z = \alpha$, we have from the second equation $y = 4\alpha$. Substituting these values into the first equation we have

$$2x - 4\alpha + 2\alpha = 0 \qquad \text{or} \qquad x = \alpha.$$

Thus the system is consistent with dependent equations, and the solutions are $(\alpha, \ 4\alpha, \ \alpha)$, where α is any real number.

21. By elimination

$$\left.\begin{array}{r} -x + 3y + 2z = 2 \\ \frac{1}{2}x - \frac{3}{2}y - z = -1 \\ -\frac{1}{3}x + y + \frac{2}{3}z = \frac{2}{3} \end{array}\right\} \xrightarrow[-3E_3]{2E_2} \left\{\begin{array}{r} -x + 3y + 2z = 2 \\ x - 3y - 2z = -2 \\ x - 3y - 2z = -2 \end{array}\right.$$

The three equations are equivalent, so the system is consistent with dependent equations. Letting $y = \alpha$ and $z = \beta$ we have $x = 3\alpha + 2\beta - 2$. The solutions are $(3\alpha + 2\beta - 2, \ \alpha, \ \beta)$, where α and β are any real numbers.

24. By elimination

$$\left.\begin{array}{r} x + y + z = 4 \\ 2x - y + 2z = 11 \\ 4x + 3y - 6z = -18 \end{array}\right\} \xrightarrow[-4E_1+E_3]{-2E_1+E_2} \left\{\begin{array}{r} x + y + z = 4 \\ -3y = 3 \\ -y - 10z = -34 \end{array}\right.$$

From the second equation we have $y = -1$. Substituting this into the third equation we have

$$-(-1) - 10z = -34 \qquad \text{or} \qquad z = \frac{35}{10} = \frac{7}{2}.$$

Substituting the values for y and z into the first equation we have

$$x - 1 + \frac{7}{2} = 4 \qquad \text{or} \qquad x = \frac{3}{2}.$$

The system is consistent with independent equations, and the solution is $\left(\frac{3}{2}, \ -1, \ \frac{7}{2}\right)$.

27. Using the method of elimination, we have

$$\left.\begin{array}{r} x - y + 4z = 1 \\ 6x + y - z = 2 \end{array}\right\} \xrightarrow{-6E_1+E_2} \left\{\begin{array}{r} x - y + 4z = 1 \\ 7y - 25z = -4 \end{array}\right.$$

Letting $z = \alpha$, we obtain $y = \frac{1}{7}(-4 + 25\alpha) = -\frac{4}{7} + \frac{25}{7}\alpha$, and $x = 1 + \left(-\frac{4}{7} + \frac{25}{7}\alpha\right) - 4\alpha = \frac{3}{7} - \frac{3}{7}\alpha$. The solutions are $\left(\frac{3}{7} - \frac{3}{7}\alpha, \ -\frac{4}{7} + \frac{25}{7}\alpha, \ \alpha\right)$ where α is any real number.

30. Using the method of elimination, we have

$$\left.\begin{array}{r} x - y + z = 0 \\ x + y - z = 0 \end{array}\right\} \xrightarrow{-E_1+E_2} \left\{\begin{array}{r} x - y + z = 0 \\ 2y - 2z = 0 \end{array}\right.$$

Letting $z = \alpha$, we obtain $y = \alpha$ and $x = \alpha - \alpha = 0$. The solutions are $(0, \ \alpha, \ \alpha)$ where α is any real number.

33. We need to balance the chemical equation $Fe_3O_4 + C \rightarrow Fe + CO$. That is, we look for positive integers x, y, z, and w so that the balanced equation is

$$x Fe_3O_4 + y C \rightarrow z Fe + w CO.$$

Then

$$
\begin{aligned}
3x &= z \\
4x &= w \qquad \text{or} \\
y &= w
\end{aligned}
\qquad
\begin{aligned}
3x + 0y - z + 0w &= 0 \\
4x + 0y + 0z - w &= 0 \\
0x + y + 0z - w &= 0.
\end{aligned}
$$

and

$$
\left.
\begin{aligned}
3x \quad - z \quad\quad &= 0 \\
4x \quad\quad - w &= 0 \\
y \quad\quad - w &= 0
\end{aligned}
\right\}
\xrightarrow[\text{operations}]{\text{elimination}}
\left\{
\begin{aligned}
x \quad - \tfrac{1}{3}z \quad\quad &= 0 \\
y \quad\quad - w &= 0 \\
\tfrac{4}{3}z - w &= 0
\end{aligned}
\right.
$$

Letting $w = \alpha$, we obtain $z = \tfrac{3}{4}\alpha$, $y = \alpha$ and $x = \tfrac{1}{4}\alpha$. Since the four variables must each be a positive integer, we can choose $\alpha = 4$. Then the balanced equation will be

$$Fe_3O_4 + 4C \rightarrow 3Fe + 4CO.$$

36. We need to balance the chemical equation $Ca_3(PO_4)_2 + H_3PO_4 \rightarrow Ca(H_2PO_4)_2$. That is, we look for positive integers x, y, and z so that the balanced equation is

$$x Ca_3(PO_4)_2 + y H_3PO_4 \rightarrow z Ca(H_2PO_4)_2.$$

Then

$$
\begin{aligned}
3x &= z \\
2x + y &= 2z \\
8x + 4y &= 8z \\
3y &= 4z
\end{aligned}
\qquad \text{or} \qquad
\begin{aligned}
3x + 0y - z &= 0 \\
2x + y - 2z &= 0 \\
8x + 4y - 8z &= 0 \\
0x + 3y - 4z &= 0
\end{aligned}
$$

and

$$
\left.
\begin{aligned}
3x \quad\quad - z &= 0 \\
2x + y - 2z &= 0 \\
8x + 4y - 8z &= 0 \\
3y - 4z &= 0
\end{aligned}
\right\}
\xrightarrow[\text{operations}]{\text{elimination}}
\left\{
\begin{aligned}
x + \tfrac{1}{2}y - z &= 0 \\
y - \tfrac{4}{3}z &= 0 \\
0 &= 0 \\
0 &= 0
\end{aligned}
\right.
$$

Letting $z = \alpha$, we obtain $y = \tfrac{4}{3}\alpha$ and $x = -\tfrac{1}{2}(\tfrac{4}{3})\alpha + \alpha = \tfrac{1}{3}\alpha$. Since the three variables must each be a positive integer, we can choose $\alpha = 3$. Then the balanced equation will be

$$Ca_3(PO_4)_2 + 4H_3PO_4 \rightarrow 3Ca(H_2PO_4)_2.$$

39. Letting $x' = \log_{10} x$ and $y' = \log_{10} y$, the system can be written

$$3x' + \ y' = 2$$
$$5x' + 2y' = 1.$$

We use the method of substitutuion. Solving the first equation for y' gives $y' = -3x' + 2$. We substitute this expression into the second equation and solve for x':

$$5x' + 2(-3x' + 2) = 1 \qquad \text{or} \qquad -x' = -3 \qquad \text{or} \qquad x' = 3.$$

Substituting this value back into $y' = -3x' + 2$ gives

$$y' = -3(3) + 2 = -7.$$

Then $x' = 3$ so $x = 10^3$, and $y' = -7$ so $y = 10^{-7}$.

42. Using

$$\cos 30° = \frac{\sqrt{3}}{2}, \qquad \cos 60° = \frac{1}{2}, \qquad \sin 30° = \frac{1}{2}, \qquad \text{and} \qquad \sin 60° = \frac{\sqrt{3}}{2},$$

we can write the system as

$$\frac{\sqrt{3}}{2} P - 0.4 \left(\frac{\sqrt{3}}{2} \right) N - \frac{1}{2} N = 0 \qquad \qquad \sqrt{3}\, P - \left(1 + 0.4\sqrt{3} \right) N = \ \ 0$$
$$\text{or}$$
$$\frac{1}{2} P - 0.4 \left(\frac{1}{2} \right) N + \frac{\sqrt{3}}{2} N - 300 = 0 \qquad \qquad P + \left(\sqrt{3} - 0.4 \right) N = 600.$$

We interchange the two equations and use the method of elimination.

$$\left. \begin{array}{l} P + \left(\sqrt{3} - 0.4 \right) N = 600 \\ \sqrt{3}\, P - \left(1 + 0.4\sqrt{3} \right) N = \ \ 0 \end{array} \right\} \xrightarrow{\sqrt{3}\, E_1 + E_2} \left\{ \begin{array}{l} P + \left(\sqrt{3} - 0.4 \right) N = \ \ 600 \\ \qquad\qquad\qquad -4N = -600\sqrt{3} \end{array} \right.$$

From the second equation we find $N = 150\sqrt{3}$. Substituting this into the first equation and solving for P we have

$$P = 600 - \left(\sqrt{3} - 0.4 \right) \left(150\sqrt{3} \right) = 600 - 450 + 60\sqrt{3} = 150 + 60\sqrt{3}.$$

Miscellaneous Applications

45. Let A_1 be the amount removed from the first tank and A_2 the amount from the second tank. Then $A_1 + A_2 = 90$. The amount of salt removed from the first tank is $\frac{50}{100} = \frac{1}{2}$ lb, and the amount from the second tank is $\frac{75}{200} = \frac{3}{8}$ lb. Since the mixture of these two amounts in $\frac{4}{9}(90) = 40$ lb, we have $\frac{1}{2} A_1 + \frac{3}{8} A_2 = 40$. Solving $A_1 + A_2 = 90$ for A_1 we obtain $A_1 = 90 - A_2$. Substituting this into $\frac{1}{2} A_1 + \frac{3}{8} A_2 = 40$, and solving for A_2 we find

$$\frac{1}{2} (90 - A_2) + \frac{3}{8} A_2 = 40 \qquad \text{or} \qquad A_2 = (360 - 4A_2) + 3A_2 = 320 \qquad \text{or} \qquad A_2 = 40.$$

Then $A_1 = 90 - 40 = 50$ and the amount removed from the first tank is 50 lb and the amount removed from the second tank is 40 lb.

48. Substituting the points $(1, 10)$, $(-1, 12)$, and $(2, 18)$ into the equation of the parabola we obtain a system of equations in the unknowns a, b, and c that can be solved by elimination:

$$\left.\begin{array}{r} a + b + c = 10 \\ a - b + c = 12 \\ 4a + 2b + c = 18 \end{array}\right\} \xrightarrow[-4E_1+E_3]{-E_1+E_2} \left\{\begin{array}{r} a + b + c = 10 \\ -2b = 2 \\ -2b - 3c = -22. \end{array}\right.$$

From the second equation we have $b = -1$. Substituting this value into the third equation gives $-3c = -24$ or $c = 8$. Finally, substituting the values of b and c into the first equation, we find $a = 10 - (-1) - 8 = 3$. Thus, the equation of the parabola is $y = 3x^2 - x + 8$.

51. Let x be the number of A's, y the number of B's, and z the number of C's. The resulting system, obtained using weighted averages, is

$$x + y + z = 40$$

$$\frac{4x + 3y + 2z}{x + y + z} = 3.125$$

$$\frac{4x + 3y}{x + y} = 3.8$$

We write this system as three linear equations in x, y, and z, and solve by elimination:

$$\left.\begin{array}{r} x + y + z = 40 \\ 0.875x - 0.125y - 1.125z = 0 \\ 0.2x - 0.8y = 0 \end{array}\right\} \xrightarrow[-0.2E_1+E_3]{-0.875E_1+E_2} \left\{\begin{array}{r} x + y + z = 40 \\ y - 2z = -35 \\ y - 0.2z = -8 \end{array}\right.$$

$$\xrightarrow{-E_2+E_3} \left\{\begin{array}{r} x + y + z = 40 \\ y - 2z = -35 \\ 1.8z = 27. \end{array}\right.$$

From the third equation we see that $z = 15$. Substituting into the second equation and solving for y we have $y = 5$. Substituting these values into the first equation we have

$$x + 5 + 15 = 40 \qquad \text{or} \qquad x = 20.$$

Thus, she received 20 A's, 5 B's, and 15 C's.

9.2 | Determinants and Cramer's Rule

3. The minors are

$$M_{11} = \begin{vmatrix} 1 & 0 \\ 0 & 5 \end{vmatrix} = 5 \qquad M_{12} = \begin{vmatrix} 2 & 0 \\ -3 & 5 \end{vmatrix} = 10 \qquad M_{13} = \begin{vmatrix} 2 & 1 \\ -3 & 0 \end{vmatrix} = 3$$

$$M_{21} = \begin{vmatrix} -7 & 8 \\ 0 & 5 \end{vmatrix} = -35 \qquad M_{22} = \begin{vmatrix} 1 & 8 \\ -3 & 5 \end{vmatrix} = 29 \qquad M_{23} = \begin{vmatrix} 1 & -7 \\ -3 & 0 \end{vmatrix} = -21$$

$$M_{31} = \begin{vmatrix} -7 & 8 \\ 1 & 0 \end{vmatrix} = -8 \qquad M_{32} = \begin{vmatrix} 1 & 8 \\ 2 & 0 \end{vmatrix} = -16 \qquad M_{33} = \begin{vmatrix} 1 & -7 \\ 2 & 1 \end{vmatrix} = 15.$$

The cofactors are

$$A_{11} = 5 \qquad\qquad A_{12} = -1(10) = -10 \qquad A_{13} = 3$$

$$A_{21} = -1(-35) = 35 \qquad A_{22} = 29 \qquad\qquad A_{23} = -1(-21) = 21$$

$$A_{31} = -8 \qquad\qquad A_{32} = -1(-16) = 16 \qquad A_{33} = 15.$$

6. $\begin{vmatrix} 0 & -1 \\ 8 & 0 \end{vmatrix} = 0(0) - (-1)(8) = 8$

9. $\begin{vmatrix} a & -b \\ b & a \end{vmatrix} = a(a) - (-b)(b) = a^2 + b^2$

12. $\begin{vmatrix} 6 & 2 & 1 \\ 0 & 3 & -4 \\ 1 & 0 & 2 \end{vmatrix} = 6\begin{vmatrix} 3 & -4 \\ 0 & 2 \end{vmatrix} - 2\begin{vmatrix} 0 & -4 \\ 1 & 2 \end{vmatrix} + 1\begin{vmatrix} 0 & 3 \\ 1 & 0 \end{vmatrix} = 6(6) - 2(4) + 1(-3) = 25$

15. $\begin{vmatrix} 5 & 9 & 1 \\ 1 & 2 & -3 \\ 0 & 0 & 0 \end{vmatrix} = 5\begin{vmatrix} 2 & -3 \\ 0 & 0 \end{vmatrix} - 9\begin{vmatrix} 1 & -3 \\ 0 & 0 \end{vmatrix} + 1\begin{vmatrix} 1 & 2 \\ 0 & 0 \end{vmatrix} = 5(0) - 9(0) + 1(0) = 0$

18. $\begin{vmatrix} 1 & 2 & 4 \\ 5 & 1 & -1 \\ 1 & 2 & 4 \end{vmatrix} = 1\begin{vmatrix} 1 & -1 \\ 2 & 4 \end{vmatrix} - 2\begin{vmatrix} 5 & -1 \\ 1 & 4 \end{vmatrix} + 4\begin{vmatrix} 5 & 1 \\ 1 & 2 \end{vmatrix} = 1(6) - 2(21) + 4(9) = 0$

21. Since

$$D = \begin{vmatrix} 2 & -1 \\ -1 & 3 \end{vmatrix} = 5 \neq 0,$$

the system has a unique solution. From

$$D_x = \begin{vmatrix} -3 & -1 \\ 19 & 3 \end{vmatrix} = 10 \qquad \text{and} \qquad D_y = \begin{vmatrix} 2 & -3 \\ -1 & 19 \end{vmatrix} = 35,$$

we find

$$x = \frac{10}{5} = 2 \qquad \text{and} \qquad y = \frac{35}{5} = 7.$$

24. Since

$$D = \begin{vmatrix} -1 & -3 \\ -2 & 6 \end{vmatrix} = -12 \neq 0,$$

the system has a unique solution. From

$$D_x = \begin{vmatrix} -7 & -3 \\ -9 & 6 \end{vmatrix} = -69 \qquad \text{and} \qquad D_y = \begin{vmatrix} -1 & -7 \\ -2 & -9 \end{vmatrix} = -5,$$

we find

$$x = \frac{-69}{-12} = \frac{23}{4} \qquad \text{and} \qquad y = \frac{-5}{-12} = \frac{5}{12}.$$

27. Since

$$D = \begin{vmatrix} 1 & 1 & -1 \\ 2 & -1 & 3 \\ 2 & 3 & 0 \end{vmatrix} = 1 \begin{vmatrix} -1 & 3 \\ 3 & 0 \end{vmatrix} - 1 \begin{vmatrix} 2 & 3 \\ 2 & 0 \end{vmatrix} + (-1) \begin{vmatrix} 2 & -1 \\ 2 & 3 \end{vmatrix}$$

$$= 1(-9) - (-6) - 8 = -11 \neq 0,$$

the system is consistent and has a unique solution. From

$$D_x = \begin{vmatrix} 5 & 1 & -1 \\ -3 & -1 & 3 \\ -4 & 3 & 0 \end{vmatrix} = 5 \begin{vmatrix} -1 & 3 \\ 3 & 0 \end{vmatrix} - 1 \begin{vmatrix} -3 & 3 \\ -4 & 0 \end{vmatrix} + (-1) \begin{vmatrix} -3 & -1 \\ -4 & 3 \end{vmatrix}$$

$$= 5(-9) - 12 - (-13) = -44,$$

$$D_y = \begin{vmatrix} 1 & 5 & -1 \\ 2 & -3 & 3 \\ 2 & -4 & 0 \end{vmatrix} = 1 \begin{vmatrix} -3 & 3 \\ -4 & 0 \end{vmatrix} - 5 \begin{vmatrix} 2 & 3 \\ 2 & 0 \end{vmatrix} + (-1) \begin{vmatrix} 2 & -3 \\ 2 & -4 \end{vmatrix}$$

$$= 1(12) - 5(-6) - (-2) = 44,$$

and

$$D_z = \begin{vmatrix} 1 & 1 & 5 \\ 2 & -1 & -3 \\ 2 & 3 & -4 \end{vmatrix} = 1 \begin{vmatrix} -1 & -3 \\ 3 & -4 \end{vmatrix} - 1 \begin{vmatrix} 2 & -3 \\ 2 & -4 \end{vmatrix} + 5 \begin{vmatrix} 2 & -1 \\ 2 & 3 \end{vmatrix}$$

$$= 1(13) - (-2) + 5(8) = 55,$$

we find

$$x = \frac{-44}{-11} = 4, \qquad y = \frac{44}{-11} = -4, \qquad \text{and} \qquad z = \frac{55}{-11} = -5.$$

30. Since

$$D = \begin{vmatrix} 2 & -1 & -2 \\ 4 & -1 & 2 \\ 2 & 3 & 8 \end{vmatrix} = 2 \begin{vmatrix} -1 & 2 \\ 3 & 8 \end{vmatrix} - (-1) \begin{vmatrix} 4 & 2 \\ 2 & 8 \end{vmatrix} + (-2) \begin{vmatrix} 4 & -1 \\ 2 & 3 \end{vmatrix}$$

$$= 2(-14) + 1(28) - 2(14) = -28 \neq 0,$$

the system is consistent and has a unique solution. From

$$D_x = \begin{vmatrix} 4 & -1 & -2 \\ -1 & -1 & 2 \\ 3 & 3 & 8 \end{vmatrix} = 4\begin{vmatrix} -1 & 2 \\ 3 & 8 \end{vmatrix} - (-1)\begin{vmatrix} -1 & 2 \\ 3 & 8 \end{vmatrix} + (-2)\begin{vmatrix} -1 & -1 \\ 3 & 3 \end{vmatrix}$$

$$= 4(-14) + 1(-14) - 2(0) = -70,$$

$$D_y = \begin{vmatrix} 2 & 4 & -2 \\ 4 & -1 & 2 \\ 2 & 3 & 8 \end{vmatrix} = 2\begin{vmatrix} -1 & 2 \\ 3 & 8 \end{vmatrix} - 4\begin{vmatrix} 4 & 2 \\ 2 & 8 \end{vmatrix} + (-2)\begin{vmatrix} 4 & -1 \\ 2 & 3 \end{vmatrix}$$

$$= 2(-14) - 4(28) - 2(14) = -168,$$

and

$$D_z = \begin{vmatrix} 2 & -1 & 4 \\ 4 & -1 & -1 \\ 2 & 3 & 3 \end{vmatrix} = 2\begin{vmatrix} -1 & -1 \\ 3 & 3 \end{vmatrix} - (-1)\begin{vmatrix} 4 & -1 \\ 2 & 3 \end{vmatrix} + 4\begin{vmatrix} 4 & -1 \\ 2 & 3 \end{vmatrix}$$

$$= 2(0) + 1(14) + 4(14) = 70,$$

we find

$$x = \frac{-70}{-28} = \frac{5}{2}, \qquad y = \frac{-168}{-28} = 6, \qquad \text{and} \qquad z = \frac{70}{-28} = -\frac{5}{2}.$$

33. $\begin{cases} -3x - 6y + 9z = 2 \\ x - y + 5z = 0 \\ x + 2y - 3z = 1 \end{cases} \Rightarrow D = \begin{vmatrix} -3 & -6 & 9 \\ 1 & -1 & 5 \\ 1 & 2 & -3 \end{vmatrix} = 0 \Rightarrow$ Cramer's rule not applicable

36. Expanding the determinant around the first row we have

$$\begin{vmatrix} x & 1 & -1 \\ 1 & x & 1 \\ -1 & x & 2 \end{vmatrix} = x\begin{vmatrix} x & 1 \\ x & 2 \end{vmatrix} - \begin{bmatrix} 1 & 1 \\ -1 & 2 \end{bmatrix} - \begin{vmatrix} 1 & x \\ -1 & x \end{vmatrix} = x(2x - x) - (2 + 1) - (x + x)$$

$$= x(x) - 3 - 2x = x^2 - 2x - 3 = (x + 1)(x - 3) = 0.$$

Thus, $x = -1$ and $x = 3$.

39. Evaluating the determinant we see that

$$\begin{vmatrix} a & b \\ a & b \end{vmatrix} = ab - ba = 0.$$

42. Evaluating the determinant we see that

$$\begin{vmatrix} x & y & 1 \\ x_1 & y_1 & 1 \\ x_2 & y_2 & 1 \end{vmatrix} = x\begin{vmatrix} y_1 & 1 \\ y_2 & 1 \end{vmatrix} - y\begin{vmatrix} x_1 & 1 \\ x_2 & 1 \end{vmatrix} + 1\begin{vmatrix} x_1 & y_1 \\ x_2 & y_2 \end{vmatrix} = 0$$

is the equation of a straight line. To see that it passes through (x_1, y_1) and (x_2, y_2) we substitute these points into the equation:

$$x_1 \begin{vmatrix} y_1 & 1 \\ y_2 & 1 \end{vmatrix} - y_1 \begin{vmatrix} x_1 & 1 \\ x_2 & 1 \end{vmatrix} + \begin{vmatrix} x_1 & y_1 \\ x_2 & y_2 \end{vmatrix} = x_1(y_1 - y_2) - y_1(x_1 - x_2) + x_1 y_2 - x_2 y_1$$

$$= x_1 y_1 - x_1 y_2 - x_1 y_1 + x_2 y_1 + x_1 y_2 - x_2 y_1 = 0$$

$$x_2 \begin{vmatrix} y_1 & 1 \\ y_2 & 1 \end{vmatrix} - y_2 \begin{vmatrix} x_1 & 1 \\ x_2 & 1 \end{vmatrix} + \begin{vmatrix} x_1 & y_1 \\ x_2 & y_2 \end{vmatrix} = x_2(y_1 - y_2) - y_2(x_1 - x_2) + x_1 y_2 - x_2 y_1$$

$$= x_2 y_1 - x_2 y_2 - x_1 y_2 + x_2 y_2 + x_1 y_2 - x_2 y_1 = 0.$$

This proves the result.

45. Evaluating the determinant and setting equal to 0 we have

$$\begin{vmatrix} -1 - \lambda & 1 & 0 \\ 1 & 2 - \lambda & 1 \\ 0 & 3 & -1 - \lambda \end{vmatrix} = -(\lambda + 1)(\lambda + 2)(\lambda - 3) = 0.$$

Thus, $\lambda = -1$, $\lambda = -2$, and $\lambda = 3$.

Miscellaneous Applications

48. Let x, y, and z be the number of ounces of food groups X, Y, and Z, respectively. The system of equations is

$$9x + 5y + 4z = 100$$
$$3x + 5y \quad\quad = 30$$
$$24x + 10y + 5z = 200.$$

With $A = \begin{bmatrix} 9 & 5 & 4 \\ 3 & 5 & 0 \\ 24 & 10 & 5 \end{bmatrix}$ and $C = \begin{bmatrix} 100 \\ 30 \\ 200 \end{bmatrix}$ we have

$$\det A = \begin{vmatrix} 9 & 5 & 4 \\ 3 & 5 & 0 \\ 24 & 10 & 5 \end{vmatrix} = -210, \qquad \det A_x = \begin{vmatrix} 100 & 5 & 4 \\ 30 & 5 & 0 \\ 200 & 10 & 5 \end{vmatrix} = -1050,$$

$$\det A_y = \begin{vmatrix} 9 & 100 & 4 \\ 3 & 30 & 0 \\ 24 & 200 & 5 \end{vmatrix} = -630, \qquad \det A_z = \begin{vmatrix} 9 & 5 & 100 \\ 3 & 5 & 30 \\ 24 & 10 & 200 \end{vmatrix} = -2100.$$

Thus

$$x = \frac{-1050}{-210} = 5, \qquad y = \frac{-630}{-210} = 3, \qquad \text{and} \qquad z = \frac{-2100}{-210} = 10.$$

9.3 | Systems of Nonlinear Equations

3. From the graph we see that the nonlinear system has two solutions.

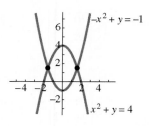

6. From the graph we see that the nonlinear system has two solutions.

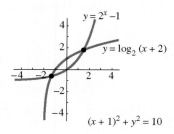

9. We set $2x - 1 = y = x^2$, so that

$$x^2 - 2x + 1 = 0 \qquad \text{or} \qquad (x - 1)^2 = 0.$$

Thus, $x = 1$, so $y = x^2 = 1$, and $(1, 1)$ is the solution of the system.

12. We solve the first equation for y, obtaining $y = x + 3$. Substituting this into the second equation we have

$$x^2 + (x + 3)^2 = 9 \qquad \text{or} \qquad 2x^2 + 6x = 0.$$

Factoring, we have $2x(x + 3) = 0$, so $x = 0$ and $x = -3$. The corresponding y-values are, respectively, $y = 0 + 3 = 3$ and $y = -3 + 3 = 0$. Thus, the solutions of the system are $(0, 3)$ and $(-3, 0)$.

15. We solve the second equation for y, obtaining $y = 1 - x$. Substituting this into the first equation we have

$$x(1 - x) = 1 \qquad \text{or} \qquad x^2 - x + 1 = 0.$$

Using the quadratic formula we see that $x^2 - x + 1 = 0$ has no real solutions, so the system has no solutions.

18. We solve the first equation for y obtaining $y = 1/x$. Substituting this into the second equation we have

$$x^2 = \frac{1}{x^2} + 2, \qquad x^4 = 1 + 2x^2, \qquad \text{or} \qquad x^4 - 2x^2 - 1 = 0.$$

This is a quadratic equation in x^2, so by the quadratic formula we have

$$x^2 = \frac{2 \pm \sqrt{4+4}}{2} = 1 \pm \sqrt{2}.$$

Since $1 - \sqrt{2}$ is negative, we have $x = \pm\sqrt{1 + \sqrt{2}}$. The corresponding y-values are $y = \pm 1/\sqrt{1 + \sqrt{2}}$. Thus, the solutions of system are

$$\left(\sqrt{1 + \sqrt{2}},\ 1/\sqrt{1 + \sqrt{2}}\right) \quad \text{and} \quad \left(-\sqrt{1 + \sqrt{2}},\ -1/\sqrt{1 + \sqrt{2}}\right).$$

21. Adding the two equations we obtain $3x^2 = 5$, so $x^2 = \frac{5}{3}$. Substituting this value into the first equation we get

$$\frac{5}{3} - y^2 = 4 \quad \text{or} \quad y^2 = \frac{5}{3} - 4 = -\frac{7}{3}.$$

Since y^2 cannot be negative, the system has no real solutions.

24. Completing the square, we write the first equation as

$$x^2 + (y^2 - 6y + 9) = -9 + 9 \quad \text{or} \quad x^2 + (y - 3)^2 = 0.$$

The only point that satisfies this equation is $(0, 3)$. Since $x = 0$ and $y = 3$ do not satisfy the second equation, $x^2 + 4x + y^2 = -1$, the system has no real solutions.

27. Taking the square root of both sides of the first equation we obtain $x - y = 2$ and $x - y = -2$. Taking the square root of both sides of the second equation we obtain $x + y = 2\sqrt{3}$ and $x + y = -2\sqrt{3}$. We thus need to solve the four sets of equations

$$
\begin{array}{llll}
x - y = 2 & x - y = -2 & x - y = 2 & x - y = -2 \\
x + y = 2\sqrt{3} & x + y = 2\sqrt{3} & x + y = -2\sqrt{3} & x + y = -2\sqrt{3}.
\end{array}
$$

Adding the first two equations and using $y = x - 2$, we find $2x = 2 + 2\sqrt{3}$ or $x = 1 + \sqrt{3}$ and $y = -1 + \sqrt{3}$. Adding the second pair of equations and using $y = x + 2$, we find $2x = -2 + 2\sqrt{3}$ or $x = -1 + \sqrt{3}$ and $y = 1 + \sqrt{3}$. Adding the third pair of equations and using $y = x - 2$, we find $2x = 2 - 2\sqrt{3}$ or $x = 1 - \sqrt{3}$. Adding the fourth pair of equations and using $y = x + 2$, we find $2x = -2 - 2\sqrt{3}$ or $x = -1 - \sqrt{3}$ and $y = 1 - \sqrt{3}$. The solutions are $\left(1 + \sqrt{3},\ -1 + \sqrt{3}\right)$, $\left(-1 + \sqrt{3},\ 1 + \sqrt{3}\right)$, $\left(1 - \sqrt{3},\ -1 - \sqrt{3}\right)$, and $\left(-1 - \sqrt{3},\ 1 - \sqrt{3}\right)$.

30. Substituting the first equation into the second we get

$$2\cos x \tan x = \sqrt{3} \quad \text{or} \quad \sin x = \frac{\sqrt{3}}{2}.$$

This implies that $x = \pi/3 + 2n\pi$ and $x = 2\pi/3 + 2n\pi$, where n is an integer. Using the first equation, $y = \cos(\pi/3 + 2n\pi) = \frac{1}{2}$ and $y = (\cos 2\pi/3 + 2n\pi) = -\frac{1}{2}$. Thus, the solutions are $\left(\pi/3 + 2n\pi,\ \frac{1}{2}\right)$ and $\left(2\pi/3 + 2n\pi,\ -\frac{1}{2}\right)$, for n an integer.

33. Substituting $\log_{10} x = y$ into the second equation we obtain

$$y^2 = 5 + 4y \qquad \text{so} \qquad y^2 - 4y - 5 = (y+1)(y-5) = 0.$$

When $y = -1$, $\log_{10} x = -1$ and $x = 10^{-1} = 0.1$. When $y = 5$, $\log_{10} x = 5$ and $x = 10^5 = 100,000$. Thus, the solutions are $(0.1, -1)$ and $(100,000, 5)$.

36. Solving the first equation for y and substituting into the second equation we have

$$7 = \log_{10} x + 5 - \log_{10}(x + 6) \qquad \text{or} \qquad 2 = \log_{10} x - \log 10(x + 6) = \log_{10} \frac{x}{x + 6}.$$

Writing this logarithmic equation as an exponential equation we get

$$\frac{x}{x + 6} = 10^2 = 100, \qquad x = 100x + 600, \qquad 99x = -600, \qquad \text{and} \qquad x = -\frac{600}{99}.$$

Since the domain of $\log_{10} x$ is all positive real numbers, the system has no real solutions.

39. Subtracting the first two equations, we see that $2x - 2y = 0$ or $x = y$. Substituting $x = y$ in the third equation, we obtain $y^2 - 3 = 0$ or $y = \pm\sqrt{3}$. Also, from either the first or second equation, we have $\lambda = 2\sqrt{3}$. Thus, the solutions are $(\sqrt{3}, \sqrt{3}, -2\sqrt{3})$ and $(-\sqrt{3}, -\sqrt{3}, 2\sqrt{3})$.

42. Writing the second equation as $x^2\lambda - 5x = x(x\lambda - 5) = 0$ we see that either $x = 0$ or $x\lambda - 5 = 0$. Since $x = 0$ implies that $-1000 = 0$, we see that x cannot be 0. Thus, there will only be a solution if $x\lambda = 5$ or $\lambda = 5/x$. Substituting this value for λ into the first equation we obtain $8x + 5y = 2xy(5/x) = 10y$. This implies that $8x = 5y$ or $y = 8x/5$. Substituting into the third equation we have

$$x^2(8x/5) - 1000 = 0, \qquad \text{so} \qquad 8x^3 = 5000 \qquad \text{and} \qquad x = 5^{4/3}.$$

Thus, $y = 8(5^{4/3})/5 = 8(5^{1/3})$ and $\lambda = 5/5^{4/3} = 5^{-1/3}$. The only solution is $(5^{4/3}, 8(5^{1/3}), 5^{-1/3})$.

Miscellaneous Applications

45. Letting the radii of the circles be r and R, we have $r + R = 8$ and $\pi r^2 + \pi R^2 = 32\pi$. The second equation is equivalent to $r^2 + R^2 = 32$. Rewriting the first equation as $R = 8 - r$ and substituting into the second equation we have

$$r^2 + (8 - r)^2 = 32 \quad \text{or} \quad r^2 + 64 - 16r + r^2 = 32$$
$$\text{or} \quad 2r^2 - 16r + 32 = 2(r^2 - 8r + 16) = 2(r - 4)^2 = 0.$$

Thus, the radii are $r = R = 4$ cm.

48. If we let the lengths of the two sides be x and y, then $y = 2x$ and $x^2 + y^2 = 20^2 = 400$. Substituting the first equation into the second we obtain

$$x^2 + (2x)^2 = 400 \qquad \text{or} \qquad 5x^2 = 400 \qquad \text{or} \qquad x^2 = 80.$$

Then $x = 4\sqrt{5}$ and $y = 8\sqrt{5}$.

9.4 | Systems of Inequalities

3. We write the inequality as $2x - y < 0$ and graph $2x - y = 0$ as a dashed line since the inequality is strict. We test the point $(1, 1)$ and see that $2(1) - 1 = 1 > 0$, so the region does not contain the point.

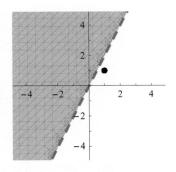

6. We write the inequality as $2x + y \le -3$ and graph $2x + y = -3$ as a solid line since the inequality is nonstrict. We test the point $(0, 0)$ and see that $2(0) + 0 = 0 > -3 > 0$, so the region satisfying the inequality is on the other side of the line.

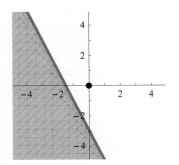

9. We first note that x is always nonnegative (so the region must be to the right of the y-axis), and then write the inequality as $y \le \sqrt{x} + 1$. Graphing $y = \sqrt{x} + 1$ as a solid curve, since the inequality is nonstrict, and testing the point $(1, 0)$, we see that $0 = \sqrt{1} + 1 = 2$, is true, so the region satisfying the inequality is below the graph of the function. As already noted, it must also be to the right of the y-axis.

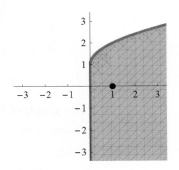

12. We write $y = 3/x$ and graph this function as a solid curve because the inequality is nonstrict. When $x < 0$, $y \le 3/x$, and when $x > 0$, $y \ge 3/x$. Since $y = 3/x$ is discontinuous at $x = 0$ we test the points $(-3, -3)$ and $(3, 3)$ separately. Substituting $x = -3$ and $y = -3$ into $y \le 3/x$ we get $-3 \le 3/(-3)$ or $-3 \le -1$, which is true, so for $x < 0$ the region satisfying the inequality is below the graph. Substituting $x = 3$ and $y = 3$ into $y \ge 3/x$ we get $3 \ge 3/3 = 1$, which is true, so for $x > 0$ the region satisfying the inequality is above the graph.

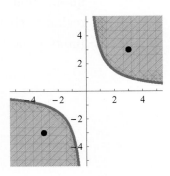

15. The graph of the first inequality is shown below in (a), where we test $(1, 0)$ obtaining $1 - 0 > 0$, which is true, so the region is below the line. We use a dashed line since the inequality is strict. The graph of the second inequality is shown in (b) where we test $(0, 0)$, obtaining

$0 + 0 > 1$, which is false, so the region is above the line. We use a dashed line since the inequality is strict. We show the intersection of the two regions in (c).

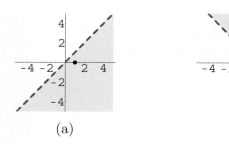

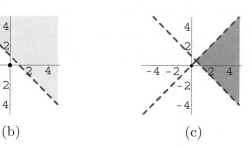

           (a)   (b)   (c)

18. The graph of the first inequality is shown below in (a), where we test $(0, 0)$ obtaining $0+0 \geq 12$, which is false, so the region is to the right of the line. The graph of the second inequality is shown in (b) where we test $(5, 0)$, obtaining $-10+0 \leq 0$, which is true, so the region is to the right of the line. Finally, the inequality $y \geq 0$ says that the region lies on or above the x-axis. This region is shown in (c). In all three cases we use a solid line because the inequalities are nonstrict. Intersecting the three regions, we obtain the graph of the system shown below in (d).

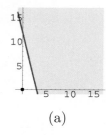
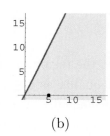
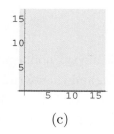
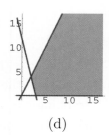

        (a)   (b)   (c)   (d)

21. The graph of the first inequality is shown below in (a), where we test $(0, 0)$ obtaining $0 < 0+2$, which is true, so the region is below the line. We use a dashed line since the inequality is strict. The graph of the second inequality is shown in (b) where the region is between the lines. Finally, the inequality $y \geq 1$ says that the region lies on or above the line $y = 1$. This region is shown in (c). In (b) and (c) the inequalities are nonstrict so the lines are solid. Intersecting the three regions, we obtain the graph of the system shown below in (d).

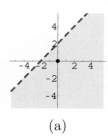
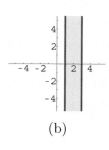
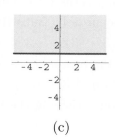
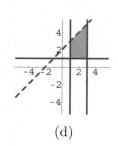

        (a)   (b)   (c)   (d)

24. The graph of the first inequality is shown below in (a), where we test $(0, 0)$ obtaining $0+0 \geq 6$, which is false, so the region is above the line. The graph of the second inequality is shown in (b) where we test $(0, 0)$, obtaining $0 - 0 \geq -6$, which is true, so the region is below the line. The graph of the third inequality is shown below in (c) where we test $(0, 0)$, obtaining $0 + 0 \le 6$, which is true, so the region is to the left of the line. In all three cases we use a solid line because the inequalities are nonstrict. Intersecting the three regions, we obtain the graph of the system shown below in (d).

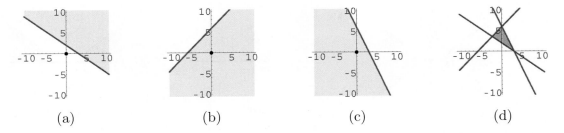

(a)   (b)   (c)   (d)

27. The graph of the first inequality is shown below in (a), where we test $(0, 0)$ obtaining $0+0 \ge 1$, which is false, so the region is outside the circle. The graph of the second inequality is shown in (b) where we test $(0, 0)$, obtaining $0 + 0 \le 1$, which is true, so the region is inside the ellipse. In both cases we use solid curves because the inequalities are nonstrict. We intersect the two regions in (c) to graph the system of inequalities.

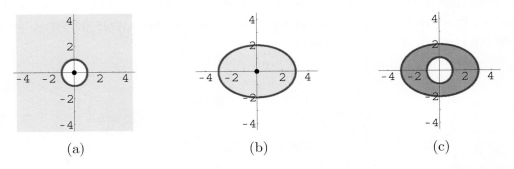

(a)   (b)   (c)

30. The graph of the first inequality is shown below in (a), where we test $(0, 0)$ obtaining $0+0 \le 4$, which is true, so the region is inside the circle. The graph of the second inequality is shown in (b) where we test $(0, 0)$, obtaining $0 \le 0 - 1$, which is false, so the region is below the parabola. In both cases we use solid curves because the inequalities are nonstrict. In (c) we show the intersection of the two regions.

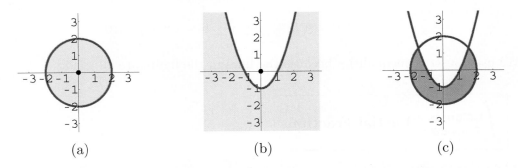

(a)   (b)   (c)

33. The graph of the first inequality is shown below in (a), where we test the points $(-7, 0)$, $(0, 0)$, and $(7, 0)$, obtaining $\frac{49}{9} - 0 \ge 1$, which is true, $0 - 0 \ge 1$, which is false, and $\frac{49}{9} - 0 \ge 1$, which is true, respectively. Thus, the region is to the left of the left-hand branch of the hyperbola, and to the right of the right-hand branch of the hyperbola. The inequality $y \geq 0$ implies that the region is above the x-axis. In both cases we use solid curves because the inequalities are nonstrict. In (c) we show the intersection of the two regions.

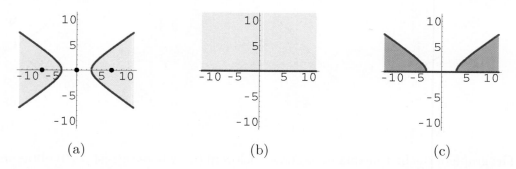

(a)   (b)   (c)

36. The graph of the first inequality is shown below in (a), where we test the point $(0,\ 1)$, obtaining $1 \geq 0$, which is true, so the region is inside of the graph of $y = x^4$. The inequality $y \leq 2$ implies that the region is below the line $y = 2$. This region is shown in (b). The inequality $x \geq -1$ implies that the region is to the right of the line $x = -1$, and the inequality $x \leq 1$ implies that the region is to the left of the line $x = 1$. These regions are shown in (c) and (d) below. In all cases we use solid curves because the inequalities are nonstrict. In (e) we show the intersection of the four regions.

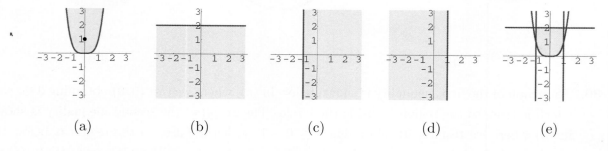

(a)   (b)   (c)   (d)   (e)

39. The region in the graph is bounded on the left by $x = 1$, on the right by $x = 5$, above by $y = 6$, and and $x + y = 10$, and below by $y = 2$. Thus, the system of inequalities is

$$1 \leq x \leq 5 \qquad 2 \leq y \leq 6 \qquad x + y \leq 10.$$

We use nonstrict inequalities because the lines in the figure are all solid.

9.5 | $\int$ **Calculus** PREVIEW   **Partial Fractions**

3. The partial fraction decomposition looks like

$$\frac{-9x + 27}{x^2 - 4x - 5} = \frac{-9x + 27}{(x - 5)(x + 1)} = \frac{A}{x - 5} + \frac{B}{x + 1}.$$

Multiplying both sides of this equality by the denominator $(x-5)(x+1)$, we obtain

(9.1) $$-9x + 27 = A(x+1) + B(x-5).$$

Since the denominator consists solely of linear factors, none of which is repeated, we can use the shortcut described on page 481 in the text. Letting $x = -1$ in (9.1) we have

$$-9(-1) + 27 = 36 = B(-1-5) = -6B,$$

so $B = -6$. Letting $x = 5$ in (9.1) we have

$$-9(5) + 27 = -18 = A(5+1) = 6A,$$

so $A = -3$. The partial fraction decomposition is

$$\frac{-9x+27}{x^2-4x-5} = -\frac{3}{x-5} - \frac{6}{x+1}.$$

6. The partial fraction decomposition looks like

$$\frac{1}{x(x-2)(2x-1)} = \frac{A}{x} + \frac{B}{x-2} + \frac{C}{2x-1}.$$

Multiplying both sides of this equality by the denominator $x(x-2)(2x-1)$, we obtain

(9.2) $$1 = A(x-2)(2x-1) + Bx(2x-1) + Cx(x-2).$$

Since the denominator consists solely of linear factors, none of which is repeated, we can use the *Shortcut Worth Knowing* described on page 481 in the text. Letting $x = 0$ in (9.2) we have

$$1 = A(-2)(-1) = 2A,$$

so $A = \frac{1}{2}$. Letting $x = 2$ in (9.2) we have

$$1 = B(2)(4-1) = 6B,$$

so $B = \frac{1}{6}$. Letting $x = \frac{1}{2}$ in (9.2) we have

$$1 = C\left(\frac{1}{2}\right)\left(\frac{1}{2}-2\right) = -\frac{3}{4}C,$$

so $C = -\frac{4}{3}$. The partial fraction decomposition is

$$\frac{1}{x(x-2)(2x-1)} = \frac{\frac{1}{2}}{x} + \frac{\frac{1}{6}}{x-2} - \frac{\frac{4}{3}}{2x-1}.$$

9. The partial fraction decomposition looks like

$$\frac{5x - 6}{(x - 3)^2} = \frac{A}{x - 3} + \frac{B}{(x - 3)^2}.$$

Multiplying both sides of this equality by the denominator $(x - 3)^2$ we obtain

(9.3) $5x - 6 = A(x - 3) + B$

(9.4)   or $5x - 6 = Ax + (-3A + B).$

Letting $x = 3$ in (9.3) we have

$$5(3) - 6 = B,$$

so $B = 9$. Then (9.4) becomes

$$5x - 6 = Ax + (9 - 3A).$$

Equating coefficients of like terms, we see that $A = 5$. The partial fraction decomposition is

$$\frac{5x - 6}{(x - 3)^2} = \frac{5}{x - 3} + \frac{9}{(x - 3)^2}.$$

12. The partial fraction decomposition looks like

$$\frac{-4x + 6}{(x - 2)^2(x - 1)^2} = \frac{A}{x - 2} + \frac{B}{(x - 2)^2} + \frac{C}{x - 1} + \frac{D}{(x - 1)^2}.$$

Multiplying both sides of this equality by the denominator $(x - 2)^2(x - 1)^2$, we obtain

(9.5) $-4x + 6 = A(x - 2)(x - 1)^2 + B(x - 1)^2 + C(x - 2)^2(x - 1) + D(x - 2)^2$

or

(9.6)
$-4x + 6 = (A + C)x^3 + (-4A + B - 5C + D)x^2 + (5A - 2B + 8C - 4D)x + (-2A + B - 4C + 4D).$

Letting $x = 1$ in (9.5) we have

$$-4(1) + 6 = 2 = A(0) + B(0) + C(0) + D(-1)^2 = D,$$

so $D = 2$. Letting $x = 2$ in (9.5) we have

$$-4(2) + 6 = -2 = A(0) + B(1)^2 + C(0) + D(0) = B,$$

so $B = -2$, Substituting the values for B and D into (9.6) we have

$$-4x + 6 = (A + C)x^3 + (-4A - 5C)x^2 + (5A + 8C - 4)x + (-2A - 4C + 6).$$

Equating like coefficients, we get

$$\begin{array}{ccc} 5A + 8C - 4 = -4 & & 5A + 8C = 0 \\ & \text{or} & \\ -2A - 4C + 6 = 6 & & A + 2C = 0. \end{array}$$

Solving this system, we find $A = 0$ and $C = 0$. The partial fraction decomposition is

$$\frac{-4x + 6}{(x - 2)^2(x - 1)^2} = \frac{-2}{(x - 2)^2} + \frac{2}{(x - 1)^2}.$$

15. The partial fraction decomposition looks like

$$\frac{6x^2 - 7x + 11}{(x-1)(x^2+9)} = \frac{A}{x-1} + \frac{Bx+C}{x^2+9}.$$

Multiplying both sides of this equality by the denominator $(x-1)(x^2+9)$ we obtain

(9.7) $$6x^2 - 7x + 11 = A(x^2+9) + (Bx+C)(x-1)$$

(9.8) or $$6x^2 - 7x + 11 = (A+B)x^2 + (-B+C)x + (9A-C).$$

Letting $x = 1$ in (9.7) we have

$$6(1)^2 - 7(1) + 11 = 10 = A(1+9) + [B(1)+C](0) = 10A,$$

so $A = 1$. Substituting $A = 1$ into (9.8) we have

$$6x^2 - 7x + 11 = (B+1)x^2 + (-B+C)x + (9-C).$$

Equating like coefficients, we get

$$B + 1 = 6, \qquad \text{so} \qquad B = 5$$

and

$$9 - C = 11, \qquad \text{so} \qquad C = -2.$$

The partial fraction decomposition is

$$\frac{6x^2 - 7x + 11}{(x-1)(x^2+9)} = \frac{1}{x-1} + \frac{5x-2}{x^2+9}.$$

18. The partial fraction decomposition looks like

$$\frac{2x^2 - x + 7}{(x-6)(x^2+x+5)} = \frac{A}{x-6} + \frac{Bx+C}{x^2+x-5}.$$

Multiplying both sides of this equality by the denominator $(x-6)(x^2+x+5)$, we obtain

(9.9) $$2x^2 - x + 7 = A(x^2+x+5) + (Bx+C)(x-6)$$

(9.10) or $$2x^2 - x + 7 = (A+B)x^2 + (A-6B+C)x + (5A-6C).$$

Letting $x = 6$ in (9.9) we have

$$2(6)^2 - 6 + 7 = 73 = A(6^2+6+5) + (6B+C)(0) = 47A,$$

so $A = \frac{73}{47}$. Substituting $A = \frac{73}{47}$ into (9.10) we have

$$2x^2 - x + 7 = \left(\frac{73}{47} + B\right)x^2 + \left(\frac{73}{47} - 6B + C\right)x + \left(5\cdot\frac{73}{47} - 6C\right).$$

Equating like coefficients, we get

$$\frac{73}{47} + B = 2 \quad \text{so} \quad B = 2 - \frac{73}{47} = \frac{21}{47}$$

and

$$5 \cdot \frac{73}{47} - 6C = 7 \quad \text{so} \quad C = \frac{1}{6}\left(\frac{365}{47} - 7\right) = \frac{1}{6}\left(\frac{36}{47}\right) = \frac{6}{47}.$$

The partial fraction decomposition is

$$\frac{2x^2 - x + 7}{(x-6)(x^2 + x + 5)} = \frac{\frac{73}{47}}{x-6} + \frac{\frac{21}{47}x + \frac{6}{47}}{x^2 + x + 5}.$$

21. The partial fraction decomposition looks like

$$\frac{x^3}{(x^2 + 2)(x^2 + 1)} = \frac{Ax + B}{x^2 + 2} + \frac{Cx + D}{x^2 + 1}.$$

Multiplying both sides of this equality by the denominator $(x^2 + 2)(x^2 + 1)$, we obtain

$$x^3 = (Ax + B)(x^2 + 1) + (Cx + D)(x^2 + 2)$$

(9.11)   or $x^3 = (A + C)x^3 + (B + D)x^2 + (A + 2C)x + (B + 2D).$

Equating like coefficients in (9.11) we get

(9.12) $$A + C = 1$$

(9.13) $$B + D = 0$$

(9.14) $$A + 2C = 0$$

(9.15) $$B + 2D = 0.$$

Subtracting the equation in (9.12) from the equation in (9.14) we get

$$(A + 2C) - (A + C) = C = 0 - 1 = -1,$$

so $C = -1$. Substituting this value into (9.12) we find $A = 2$. Next, subtracting the equation in (9.13) from the equation in (9.15) we get $D = 0$. Substituting this value into (9.13) we find $B = 0$. Thus, the partial fraction decomposition is

$$\frac{x^3}{(x^2 + 2)(x^2 + 1)} = \frac{2x}{x^2 + 2} - \frac{x}{x^2 + 1}.$$

24. The partial fraction decomposition looks like

$$\frac{2x^2}{(x-2)(x^2 + 4)^2} = \frac{A}{x-2} + \frac{Bx + C}{x^2 + 4} + \frac{Dx + E}{(x^2 + 4)^2}.$$

Multiplying both sides of this equality by the denominator $(x - 2)(x^2 + 4)^2$ we obtain

(9.16) $$2x^2 = A(x^2 + 4)^2 + (Bx + C)(x - 2)(x^2 + 4) + (Dx + E)(x - 2)$$

or

$$2x^2 = (A + B)x^4 + (-2B+C)x^3 + (8A + 4B - 2C + D)x^2$$

(9.17)
$$+ (-8B + 4C - 2D + E)x + (16A - 8C - 2E).$$

Letting $x = 2$ in (9.16) we have

$$2(2)^2 = 8 = 64A + (2B + C)(0)(8) + (2D + E)(0) = 64A,$$

so $A = \frac{1}{8}$. Substituting $A = \frac{1}{8}$ into (9.17), we get

$$2x^2 = \left(B + \frac{1}{8} \right) x^4 + (-2B + C)x^3 + (4B - 2C + D + 1)x^2$$

$$+ (-8B + 4C - 2D + E)x + (-8C - 2E + 2).$$

Equating like coefficients, we get

$$B + \frac{1}{8} = 0 \qquad \text{so} \qquad B = -\frac{1}{8}$$

$$-2B + C = 0 \qquad \text{so} \qquad C = 2B = 2\left(-\frac{1}{8} \right) = -\frac{1}{4}$$

$$4B - 8C + D + 1 = 2 \qquad \text{so} \qquad D = 1 - 4B + 2C = 1 + \frac{1}{2} - \frac{1}{2} = 1$$

$$-8B + 4C - 2D + E = 0 \qquad \text{so} \qquad E = 8B - 4C + 2D = -1 + 1 + 2 = 2.$$

Thus, the partial fraction decomposition is

$$\frac{2x^2}{(x - 2)(x^2 + 4)^2} = \frac{\frac{1}{8}}{x - 2} - \frac{\frac{1}{8}x + \frac{1}{4}}{x^2 + 4} + \frac{x + 2}{(x^2 + 4)^2}.$$

27. Using polynomial long division, we have

$$
\begin{array}{r}
\frac{1}{2} \\
2x^2 + 5x + 2 \overline{\smash{\big)}\ x^2 - 4x + 1} \\
\underline{x^2 + \tfrac{5}{2}x + 1} \\
-\tfrac{13}{2}x + 0
\end{array}
$$

so

$$\frac{x^2 - 4x + 1}{2x^2 + 5x + 2} = \frac{1}{2} - \frac{\frac{13}{2}x}{2x^2 + 5x + 2}.$$

The partial fraction decomposition of the proper fraction looks like

$$-\frac{\frac{13}{2}x}{2x^2 + 5x + 2} = -\frac{\frac{13}{2}x}{(x + 2)(2x + 1)} = \frac{A}{x + 2} + \frac{B}{2x + 1}.$$

Multiplying both sides of this equality by the denominator $(x + 2)(2x + 1)$ we obtain

(9.18)
$$-\frac{13}{2}x = A(2x + 1) + B(x + 2).$$

Since the denominator consists solely of linear factors, none of which is repeated, we can use the *Shortcut Worth Knowing* described on page 481 in the text. Letting $x = -\frac{1}{2}$ in (9.18), we have

$$-\frac{13}{2}\left(-\frac{1}{2}\right) = \frac{13}{4} = A(0) + B\left(-\frac{1}{2} + 2\right) = \frac{3}{2}B,$$

so $B = \frac{13}{6}$. Letting $x = -2$ in (9.18), we have

$$-\frac{13}{2}(-2) = 13 = A[2(-2) + 1] + B(0) = -3A,$$

so $A = -\frac{13}{3}$. The partial fraction decomposition is

$$\frac{x^2 - 4x + 1}{2x^2 + 5x + 2} = \frac{1}{2} - \frac{\frac{13}{3}}{x + 2} + \frac{\frac{13}{6}}{2x + 1}.$$

30. Using polynomial long division, we have

$$
\begin{array}{r}
1 \\
x^3 + 3x^2 + 3x + 1 \overline{\smash{)}\, x^3 + x^2 - x + 1} \\
\underline{x^3 + 3x^2 + 3x + 1} \\
-2x^2 - 4x + 0
\end{array}
$$

so

$$\frac{x^3 + x^2 - x + 1}{x^3 + 3x^2 + 3x + 1} = 1 + \frac{-2x^2 - 4x}{(x + 1)^3}.$$

The partial fraction decomposition of the proper fraction looks like

$$\frac{-2x^2 - 4x}{(x + 1)^3} = \frac{A}{x + 1} + \frac{B}{(x + 1)^2} + \frac{C}{(x + 1)^3}.$$

Multiplying both sides of this equality by the denominator $(x + 1)^3$ we obtain

(9.19) $$-2x^2 - 4x = A(x + 1)^2 + B(x + 1) + C$$

(9.20) or $$-2x^2 - 4x = Ax^2 + (2A + B)x + (A + B + C).$$

Letting $x = -1$ in (9.19), we find

$$-2(-1)^2 - 4(-1) = 2 = A(0) + B(0) + C,$$

so $C = 2$. Equation (9.20) then becomes

$$-2x^2 - 4x = Ax^2 + (2A + B)x + (A + B + 2).$$

Equating coefficients of like terms, we have $A = -2$, $A + B + 2 = 0$, and $-2 + B + 2 = B = 0$. Thus, the partial fraction decomposition of the improper fraction is

$$\frac{x^3 + x^2 - x + 1}{x^3 + 3x^2 + 3x + 1} = 1 + \frac{-2}{x + 1} + \frac{2}{(x + 1)^3}.$$

Chapter 9 Review Exercises

A. Fill in the Blanks

3. Since

$$\begin{vmatrix} x & 1 & 1 \\ 1 & 1 & x \\ 1 & x & 1 \end{vmatrix} = x \begin{vmatrix} 1 & x \\ x & 1 \end{vmatrix} + (-1) \begin{vmatrix} 1 & x \\ 1 & 1 \end{vmatrix} + 1 \begin{vmatrix} 1 & 1 \\ 0 & 1 \end{vmatrix} = x(1 - x^2) - (1 - x) + 1(x - 1)$$

$$= x(1 + x)(1 - x) - 2(1 - x) = (1 - x)(x^2 + x - 2)$$

$$= (1 - x)(x - 1)(x + 2) = 0,$$

we see that $x = 1$ or $x = -2$.

6. From the graphs of the two curves there is a single point of intersection, probably at $(1, 0)$ Substituting $x = 1$ and $y = 0$ into the two equations, we see that this is indeed the solution of the system.

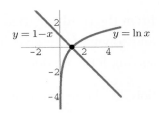

9. Substituting $(1, 1)$ and $(2, 1)$ into $y = ax^2 + bx$ we have

$$\begin{array}{cc} a(1^2) + b(1) = 1 & \quad a + b = 1 \\ a(2^2) + b(2) = 1 & \text{or} \quad 4a + 2b = 1. \end{array}$$

To solve this system we multiply the first equation by 2 and subtract the second equation from the first.

$$\begin{array}{c} 2a + 2b = 2 \\ 4a + 2b = 1 \end{array} \implies -2a = 1 \implies a = -\frac{1}{2}.$$

Since $a + b = 1$, we have $b = \frac{3}{2}$. Thus, $y = -\frac{1}{2}x + \frac{3}{2}$.

B. True/False

3. This is true. When $m \neq 0$, $y = mx$ is a straight line through the origin, and when $k > 0$, $x^2 + y^2 = k$ is a circle centered at the origin. The line must intersect the cicle in exactly two points. Instead of this geometric argument, an algebraic one can also be given.

6. This is false, since $4(1) - 3(-2) + 5 = 15$, which is not less than or equal to 0.

9. This is true. Solving the second equation for y gives $y = x^2 - 5$. Substituting this into the first equation and solving for x we have

$$x^2 + (x^2 - 5)^2 = 25$$

$$x^2 + x^4 - 10x^2 + 25 = 25$$

$$x^4 - 9x^2 = 0$$

$$x^2(x^2 - 9) = 0.$$

Thus, $x = 0$, $x = -3$, and $x = 3$. The corresponding y-values are -5, 4, and 4. The solutions of the system are $(0, -5)$, $(-3, 4)$, and $(3, 4)$.

C. Review Exercises

3. We interchange the first two equations and then use elimination.

$$\left.\begin{array}{rcl} x + y + z &=& -2 \\ 2x + y - z &=& 7 \\ 4x + 2y + 2z &=& -6 \end{array}\right\} \xrightarrow[\substack{-4E_1+E_3}]{-2E_1+E_2} \left\{\begin{array}{rcl} x + y + z &=& -2 \\ -y - 3z &=& 11 \\ -2y - 2z &=& 2 \end{array}\right.$$

$$\xrightarrow{-2E_2+E_3} \left\{\begin{array}{rcl} x + y + z &=& -2 \\ -y - 3z &=& 11 \\ 4z &=& -20 \end{array}\right.$$

From the third equation we see that $z = -5$. Substituting this value into the second equation and solving for y gives $y = 4$. Finally, substituting these values into the first equation and solving for x we obtain $x = -1$. The solution of the system is $(-1, 4, -5)$.

6. Solving the second equation for y and substituting into the first equation we get

$$101 \cdot 10^x = 10^x + 10^{-x}$$

$$100 \cdot 10^x = 10^{-x}$$

$$10^{2x} = \frac{1}{100} = 10^{-2}$$

$$2x = = -2$$

$$x = -1.$$

Thus $x = -1$ and $y = 10^{-1} = \frac{1}{10}$. The solution of the system is $\left(-1, \frac{1}{10}\right)$.

9. We multiply the first equation by -4 and add the equations. This gives $y^2 - 4y + 4 = 0$ or $(y - 2)^2 = 0$. Thus, $y = 2$ and $\log_{10} x = 2$ so $x = 10^2 = 100$. The solution of the system is $(100, 2)$.

12. We first note that because of the second equation, neither x nor y can be 0. Writing the second equation as $y = 1/x$ and squaring both sides we have $y^2 = 1/x^2$. Then substituting this into the first equation we find $x^2 + 1/x^2 = 4$. Multiplying both sides by x^2 we obtain $x^4 + 1 = 4x^2$ or $x^4 - 4x^2 + 1 = 0$. Since this does not factor we apply the quadratic formula to find

$$x^2 = \frac{4 \pm \sqrt{16 - 4}}{2} = 2 \pm \sqrt{3}.$$

Thus, $x = \sqrt{2 \pm \sqrt{3}}$. From $y^2 = 4 - x^2$ we have $y^2 = 4 - (2 \pm \sqrt{3}) = 2 \mp \sqrt{3}$, so $y = \sqrt{2 \mp \sqrt{3}}$. The four solutions of the system are $\left(\sqrt{2 \pm \sqrt{3}}, \sqrt{2 \mp \sqrt{3}}\right)$.

15. If x is the unit's digit and y the ten's digit of the old number, then $x = 1 + 3y$, and the old number is

$$x + 10y = (1 + 3y) + 10y = 1 + 13y.$$

When the digits are reversed, the new number is

$$y + 10x = y + 10(1 + 3y) = 10 + 31y.$$

Since the new number is 45 more than the old number,

$$10 + 31y = 45 + (1 + 13y) \qquad \text{or} \qquad 18y = 36.$$

Thus, $y = 2$ so $x = 1 + 3(2) = 7$, and the old number is 27.

18. We assume that the y-intercept of the line is $(0, 4)$. Then the slope of the line is $m = (6 - 4)/(5 - 0) = \frac{2}{5}$, so the equation of the line is $y = \frac{2}{5}x + 4$. The equation of the parabola is $y = x^2 - 4x + 1$, so we can find the points of intersection by solving

$$x^2 - 4x + 1 = \frac{2}{5}x + 4 \qquad \text{or} \qquad 5x^2 - 20x + 5 = 2x + 20.$$

This leads to

$$5x^2 - 22x - 15 = (5x + 3)(x - 5) = 0.$$

Thus, $x = -\frac{3}{5}$ and $y = \frac{2}{5}\left(-\frac{3}{5}\right) + 4 = -\frac{94}{25}$. When $x = 5$, $y = 6$, and $(5, 6)$ is a point of intersection of the line and the parabola, as we already knew. Thus, P is $\left(-\frac{3}{5}, \frac{94}{25}\right)$.

21. The partial fraction decomposition looks like

$$\frac{x^2}{(x^2 + 4)^2} = \frac{Ax + B}{x^2 + 4} + \frac{Cx + D}{(x^2 + 4)^2}.$$

Multiplying both sides of this equality by $(x^2 + 4)^2$, we obtain

$$x^2 = (Ax + B)(x^2 + 4) + Cx + D \quad \text{or} \quad x^2 = Ax^3 + Bx^2 + (4A + C)x + 4B + D.$$

Equating coefficients of like powers of x we get

$$A = 0$$
$$B = 1$$
$$4A + C = 0$$
$$4B + D = 0.$$

From this we easily see that $A = 0$, $B = 1$, $C = 0$ and $D = -4$. Thus

$$\frac{x^2}{(x^2 + 4)^2} = \frac{1}{x^2 + 4} - \frac{4}{(x^2 + 4)^2}.$$

24. The graph of the first inequality is shown below in (a), where we test $(0, 0)$ obtaining $0+0 \le 4$, which is true, so the region is below the line. The graph of the second inequality is shown in (b) where we test $(0, 0)$, obtaining $0 - 0 \ge -6$, which is true, so the region is below the line. The graph of the third inequality is shown below in (c) where we test $(0, 0)$, obtaining $0 - 0 \le 12$, which is true, so the region is to the above the line. In all three cases we use a solid line because the inequalities are nonstrict. Intersecting the three regions, we obtain the graph of the system shown below in (d).

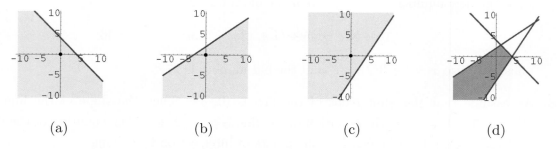

(a)   (b)   (c)   (d)

27. The graph of the first inequality is shown below in (a), where we test $(0, 0)$ obtaining $0+0 \le 4$, which is true, so the region is on and inside the circle. We complete the square in the second inequality and write it as $x^2 + (y - 2)^2 \le 4$. We test $(0, 2)$, obtaining $0 + 0 \le 4$, which is true, so the region is on and inside this circle as shown in (b) below. In both cases we use solid curves because the inequalities are nonstrict. In (c) we show the intersection of the two regions.

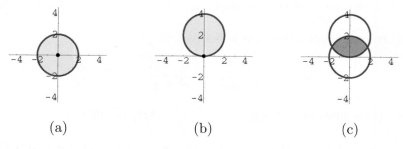

(a)   (b)   (c)

30. (a) The graph of $\begin{cases} y \ge 9 - x^2 \\ y \le 4 - x^2 \end{cases}$ shows that the system has no solution.

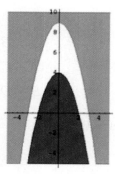

(b) The point $(1, \ 9)$ is above the point $(1, \ 8)$ on the graph of $y = 9 - x^2$ and hence in the

solution of $\begin{cases} y \geq 9 - x^2 \\ y \geq 4 - x^2 \end{cases}$.

See the figure.

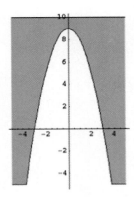

33. The shaded region shown in the graph is below both the line $x + y = 2$ and the parabola $y = x^2$. Thus, the system of inequalities is

$$x + y \leq 2$$
$$y \leq x^2$$

We use nonstrict inequalities because the lines in the figure are all solid.

36. is (c)

39. is (b)

42. The system $\begin{cases} y = x^4 \\ y = x \end{cases}$ has four solutions. From

$x^4 - x = 0 \Rightarrow x(x^3 - 1) = 0$ or $x(x - 1)(x^2 + x + 1) = 0$

we get $x = 0, x = 1, x = \frac{1}{2}(-1 + \sqrt{3}i), \ x = \frac{1}{2}(-1 - \sqrt{3}i)$,

So four solutions are $(0, \ 0)$, $(1, \ 1)$, $\left(\frac{1}{2}(-1 + \sqrt{3}i), \frac{1}{2}(-1 + \sqrt{3}i)\right)$,

and $\left(\frac{1}{2}(-1 - \sqrt{3}i), \frac{1}{2}(-1 - \sqrt{3}i)\right)$. The two real solutions $(0, \ 0)$ and $(1, \ 1)$ correspond to the points of intersection of the graphs.

Chapter 10

Sequences and Series

10.1 | Sequences

In Problems 1–10 we let n take on the values 1, 2, 3, 4, and 5.

3. $\frac{1}{2}(1)(1+1)$, $\frac{1}{2}(2)(2+1)$, $\frac{1}{2}(3)(3+1)$, $\frac{1}{2}(4)(4+1)$, $\frac{1}{2}(5)(5+1)$ or   1, 3, 6, 10, 15

6. $\frac{1+1}{1+2}$, $\frac{2+1}{2+2}$, $\frac{3+1}{3+2}$, $\frac{4+1}{4+2}$, $\frac{5+1}{5+2}$,   or $\frac{2}{3}$, $\frac{3}{4}$, $\frac{4}{5}$, $\frac{5}{6}$, $\frac{6}{7}$

9. $\frac{1+(-1)^1}{1+4(1)}$, $\frac{2+(-1)^2}{1+4(2)}$, $\frac{3+(-1)^3}{1+4(3)}$, $\frac{4+(-1)^4}{1+4(4)}$, $\frac{5+(-1)^5}{1+4(5)}$,   or   0, $\frac{1}{3}$, $\frac{2}{13}$, $\frac{5}{17}$, $\frac{4}{21}$

12. $\sqrt{1}$, $\frac{1}{2}$, $\sqrt{3}$, $\frac{1}{4}$, $\sqrt{5}$, $\frac{1}{6}$ or   1, $\frac{1}{2}$, $\sqrt{3}$, $\frac{1}{4}$, $\sqrt{5}$, $\frac{1}{6}$

15. $a_1 = 3$, $a_2 = \frac{(-1)^2}{a_1} = \frac{1}{3}$, $a_3 = \frac{(-1)^3}{a_2} = \frac{-1}{1/3} = -3$, $a_4 = \frac{(-1)^4}{a_3} = \frac{1}{-3} = -\frac{1}{3}$

$a_5 = \frac{(-1)^5}{a_4} = -\frac{1}{-1/3} = 3$

18. $a_1 = 2$, $a_2 = \frac{1}{3}(2)a_1 = \frac{4}{3}$, $a_3 = \frac{1}{3}(3)a_2 = \frac{4}{3}$, $a_4 = \frac{1}{3}(4)a_3 = \frac{16}{9}$, $a_5 = \frac{1}{3}(5)a_4 = \frac{80}{27}$

21. $a_1 = 7$, $a_2 = a_1 + 2 = 9$, $a_3 = a_2 + 2 = 11$, $a_4 = a_3 + 2 = 13$, $a_5 = a_4 + 2 = 15$

24. Since $a_2 - a_1 = \frac{1}{8} - \frac{1}{16} = \frac{1}{16}$ and $a_3 - a_2 = \frac{1}{4} - \frac{1}{8} = \frac{1}{8}$, the differences are not equal and the sequence is geometric with common ratio $r = a_2/a_1 = \frac{1}{8}/\frac{1}{16} = 2$. From (7) in the text, the general term of the sequence is $a_n = \frac{1}{16}(2)^{n-1} = 2^{-4}2^{n-1} = 2^{n-5}$. By (6) in the text, the recursion formula is $a_{n+1} = a_n(2) = 2a_n$, with $a_1 = \frac{1}{16}$.

27. Since $a_2 - a_1 = -9 - 2 = -11$ and $a_3 - a_2 = -20 - (-9) = -11$, the differences are equal and the sequence is arithmetic with common difference $d = -11$. From (5) in the text, the general term of the sequence is $a_n = 2 + (n-1)(-11) = 13 - 11n$. By (4) in the text, the recursion formula is $a_{n+1} = a_n - 11$, with $a_1 = 2$.

298

30. Since $a_2 - a_1 = 7x - 4x = 3x$ and $a_3 - a_2 = 10x - 7x = 3x$, the differences are equal and the sequence is arithmetic with common difference $d = 3x$. From (5) in the text, the general term of the sequence is $a_n = 4x + (n-1)(3x) = x(1 + 3n)$. By (4) in the text, the recursion formula is $a_{n+1} = a_n + 3x$, with $a_1 = 4x$.

33. This is an arithmetic sequence with common difference $d = 6$ and first term $a_1 = -1$. By (5) in the text, $a_n = -1 + (n-1)(6) = 6n - 7$. Letting $n = 20$, we have $a_{20} = 6(20) - 7 = 113$.

36. This is a geometric sequence with $r = \frac{1}{128} / \frac{1}{1024} = 8$ and first term $a_1 = \frac{1}{1024}$. By (7) in the text,

$$a_n = \frac{1}{1024}(8)^{n-1} = \frac{1}{2^{10}}(2^3)^{n-1} = 2^{3n-13}.$$

Letting $n = 8$, we have $a_8 = 2^{3(8)-13} = 2^{11} = 2048$.

39. The difference between the first and third terms of an arithmetic sequence is $2d = a_3 - a_1 = 323 - 357 = -34$. thus, $d = -17$ is the common difference. Using (5) in the text, $a_7 = a_1 + (7-1)d = 357 + 6(-17) = 255$.

42. In a geometric sequence, $a_5 = a_4 r = (a_3 r)r = a_3 r^2$, so $r^2 = a_5/a_3 = 64$ and r is ± 8. When $r = 8$,

$$1 = a_2 = a_1 r = a_1(8) = 8a_1,$$

so $a_1 = \frac{1}{8}$. In this case the series is $\frac{1}{8}$, 1, 8, 64, 512, ... When $r = -8$,

$$1 = a_2 = a_1 r = a_1(-8) = -8a_1,$$

so $a_1 = -\frac{1}{8}$. In this case the series is $-\frac{1}{8}$, 1, -8, 64, -512, ...

45. We identify $P = 450, n = 8$, and $A_8 = 750$. then, by (8) in the text,

$$750 = A_8 = P(1+r)^8 = 450(1+r)^8,$$

so $(1+r)^8 = \frac{750}{450} = \frac{5}{3}$ and $1 + r = (\frac{5}{3})^{1/8} \approx 1.0659$. Thus, the annual rate of interest is $r = 1.0659 - 1 = 0.0659$ or $r = 6.59\%$.

Miscellaneous Applications

48. From Problem 47, the amount they set aside in the nth year is given by $a_n = 5 + (n-1)10 = 10n - 5$.

51. If a person has two parents, each of whom has two parents, the first person has four grandparents. Each grandparent has two parents, who are great-grandparents of the first person. Thus, the person has $2(4) = 8$ great-grandparents. This is forming the geometric sequence

2, 4, 8, ... with first term $a_1 = 2$ and common ratio $r = 2$. The number of great-great-great-grandparents corresponds to the fifth term in the sequence. This is

$$a_5 = a_1 r^{5-1} = 2(2)^4 = 32,$$

so every person has 32 great-great-great-grandparents.

10.2   Series

3. $\displaystyle\sum_{k=0}^{5}(k-k^2) = (0-0^2)+(1-1^2)+(2-2^2)+(3-3^2)+(4-4^2)+(5-5^2) = 0+0-2-6-12-20 = -40$

6. $\displaystyle\sum_{k=0}^{3}(1-k)^3 = (1-0)^3 + (1-1)^3 + (1-2)^3 + (1-3)^3 = 1+0-1-8 = -8$

9. $\displaystyle\sum_{k=0}^{3}(-1)^n = 1-1+1-1 = 0.$

12. $\displaystyle\frac{1}{2} + \frac{4}{3} + \frac{9}{4} + \frac{16}{5} + \frac{25}{6} + \frac{36}{7} = \sum_{k=1}^{6}\frac{k^2}{k+1}$

15. Here, we note that the denominators are powers of 2 and the numerators are each 1 more that the corresponding denominator. Thus,

$$\frac{3}{2} + \frac{5}{4} + \frac{9}{8} + \frac{17}{16} + \frac{33}{32} = \sum_{k=1}^{5}\frac{2^k+1}{2^k} = \sum_{k=1}^{5}\left(1 + \frac{1}{2^k}\right).$$

18. The number of terms is $n = 6$, the first term is $a_1 = 131$, and the sixth term is $a_6 = 31$. Then, by formula (4) in the text,

$$S_6 = 6\left(\frac{131+31}{2}\right) = 486.$$

21. The first term of the series is $a_1 = 12$ and the last term is $a_n = -100$. To find n we note that the common difference is $d = -7$, and solve $a_1 + (n-1)d = 12 + (n-1)(-7) = -100$ for n. This gives $n = 17$. Then, from (4) in the text,

$$S_{17} = 17\left(\frac{12-100}{2}\right) = -748.$$

24. The number of terms is $n = 6$, the first term is $a_1 = 7$, and the common ratio is $r = 2$. Then, by formula (8) in the text,

$$S_6 = \frac{7(1-2^6)}{1-2} = 441.$$

27. The number of terms is $n = 8$, the first term is $a_1 = 1$, and the common ratio is $r = -\frac{1}{2}$. Then, by formula (8) in the text,

$$S_8 = \frac{1[1 - (-\frac{1}{2})^8]}{1 - (-\frac{1}{2})} = \frac{85}{128}.$$

30. From formula (4) in the text we identify $n = 8$, $a_1 = 4$, and $S_8 = 86$. Solving

$$86 = 8\left(\frac{4 + a_8}{2}\right)$$

for a_8, we find $a_8 = \frac{35}{2}$. Since $a_8 = a_1 + (8 - 1)d$, we have $d = \frac{1}{7}(\frac{35}{2} - 4) = \frac{27}{14}$.

33. The first term of the sequence is $a = 2$ and the common ratio is $r = 2$. Solving

$$8190 = \frac{2(1 - 2^n)}{1 - 2} = -2 + 2^{n+1}$$

for n we obtain $n = \log_2 8192 - 1 = 12$.

36. This is an arithmetic sequence with common difference $d = 1$, first term $a_1 = 1$ and nth term $a_n = n$. By formula (4) in the text,

$$S_n = 1 + 2 + 3 + \cdots + n = n\left(\frac{1 + n}{2}\right) = \frac{1}{2}n(n + 1).$$

39. From Problem 36, with $n = 1,000$, we have

$$S_{1,000} = \frac{1}{2}(1,000)(1,001) = 500,500.$$

Miscellaneous Applications

42. The nth term in Problem 41 is

$$a_n = [5 + (n - 1)10](12) = (10n - 5)(12),$$

so the total amount at the end of the nth year is

$$S_n = n\left(\frac{5(12) + (10n - 5)(12)}{2}\right) = 60n^2.$$

45. The amount accumulated after n payments is

$$S = P + P(1 + r) + P(1 + r)^2 + \cdots + P(1 + r)^{n-1}.$$

This is a geometric series with first term $a = P$ and comon ratio $1 + r$. By formula (8) in the text

$$\frac{P[1 - (1 + r)^n]}{1 - (1 + r)} = \frac{P[1 - (1 + r)^n]}{-r} = P\left[\frac{(1 + r)^n - 1}{r}\right].$$

48. Since 20% of 10 is 2, initially 2 kg of salt is removed, leaving 8 kg of salt. The next filtration removes 20% of 8 kg or 1.6 kg of salt, leaving $9 - 1.6 = 6.4$ kg of salt. The third filtration removes 20% of 6.4 kg or 1.28 kg of salt, leaving $6.4 - 1.28 = 5.12$ kg of salt. Since

$$\frac{1.6}{2} = \frac{1.28}{1.6} = 0.8,$$

the sequence $2, 1.6, 1.28, \ldots$ is a geometric sequence with first term 2 and common ratio $r = 0.8$. The sum of the first 10 terms is

$$S_{10} = \frac{2\left(1 - 0.8^{10}\right)}{1 - 0.8} = 10\left(1 - 0.8^{10}\right) \approx 8.93 \text{ kg.}$$

51. This is a geometric series with $a = 1$, $r = 2$, and $n = 64$. By (8) in text

$$S_{64} = \frac{1(1 - 2^{64})}{1 - 2} = 2^{64} - 1 \approx 1.84 \times 10^{19}.$$

Since the average bushel contains 10^6 grains of wheat, the number of bushels of wheat owed the peasant is

$$\frac{1.84 \times 10^{19}}{10^6} = 1.84 \times 10^{13} \approx 1.84 \times 10,000,000,000,000.$$

This is about 1.84 trillion bushels, which has to be way more than the king could possibly pay. Once the king finds out how much he really owes, the peasant would do well to make himself very scarce.

10.3   Mathematical Induction

3. Let $S(n)$ be the statement $1^2 + 2^2 + 3^2 + \cdots + n^2 = \frac{1}{6}n(n+1)(2n+1)$. When $n = 1$ we have $1^2 = \frac{1}{6}(1)(2)(3) = 1$, which is true. Now assume that

$$S(k): 1^2 + 2^2 + 3^2 + \cdots + k^2 = \frac{1}{6}k(k+1)(2k+1)$$

is true. We need to show that

$$S(k+1): 1^2 + 2^2 + 3^2 + \cdots + k^2 + (k+1)^2 = \frac{1}{6}(k+1)[(k+1)+1][2(k+1)+1]$$

$$= \frac{1}{6}(k+1)(k+2)(2k+3)$$

is true. This follows from

$$(1^2 + 2^2 + 3^2 + \cdots + k^2) + (k+1)^2 = \frac{1}{6}k(k+1)(2k+1) + (k+1)^2$$

$$= (k+1)\left[\frac{1}{6}k(2k+1) + k + 1\right]$$

$$= \frac{1}{6}(k+1)[k(2k+1) + 6k + 6]$$

$$= \frac{1}{6}(k+1)(2k^2 + 7k + 6)$$

$$= \frac{1}{6}(k+1)(k+2)(2k+3)].$$

Thus, $S(k+1)$ is true, and by the Principle of Mathematical Induction, $S(n)$ is true for every positive integer n.

6. Let $S(n)$ be the statement $\displaystyle\sum_{i=1}^{n}(4i-5) = n(2n-3)$. When $n = 1$ we have $4(1) - 5 = 1[2(1) - 3]$ or $-1 = -1$, which is true. Now assume that

$$S(k)\colon \sum_{i=1}^{k}(4i-5) = k(2k-3)$$

is true. We need to show that

$$S(k+1)\colon \sum_{i=1}^{k+1}(4i-5) = (k+1)[2(k+1) - 3] = (k+1)(2k-1)$$

is true. This follows from

$$\sum_{i=1}^{k}(4i-5) + [4(k+1) - 5] = k(2k-3) + [4(k+1) - 5]$$

$$= 2k^2 - 3k + 4k + 4 - 5$$

$$= 2k^2 + k - 1$$

$$= (k+1)(2k-1).$$

Thus, $S(k+1)$ is true, and by the Principle of Mathematical Induction, $S(n)$ is true for every positive integer n.

9. Let $S(n)$ be the statement $1 + 4 + 4^2 + \cdots + 4^{n-1} = \frac{1}{3}(4^n - 1)$. When $n = 1$ we have $1 = \frac{1}{3}(4^1 - 1) = 1$, which is true. Now assume that

$$S(k)\colon 1 + 4 + 4^2 + \cdots + 4^{k-1} = \frac{1}{3}(4^k - 1)$$

is true. We need to show that

$$S(k+1)\colon 1+4+4^2+\cdots+4^{(k+1)-1} = 1+4+4^2+\cdots+4^k = \frac{1}{3}(4^{k+1}-1)$$

is true. This follows from

$$(1+4+4^2+\cdots+4^{k-1})+4^k = \frac{1}{3}(4^k-1)+4^k$$

$$= \frac{1}{3}(4^k-1+3\cdot4^k)$$

$$= \frac{1}{3}(4\cdot4^k-1)$$

$$= \frac{1}{3}(4^{k+1}-1).$$

Thus, $S(k+1)$ is true, and by the Principle of Mathematical Induction, $S(n)$ is true for every positive integer n.

12. Let $S(n)$ be the statement n^2+n is divisible by 2. When $n=1$ we have $1^2+1=2$ is divisible by 3, which is true. Now assume that

$$S(k)\colon k^2+k \text{ is divisible by } 2.$$

We need to show that

$$S(k+1)\colon (k+1)^2+(k+1) \text{ is divisible by } 2.$$

Consider

$$(k+1)^2+(k+1) = k^2+2k+1+k+1 = \underbrace{(k^2+k)}_{\text{use } S(k)}+\underbrace{(2k+2)}_{\text{divisible by } 2}.$$

Thus, $(k+1)^2+(k+1)$ is divisible by 2, and $S(k+1)$ is true. Then, by the Principle of Mathematical Induction, $S(n)$ is true for every positive integer n.

15. Let $S(n)$ be the statement 7 is a factor of $3^{2n}-2^n$. When $n=1$ we have $3^{2(1)}-2^1 = 9-2 = 7$, of which 7 is a factor, so $S(1)$ is true. Now assume that

$$S(k)\colon 7 \text{ is a factor of } 3^{2k}-2^k.$$

We need to show that

$$S(k+1)\colon 7 \text{ is a factor of } 3^{2(k+1)}-2^{k+1}.$$

Consider

$$3^{2(k+1)} - 2^{k+1} = 3^{2k+2} - 2^{k+1}$$
$$= 3^2 \cdot 3^{2k} - 2^1 \cdot 2^k$$
$$= 9 \cdot 3^{2k} + (7 - 9) \cdot 2^k$$
$$= 9 \cdot 3^{2k} - 9 \cdot 2^k + 7 \cdot 2^k$$
$$= 9 \underbrace{\left(3^{2k} - 2^k\right)}_{\text{use } S(k)} + \underbrace{7 \cdot 2^k}_{\text{divisible by 7}}.$$

Since 7 is a factor of $3^{2(k+1)} - 2^{k+1}$, $S(k+1)$ is true, and by the Principle of Mathematical Induction, $S(n)$ is true for every positive integer n.

18. Let $S(n)$ be the statement $2n \leq 2^n$. When $n = 1$ we have $2(1) \leq 2^1$ or $2 \leq 2$, so $S(1)$ is true. Now assume that

$$S(k): \ 2k \leq 2^k.$$

We need to show that

$$S(k+1): \ 2(k+1) \leq 2^{k+1}.$$

Since $2k \leq 2^k$, by assumption, and $2 \leq 2^k$, for $k \geq 1$, we have

$$2(k+1) = 2k + 2 \leq 2^k + 2 \leq 2^k + 2^k = 2 \cdot 2^k = 2^{k+1}.$$

Thus, $S(k+1)$ is true, and by the Principle of Mathematical Induction, $S(n)$ is true for every positive integer n.

10.4 │ The Binomial Theorem

3. $\dfrac{2!}{5!} = \dfrac{2!}{5 \cdot 4 \cdot 3 \cdot 2!} = \dfrac{1}{5 \cdot 4 \cdot 3} = \dfrac{1}{60}$

6. $0!5! = 1 \cdot 5 \cdot 4 \cdot 3 \cdot 2 \cdot 1 = 120$

9. $\dbinom{7}{6} = \dfrac{7!}{6!(7 - 6)!} = \dfrac{7 \cdot 6!}{6! \, 1!} = 7$

12. $\dbinom{4}{0} = \dfrac{4!}{0!(4 - 0)!} = \dfrac{4!}{4!} = 1$

15. $\dfrac{n!(n+1)!}{(n+2)!(n+3)!} = \dfrac{n!(n+1)!}{(n+2)(n+1)n!(n+3)(n+2)(n+1)!} = \dfrac{1}{(n+3)(n+2)^2(n+1)}$

18. $7 \cdot 6 \cdot 5 \cdot 4 \cdot 3 \cdot 2 \cdot 1 = 7!$

21. $(4 \cdot 3 \cdot 2 \cdot 1)(5 \cdot 4 \cdot 3 \cdot 2 \cdot 1) = 4!\, 5!$

24. $10 \cdot 9 \cdot 8 = \dfrac{10 \cdot 9 \cdot 8 \cdot 7 \cdot 6 \cdot 5 \cdot 4 \cdot 3 \cdot 2 \cdot 1}{7 \cdot 6 \cdot 5 \cdot 4 \cdot 3 \cdot 2 \cdot 1} = \dfrac{10!}{7!}$

27. True, since $5 \cdot 4! = 5(4 \cdot 3 \cdot 2 \cdot 1) = 5!$.

30. False, since

$$\frac{8!}{4} = \frac{8 \cdot 7 \cdot 6 \cdot 5 \cdot 4 \cdot 3 \cdot 2 \cdot 1}{4} \neq 2.$$

33. Using (5) in the text with $n = 2$ we have

$$(x^2 - 5y^4)^2 = (x^2)^2 + 2(x^2)^1(-5y^4)^1 + (-5y^4)^2 = x^4 - 10x^2y^4 + 25y^8.$$

36. Using (5) in the text with $n = 4$ we have

$$(x^{-2} + 1)^4 = (x^{-2})^4 + 4(x^{-2})^3(1)^1 + \frac{4 \cdot 3}{2 \cdot 1}(x^{-2})^2(1)^2 + \frac{4 \cdot 3 \cdot 2}{3 \cdot 2 \cdot 1}(x^{-2})^1(1)^3 + 1^4$$

$$= x^{-8} + 4x^{-6} + 6x^{-4} + 4x^{-2} + 1.$$

39. Using (5) in the text with $n = 5$ we have

$$(x^2 + y^2)^5 = (x^2)^5 + 5(x^2)^4(y^2)^1 + \frac{5 \cdot 4}{2 \cdot 1}(x^2)^3(y^2)^2 + \frac{5 \cdot 4 \cdot 3}{3 \cdot 2 \cdot 1}(x^2)^2(y^2)^3$$

$$+ \frac{5 \cdot 4 \cdot 3 \cdot 2}{4 \cdot 3 \cdot 2 \cdot 1}(x^2)^1(y^2)^4 + (y^2)^5$$

$$= x^{10} + 5x^8y^2 + 10x^6y^4 + 10x^4y^6 + 5x^2y^8 + y^{10}.$$

42. Using (5) in the text with $n = 4$ we have

$$(x + y + z)^4 = [(x + y) + z]^4$$

$$= (x + y)^4 + 4(x + y)^3z^1 + \frac{4 \cdot 3}{2 \cdot 1}(x + y)^2z^2 + \frac{4 \cdot 3 \cdot 2}{3 \cdot 2 \cdot 1}(x + y)^1z^3 + z^4$$

$$= x^4 + 4x^3y + \frac{4 \cdot 3}{2 \cdot 1}x^2y^2 + \frac{4 \cdot 3 \cdot 2}{3 \cdot 2 \cdot 1}xy^3 + y^4$$

$$+ 4\left(x^3 + 3x^2y + \frac{3 \cdot 2}{2 \cdot 1}xy^2 + y^3\right)z + \frac{4 \cdot 3}{2 \cdot 1}(x^2 + 2xy + y^2)z^2 + 4(x + y)z^3 + z^4$$

$$= x^4 + +4x^3y + 6x^2y^2 + 4xy^3 + y^4 + 4x^3z + 12x^2yz$$

$$+ 12xy^2z + 4y^3z + 6x^2z^2 + 12xyz^2 + 6y^2z^2 + 4xz^3 + 4yz^3 + z^4$$

45. The sixth term of $(a + b)^6$ corresponds to $r = 5$ since $r + 1 = 6$ implies $r = 5$. Identifying $n = 6$ and $r = 5$ we have $n - r + 1 = 6 - 5 + 1 = 2$, so the sixth term is

$$\frac{6 \cdot 5 \cdot 4 \cdot 3 \cdot 2}{5 \cdot 4 \cdot 3 \cdot 2 \cdot 1} \, ab^5 = 6ab^5.$$

48. The third term of $(x - y)^5$ corresponds to $r = 2$ since $r + 1 = 3$ implies $r = 2$. Identifying $n = 5$, $r = 2$, $a = x$, and $b = -5$ we have $n - r + 1 = 5 - 2 + 1 = 4$, so the third term is

$$\frac{5 \cdot 4}{2 \cdot 1} \, x^3(-5)^2 = 250x^3.$$

51. The tenth term of $(x + y)^{14}$ corresponds to $r = 9$ since $r + 1 = 10$ implies $r = 9$. Identifying $n = 14$, $r = 9$, $a = x$, and $b = y$ we have $n - r + 1 = 14 - 9 + 1 = 6$, so the tenth term is

$$\frac{14 \cdot 13 \cdot 12 \cdot 11 \cdot 10 \cdot 9 \cdot 8 \cdot 7 \cdot 6}{9 \cdot 8 \cdot 7 \cdot 6 \cdot 5 \cdot 4 \cdot 3 \cdot 2 \cdot 1} \, x^5 y^9 = 2002 x^5 y^9.$$

54. The ninth term of $(3 - z)^{10}$ corresponds to $r = 8$ since $r + 1 = 9$ implies $r = 8$. Identifying $n = 10$, $r = 8$, $a = 3$, and $b = -z$ we have $n - r + 1 = 10 - 8 + 1 = 3$, so the ninth term is

$$\frac{10 \cdot 9 \cdot 8 \cdot 7 \cdot 6 \cdot 5 \cdot 4 \cdot 3}{8 \cdot 7 \cdot 6 \cdot 5 \cdot 4 \cdot 3 \cdot 2 \cdot 1} \, (3)^2(-z)^8 = 405z^8.$$

57. Using Pascal's triangle, the sum of the first four terms is

$$1^5 + 5(1)^4(-0.01) + 10(1)^3(-0.01)^2 + 10(1)^2(-0.01)^3 = 1 - 0.05 + 0.001 - 0.00001$$
$$= 0.95099.$$

A calculator gives $(1 - 0.01)^5 = 0.99^5 = 0.9509900499$.

10.5 | Principles of Counting

3.

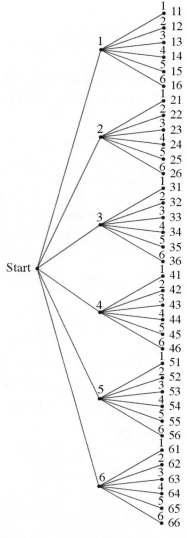

6. There are $6 \cdot 4 \cdot 2 = 48$ different stereo systems.

9. $P(6,3) = \dfrac{6!}{(6-3)!} = \dfrac{6!}{3!} = 6 \cdot 5 \cdot 4 = 120$

12. $P(4,0) = \dfrac{4!}{(4-0)!} = \dfrac{4!}{4!} = 1$

15. $P(8,6) = \dfrac{8!}{(8-6)!} = \dfrac{8!}{2!} = 8 \cdot 7 \cdot 6 \cdot 5 \cdot 4 \cdot 3 = 20{,}160$

18. $C(4,1) = \dfrac{4!}{3!1!} = \dfrac{4}{1} = 4$

21. $C(13,11) = \dfrac{13!}{2!11!} = \dfrac{13 \cdot 12}{2} = 78$

24. $C(7,4) = \dfrac{7!}{3!4!} = \dfrac{7 \cdot 6 \cdot 5}{3 \cdot 2} = 35$

Miscellaneous Applications

27. (a) Since the letters are all distinct, there are $P(7,7) = \dfrac{7!}{(7-7)!} = \dfrac{7!}{0!} = 5040$ possible "words."

(b) There are $P(7,5) = \dfrac{7!}{(7-5)!} = \dfrac{7!}{2!} = 2520$ possible 5-letter "words."

30. The number of ways is $C(10,3) = \dfrac{10!}{7!3!} = \dfrac{10 \cdot 9 \cdot 8}{3 \cdot 2 \cdot 1} = 120$.

33. Since order matters, they can be chosen in $P(10,2) = \dfrac{10!}{8!} = 90$ different ways.

36. There are 8 teams that could come in first; then 7 that could come in second; and then 6 that could come in third. Thus, there are a total $8 \cdot 7 \cdot 6 = 336$ different ways that first, second, and third place could be determined. This is the same as the number of permutations of 8 things taken 3 at a time. That is, $P(8,3) = 8!/(8-3)! = 8!/5! = 336$.

39. (a) Without repetition, 6 colors can be put in the first slot, 5 in the second slot, 4 in the third slot, and 3 in the fourth slot. Thus, there are $6 \cdot 5 \cdot 4 \cdot 3 = 360$ possible codes.

(b) Since any of 6 colors can go in each of the 4 positions, there are $6^4 = 1296$ possible colors.

(c) In this case there are 7 choices for each slot and $7^4 = 2401$ possible codes.

42. (a) If four bulbs are defective then there is $C(4,4) = \dfrac{4!}{0!4!} = 1$ way in which four bulbs can be chosen so that all four are defective.

(b) In this case, there are $24 - 4 = 20$ good bulbs, so there are $C(20,4) = \dfrac{20!}{16!4!} = 4845$ ways in which four bulbs can be chosen so that all four are defective.

(c) The number of ways in which two good bulbs can be chosen from the the total of 20 good bulbs is $C(20,2) = \dfrac{20!}{18!2!} = 190$, and the number of ways in which two defective bulbs can be chosen from the total of four defective bulbs is $C(4,2) = \dfrac{4!}{2!2!} = 6$. By the Fundamental Counting Principle then there is a total of $190 \cdot 6 = 1140$ ways in which two good bulbs and 2 defective bulbs can be chosen.

(d) The number of ways in which three good bulbs can be chosen from the the total of 20 good bulbs is $C(20,3) = \dfrac{20!}{17!3!} = 1140$, and the number of ways in which one defective bulb can be chosen from the total of four defective bulbs is $C(4,1) = \dfrac{4!}{3!1!} = 4$. By the Fundamental Counting Principle then there is a total of $1140 \cdot 4 = 4560$ ways in which three good bulbs and 1 defective bulbs can be chosen.

10.6   Introduction to Probability

3. The sample space is $S = \{1H, 2H, 3H, 4H, 5H, 6H, 1T, 2T, 3T, 4T, 5T, 6T\}$.

6. The sample space is the set S of 52 cards, and the event E is the set of 13 hearts. Thus by (4) in the text,

$$P(E) = \frac{n(E)}{n(S)} = \frac{13}{52} = \frac{1}{4}.$$

9. The sample space is the set S of the 36 possible pairs of numbers on two dice, and the event E is the set consisting of the single pair of number 1 and 1. Thus by (4) in the text,

$$P(E) = \frac{n(E)}{n(S)} = \frac{1}{36}.$$

12. The sample space is $S = \{HHH, HHT, HTH, HTT, THH, THT, TTH, TTT\}$, and the event E is the set consisting of the elements HHH THT, and TTH. Thus by (4) in the text,

$$P(E) = \frac{n(E)}{n(S)} = \frac{3}{8}.$$

15. There are $C(13,5)$ ways to choose 5 cards from one suit, and there are 4 suits, so

$$P(E) = \frac{4C(13,5)}{C(52,5)} = \frac{4 \cdot 1287}{52!/47!5!} \approx 0.00198.$$

18. The complementary event E' is the event of obtaining none of the 12 face cards. Thus,

$$P(E) = 1 - \frac{C(40,5)}{C(52,5)} = 1 - \frac{40!/35!5!}{52!/47!5!} = 1 - \frac{47!40!}{52!35!} = 1 - \frac{40 \cdot 39 \cdot 38 \cdot 37 \cdot 36}{52 \cdot 51 \cdot 50 \cdot 49 \cdot 48} \approx 1 - 0.253 = 0.747.$$

Miscellaneous Applications

21. The complementary event E' is the event of having no girls (or all boys). In this case $n(S) = 2^4 = 16$ and $n(E') = 1$, so

$$P(E) = 1 - \frac{1}{24} = 1 - \frac{1}{16} = \frac{15}{16}.$$

24. The number of ways to choose 6 people (without ordering) is $C(4 + 7 + 8, 6)$. Thus, $n(S) = C(19, 6) = 27,132$. Since all 4 administrators are to be chosen and no faculty are to be chosen, there must be 2 staff members chosen. Because there are exactly 4 administrators, there is only one way to chose all four administrators, but there are $C(8, 2) = 28$ ways to chose the 2 staff members. The probability that the committee is constituted with 4 administrators and 2 staff members is

$$P(E) = \frac{n(E)}{n(S)} = \frac{28}{27,132} = \frac{1}{969} \approx 0.001032.$$

27. There are $6^2 = 36$ ways to roll two dice. There are 6 ways to roll 7 (16, 25, 34, 43, 52, 61) and 2 ways to roll 11 (56, 65). The probability of winning is then $\frac{6+2}{36} = \frac{8}{36} = \frac{2}{9}$.

30. Since the events of earning an A and earning a B are mutually exclusive, the probability of an A or a B is $\frac{3}{10} + \frac{2}{5} = \frac{7}{10} = 0.7$.

33. The probability that it will not rain is $100\% - 40\% = 60\%$.

36. We are given

$$P(\text{female}) = 30\%, \qquad P(\text{plumber}) = 70\%, \qquad \text{and} \qquad P(\text{female and plumber}) = 40\%.$$

Thus, the probability that the worker is either female or a plumber is

$$P(\text{female or plumber}) = P(\text{female}) + P(\text{plumber}) - P(\text{female and plumber})$$
$$= 30\% + 70\% - 40\% = 60\%.$$

10.7 | ∫ **Calculus PREVIEW** **Convergence of Sequences and Series**

3. Since $5n + 6$ increases without bound as $n \to \infty$,

$$\lim_{n \to \infty} \frac{1}{5n + 6} = 0,$$

and the sequence converges.

6. Dividing the numerator and denominator by n we have

$$\lim_{n \to \infty} \frac{n}{1 - 2n} = \lim_{n \to \infty} \frac{n/n}{1/n - 2n/n} = \lim_{n \to \infty} \frac{1}{1/n - 2} = -\frac{1}{2},$$

so the sequence converges.

9. Dividing the numerator and denominator by n we have

$$\frac{n^2 - 1}{2n} = \frac{n^2/n - 1/n}{2n/n} = \frac{n}{2}.$$

Since the terms of $\{n/2\}$ increase without bound as $n \to \infty$, the sequence diverges.

12. Dividing the numerator and denominator by n (n^2 under the square root sign) we have

$$\lim_{n \to \infty} \frac{n}{\sqrt{n + 1}} = \lim_{n \to \infty} \frac{n/n}{\sqrt{n/n^2 + 1/n^2}} = \lim_{n \to \infty} \frac{1}{\sqrt{1/n + 1/n^2}}.$$

Since the terms in the denominator approach 0, the terms of the fraction increase without bound. Thus, the sequence diverges to infinity.

15. Since $\lim\limits_{n\to\infty} 2^{-n} = 0$ and $\lim\limits_{n\to\infty} 4^{-n} = 0$, the sequence converges to $\frac{5}{6}$.

18. Since

$$4 + \frac{3^n}{2^n} = 4 + \left(\frac{3}{2}\right)^n,$$

and $\frac{3}{2} > 1$, the terms of the sequence increase without bound and the sequence diverges to infinity.

21. Written as an infinite geometric series, $0.222\ldots$ is the same as

$$\frac{2}{10} + \frac{2}{10^2} + \frac{2}{10^3} + \cdots = \sum_{k=1}^{\infty} \frac{2}{10^k}.$$

Identifying $a = \frac{2}{10}$ and $r = \frac{1}{10}$, we have

$$0.222\ldots = \frac{a}{1-r} = \frac{2/10}{1-1/10} \cdot \frac{10}{10} = \frac{2}{10-1} = \frac{2}{9}.$$

24. Written as an infinite geometric series, $0.393939\ldots$ is the same as

$$\frac{39}{100} + \frac{39}{100^2} + \frac{39}{100^3} + \cdots = \sum_{k=1}^{\infty} \frac{39}{100^k}.$$

Identifying $a = \frac{39}{100}$ and $r = \frac{1}{100}$, we have

$$0.393939\ldots = \frac{a}{1-r} = \frac{39/100}{1-1/100} \cdot \frac{100}{100} = \frac{39}{100-1} = \frac{39}{99} = \frac{13}{33}.$$

27. The first term is 2 and the common ratio is $r = 1/2 < 1$, so by (12) in the text, the infinite geometric series converges, and its sum is

$$S = \frac{2}{1-1/2} \cdot \frac{2}{2} = \frac{4}{2-1} = 4.$$

30. The first term is 1 and the common ratio is $r = 0.1/1 = 0.1 < 1$, so by (12) in the text, the infinite geometric series converges, and its sum is

$$S = \frac{1}{1-0.1} = \frac{1}{0.9} \cdot \frac{10}{10} = \frac{10}{9}.$$

33. The common ratio is

$$r = \frac{1/(\sqrt{3}-\sqrt{2})^1}{(\sqrt{3}-\sqrt{2})^0} = \frac{1}{\sqrt{3}-\sqrt{2}} \approx 3.15 \geq 1,$$

so the series diverges.

36. The common ratio is

$$r = \frac{\pi^2(1/3)^1}{\pi^1(1/3)^0} = \frac{\pi}{3} > 1,$$

so the series diverges.

Miscellaneous Applications

39. Initially the ball travels 15 feet. Then it bounces and rises and falls a total of $2\left[\frac{2}{3}(15)\right]$ feet. After the next bounce it rises and falls a total of $2\left[\left(\frac{2}{3}\right)^2(15)\right]$ feet. After the third bounce it rises and falls a total of $2\left[\left(\frac{2}{3}\right)^3(15)\right]$ feet. Continuing in this fashion, after the nth bounce it rises and falls a total of $2\left[\left(\frac{2}{3}\right)^n(15)\right]$ feet. The total distance in feet that the ball travels is then modeled by 15 plus a geometric series:

$$15 + 2\left[\frac{2}{3}(15)\right] + 2\left[\left(\frac{2}{3}\right)^2(15)\right] + 2\left[\left(\frac{2}{3}\right)^3(15)\right] + \cdots + 2\left[\left(\frac{2}{3}\right)^n(15)\right] + \cdots = 15 + \sum_{n=1}^{\infty} 2\left[\left(\frac{2}{3}\right)^n(15)\right].$$

The first term of the series is $a = 2\left[\frac{2}{3}(15)\right] = 20$ and the common ratio is $r = \frac{2}{3}$. Thus, the total distance is

$$S = 15 + \frac{20}{1 - 2/3} \cdot \frac{3}{3} = 15 + \frac{60}{3 - 2} = 75 \text{ feet.}$$

Chapter 10 Review Exercises

A. Fill in the Blanks

3. The fifth term of the sequence $\left\{\sum_{k=1}^{n}\frac{1}{k}\right\}$ is $\sum_{k=1}^{5}\frac{1}{k} = \frac{1}{1} + \frac{1}{2} + \frac{1}{3} + \frac{1}{4} + \frac{1}{5} = \frac{137}{60}$.

6. The sequence $\left\{8 - \frac{n}{2}\right\}$ is arithmetic:

$$\left(8 - \tfrac{1}{2}\right), \ \left(8 - \tfrac{2}{2}\right), \ \left(8 - \tfrac{3}{2}\right), \ \ldots \ \text{ or } 7.5, \ 7, \ 6.5, \ \ldots$$

The common difference is

$$d = 8 - \frac{n+1}{2} - \left(8 - \frac{n}{2}\right) = -\tfrac{1}{2}$$

9. This is an infinite geometric series with first term 3 and common ration $-\frac{1}{3}$, so

$$3 - 1 + \frac{1}{3} - \frac{1}{9} + \cdots = \frac{3}{1 - (-1/3)} = \frac{3}{4/3} = \frac{9}{4}.$$

12. $\displaystyle\sum_{k=1}^{\infty} 3\left(\tfrac{1}{2}\right)^{k-1} = \frac{3}{1 - \tfrac{1}{2}} = 6$

15. Since the sequence is arithmetic, it follows from (4) of Section 10.2 that the sum of the first three hundred terms is

$$S_{300} = 300\left(\frac{1+300}{2}\right) = 45{,}150.$$

18. $C(5,\ 3)/C(8,\ 3) = \frac{5}{28}$.

21. If E_1 and E_2 are mutually exclusive, then $P(E_1 \cup E_2) = P(E_1) + P(E_2) = \frac{1}{5} + \frac{1}{3} = \frac{8}{15}$.

B. True/False

3. True; since $(n-1)!n = n(n-1)! = n!$.

6. False; $(a+b)^{100}$ contains terms with $a^0,\ a^1,\ \ldots,\ a^{100}$, and this represents 101 terms.

9. True; since $\displaystyle\sum_{k=1}^{5} \ln k = \ln 1 + \ln 2 + \ln 3 + \ln 4 + \ln 5 = \ln(1\cdot 2\cdot 3\cdot 4\cdot 5) = \ln 120$.

12. True; $P(n,n) = \dfrac{n!}{(n-n)!} = \dfrac{n!}{0!} = \dfrac{n!}{1} = n!$.

15. Since

$$a_2 = 2a_1 + 1 = -2 + 1 = -1$$
$$a_3 = 2a_2 + 1 = -2 + 1 = -1$$
$$a_4 = 2a_3 + 1 = -2 + 1 = -1$$
$$a_5 = 2a_4 + 1 = -2 + 1 = -1$$
$$\vdots$$

this is true.

18. The common ratio in sequence satisfies $|r| < 1$, hence the sequence converges to 0. False.

C. Review Exercises

3. Letting n take on the values 1, 2, 3, 4, and 5 we have $-1,\ 2,\ -3,\ 4,\ -5$.

6. If $a_1 = 3$ and $a_3 = 11$, then the common difference d is determined by $2d = a_3 - a_1 = 11 - 3 = 8$, so $d = 4$. By (5) in Section 10.1 in the text

$$a_{17} = 3 + (17-1)4 = 67.$$

9. Identifying $a_1 = 2$, $r = -1/2$, and $n = 10$ we have

$$S_{10} = \frac{2\left[1 - \left(-\frac{1}{2}\right)^{10}\right]}{1 - \left(-\frac{1}{2}\right)} = \frac{4}{3}\left(1 - \frac{1}{1024}\right) = \frac{341}{256}.$$

12. We first note that $1 + 3 + 5 + \cdots + (2n - 1)$ is the sum of an arithmetic progression with first term $a_1 = 1$, nth term $a_n = 2n - 1$, and common difference $d = 2$. Using formula (4) in Section 10.2 we have $D = S_n = n[1 + (2n - 1)]/2 = n^2$.

15. The statement $S(1)$, $1^2(1 + 1)^2 = 4$ is divisible by 4, is true. Assume that $S(k)$,

$$k^2(k + 1)^2 \text{ is divisible by } 4,$$

is true. Then

$$(k + 1)^2(k + 2)^2 = (k + 1)^2(k^2 + 4k + 4) = k^2(k + 1)^2 + 4(k + 1)^3$$

is divisible by 4 whenever $k^2(k + 1)^2$ is divisible by 4. Thus, $S(k + 1)$ is true. The proof is complete by mathematical induction.

18. The statement $S(1)$, 9 is a factor of $10^{1+1} - 9(1) - 10 = 100 - 9 - 10 = 81$, is true. Assume that $S(n)$,

$$9 \text{ is a factor of } 10^{n+1} - 9n - 10$$

is true. Then $S(n + 1)$, 9 is a factor of

$$10^{(n+1)+1} - 9(n + 1) - 10 = 10^{n+2} - 9n - 9 - 10 = 10(10^{n+1}) - 9n - 19$$

$$= 10(10^{n+1} - 9n - 10) + 81n + 81$$

$$= 10(10^{n+1} - 9n - 10) + 9(9n + 9)$$

is true. The proof is complete by mathematical induction.

21. $\dfrac{6!}{4! - 3!} = \dfrac{6 \cdot 5 \cdot 4 \cdot 3 \cdot 2 \cdot 1}{24 - 6} = \dfrac{6 \cdot 5 \cdot 4 \cdot 3 \cdot 2 \cdot 1}{18} = 40$

24. $P(9, 6) = \dfrac{9!}{(9 - 6)!} = \dfrac{9!}{3!} = 9 \cdot 8 \cdot 7 \cdot 6 \cdot 5 \cdot 4 = 60{,}480$

27. $(a + 4b)^4 = a^4 + 4a^3(4b) + 6a^2(4b)^2 + 4a(4b)^3 + (4b)^4 = a^4 + 16a^3b + 96a^2b^2 + 256ab^3 + 256b^4$

30. $[4 - (a + b)]^3 = 4^3 - 3 \cdot 4^2(a + b) + 3 \cdot 4(a + b)^2 - (a + b)^3$

$$= 64 - 48a - 48b + 12a^2 + 24ab + 12b^2 - a^3 - 3a^2b - 3ab^2 - b^3$$

33. The fifth term corresponds to $r = 4$ in (9) in Section 10.4 of the text.

$$\binom{10}{4}(xy^2)^6(z^3)^4 = \frac{10!}{4!6!}x^6y^{12}z^{12} = 210x^6y^{12}z^{12}$$

36. We first write the sum as $\displaystyle\sum_{k=0}^{n} \binom{n}{k}(x^2)^{n-k}(-4)^k$. Then by the equation in magenta between (9) and (10) on page 511 in the text we see that the sum is the binomial expansion of $(x^2 - 4)^n$. Hence

$$(x^2 - 4)^n = 0 \quad \text{or} \quad x = \pm 2.$$

39. Since $n! = n(n-1)(n-2)\cdots 3\cdot 2\cdot 1$ grows much faster that $2^n = 2\cdot 2\cdot 2\cdots 2\cdot 2\cdot 2$, it seems likely that the sequence $\{2^n/n!\}$ converges to 0.

42. There will be $4\cdot 4\cdot 4 = 4^3 = 64$ distinct three digit numbers. These are

222, 224, 226, 228, 242, 244, 246, 248, 262, 264, 266, 268, 282, 284, 286, 288,

422, 424, 426, 428, 442, 444, 446, 448, 462, 464, 466, 468, 482, 484, 486, 488,

622, 624, 626, 628, 642, 644, 646, 648, 662, 664, 666, 668, 682, 684, 686, 688,

822, 824, 826, 828, 842, 844, 846, 848, 862, 864, 866, 868, 882, 884, 886, 888.

45. Since order does not matter, there are $C(10,3) = \dfrac{10!}{7!3!} = 120$ different pizzas.

48. Order does not matter here, so there are $C(9,3) = \dfrac{9!}{3!(9-3)!} = \dfrac{9!}{3!6!} = 84$ ways to order 3 seeds.

51. $\{(1,1), (1,2), (1,3), (1,4), (1,5), (2,1), (2,2), (2,3), (2,4), (2,5), (3,1), (3,2), (3,3), (3,4),$ $(3,5), (4,1), (4,2), (4,3), (4,4), (4,5), (5,1), (5,2), (5,3), (5,4), (5,5)\}$

54. We first note that there are $C(100,5) = 75,287,520$ ways to choose 5 of the pens, so $n(S) = C(100,5)$.

  (a) The number of ways to choose 5 of the 90 pens that will write the first time is $C(90,5) = 43,949,268$. Thus the probability of choosing 5 pens that will write the first time is

$$P(E) = \frac{C(90,5)}{C(100,5)} \approx 0.584.$$

  (b) Since $100 - 90 = 10$ pens won't write the first time, the probability that none of them will write the first time is

$$P(E_1) = \frac{C(10,5)}{C(100,5)} \approx 3.347 \times 10^{-6}.$$

  (c) Since at least one of them writing the first time is the complement of none of them writing the first time is

$$P(E_1') = 1 - P(E_1) \approx 1 - 3.347 \times 10^{-6} = 0.999\,997.$$

57. There are $C(12,2) = \dfrac{12!}{10!2!} = 66$ ways to choose 2 of the 12 socks.

  (a) There are $C(8,2) = \dfrac{8!}{6!2!} = 28$ ways to choose 2 black socks. The probability of this is $\dfrac{28}{66} = \dfrac{14}{33}$.

(b) There are $C(4,2) = \dfrac{4!}{2!2!} = 6$ ways to choose 2 white socks. The probability of this is $\dfrac{6}{66} = \dfrac{1}{11}$.

(c) A matching pair (assuming all the black socks are the same and all the white socks are the same) can happen in $28 + 6 = 34$ ways. The probability is $\dfrac{34}{66} = \dfrac{17}{33}$.

60. The area of the first triangle is $A_1 = \dfrac{1}{2}(1 \cdot 1) = \dfrac{1}{2}$. The area of the second triangle is $A_2 = \dfrac{1}{2}\left(\dfrac{1}{2} \cdot 1\right) = \dfrac{1}{4}$. The area of the third triangle is $A_3 = \dfrac{1}{2}\left(\dfrac{1}{4} \cdot 1\right) = \dfrac{1}{8}$. In general, the area of the nth triangle is $A_n = \dfrac{1}{2}\left(\dfrac{1}{2^{n-1}} \cdot 1\right) = \dfrac{1}{2^n}$. This sequence is geometric with first term $A_1 = \dfrac{1}{2}$ and common ratio $r = \dfrac{1}{2}$. Thus, by Theorem 10.7.1 in Section 10.7 of the text, the sum of the infinite series is

$$A_1 + A_2 + A_3 + \cdots = \frac{1/2}{1 - 1/2} = \frac{1/2}{1/2} = 1.$$

Part V
Final Examination Answers

Part A. Fill in the Blanks

1. $2\left(x+\frac{3}{2}\right)^2+\frac{1}{2}$

2. 12

3. $(-5,-3]\cup[0,3]\cup(5,\infty)$

4. $-a+3$

5. ±16

6. fourth

7. **(a)** $(1,-7)$; **(b)** $(-1,7)$; **(c)** $(-1,-7)$

8. $-27;\ 3$

9. $f(x)=x\left(x-1-\sqrt{7}\right)\left(x-1+\sqrt{7}\right)$

10. -1 and -2

11. $\pi/4$

12. -4π

13. $-\frac{24}{25}$

14. -4

15. $150°$

16. $\ln 48$

17. $x=-\frac{5}{2}$

18. $(-\infty,0)\cup(2,\infty)$

19. $67,780$

20. -25

Part B. True/False

1. false

2. false

3. false

4. false

5. true

6. false

7. true

8. true

9. false

10. true

11. false

12. false

13. true

14. false

15. true

16. true

17. true

18. false

19. true

20. true

Part C.   Exercises

1. (i) and (a);   (ii) and (c);   (iii) and (d);
(iv) and (b)

2. $(-\infty, -2) \cup (-\frac{8}{3}, \infty)$

3. yes

4. second and fourth

5. (c)

6. (i) and (d);   (ii) and (c);   (iii) and (b);
(iv) and (a)

7. $(0, 10)$

8. $(-2, 0) \cup (0, \infty)$

9. $y = -\frac{5}{8}x$

10. $f(x) = -10x^2 + 20x - 6$

11. $\dfrac{4}{\sqrt{x^6 + 4} + x^3}$

12. $f(x) = 5x^{8/3} - 4x^{13/6} + 8x^{-1/3}$

13. $f(x) = \dfrac{13}{x} + \dfrac{6}{x^2} - \dfrac{6}{x - 1}$

14. $f(x) = \sec^2 x - \sec x \tan x$

15. $f(x) = x^3$

16. $f(x) = \begin{cases} x^2 - 3x, & x \le 0 \text{ and } x \ge 3 \\ -x^2 + 3x, & 0 < x < 3 \end{cases}$

17. $x = 0, \; x = -\sqrt{\frac{8}{3}}, \; x = \sqrt{\frac{8}{3}}$

18. $x = -\pi, \; x = -\pi/3, \; x = 0, \; x = \pi,$
$x = \pi/3$

19. $\dfrac{15}{(2x + 2h + 5)(2x + 5)}$

20. $-3x^2 - 3xh - h^2 + 20x + 10h$

21. $8; \; [0, 6]$

22. $-1/\sqrt{6}$

23. 0

24. $x = e^{n\pi}, \; n$ an integer

25. $(3, 0), \; (0, -6), \; (0, -2)$

26. $(-1, 5); \; (-2, 5), \; (0, 5);$
$(-1 - \sqrt{2}, 5), \; (-1 + \sqrt{2}, 5);$
$(-1, 4), \; (-1, 6)$

27. $(-1, 3)$

28. approximately $46.6°$

29. 1.36 g

30. $0, \; \pi/4, \; \pi/2, \; 3\pi/4, \; \pi, \; 5\pi/4, \; 3\pi/2, \; 7\pi/4, \; 2\pi$

31. $\sin^2 x + \cos^2 x = 1,$
$1 + \tan^2 x = \sec^2 x,$
$1 + \cot^2 x = \csc^2 x$

32. $\sqrt{3}/2$

33. $(0, 1)$

34. $(0, \log_4 3)$ or $(0, \ln 3/\ln 4), \; (-2, 0),$
$y = -3$

35. reflect the graph of $\ln x$ in the
y-axis

36. $y = \frac{1}{3}x + \frac{4}{3}, \quad y = -\frac{1}{3}x + \frac{14}{3}$

37.

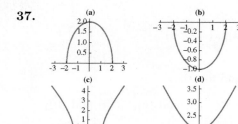

38. an ellipse with focus at the origin, axis along the y-axis, then rotated clockwise $3\pi/4$ radians about the origin

39. $\left(\frac{1}{3}(11-\alpha),\ \frac{4}{3}(1+\alpha),\ \alpha\right),\ \alpha$ real intersection of two planes

40. $0, -4, 4$

41. $x = \pm 3,\ y = 0,\ \lambda = \mp\frac{2}{3}$
$x = -2,\ y = \pm\sqrt{5},\ \lambda = 1$

42.

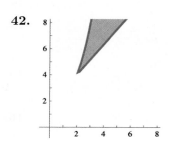

43. 1

44. 255

45. convergent

46. 256

47. $n - 3$

48. $n^2 - 1$

49. $(-1)^{n+1}10^{4-n}$

50. $\dfrac{1}{6n - 5}$

51. $\frac{d}{9}$

52. $8{,}000{,}000$